Freshwater Ecosystems

REVITALIZING EDUCATIONAL PROGRAMS IN LIMNOLOGY

Committee on Inland Aquatic Ecosystems

Water Science and Technology Board

Commission on Geosciences, Environment, and Resources

NATIONAL ACADEMY PRESS
Washington, D.C.   1996

NATIONAL ACADEMY PRESS • 2101 CONSTITUTION AVE., NW • WASHINGTON, DC 20418

NOTICE: The project that is the subject of this report was approved by the Governing Board of the National Research Council, whose members are drawn from the councils of the National Academy of Sciences, the National Academy of Engineering, and the Institute of Medicine. The members of the committee responsible for the report were chosen for their special competencies and with regard for appropriate balance.

This report has been reviewed by a group other than the authors according to procedures approved by a Report Review Committee consisting of members of the National Academy of Sciences, the National Academy of Engineering, and the Institute of Medicine.

Support for this project was provided by National Research Council internal funds, U.S. Environmental Protection Agency cooperative agreement number CR 823885-01, National Science Foundation grant number DEB-9224974, and the Johnson Foundation, Inc.

Library of Congress Cataloging-in-Publication Data

Freshwater ecosystems: revitalizing educational programs in limnology / Committee on Inland Aquatic Ecosystems, Water Science and Technology Board, Commission on Geosciences, Environment, and Resources.
    p. cm.
    Includes bibliographical references and index.
    ISBN 0-309-05443-5 (cloth)
    1. Limnology—Study and teaching (Higher)—United States.
  I. National Research Council (U.S.). Committee on Inland Aquatic Ecosystems.
    QH104.F72   1996
    574.5′2632′071173—dc20   96-25348

Cover art by Edith Socolow, Gloucester, Massachusetts. This work, titled "Spring Thaw," is from her "Window Series."

Printed in the United States of America

The National Academy of Sciences is a private, nonprofit, self-perpetuating society of distinguished scholars engaged in scientific and engineering research, dedicated to the furtherance of science and technology and to their use for the general welfare. Upon the authority of the charter granted to it by the Congress in 1863, the Academy has a mandate that requires it to advise the federal government on scientific and technical matters. Dr. Bruce Alberts is president of the National Academy of Sciences.

The National Academy of Engineering was established in 1964, under the charter of the National Academy of Sciences, as a parallel organization of outstanding engineers. It is autonomous in its administration and in the selection of its members, sharing with the National Academy of Sciences the responsibility for advising the federal government. The National Academy of Engineering also sponsors engineering programs aimed at meeting national needs, encourages education and research, and recognizes the superior achievements of engineers. Dr. William A. Wulf is interim president of the National Academy of Engineering.

The Institute of Medicine was established in 1970 by the National Academy of Sciences to secure the services of eminent members of appropriate professions in the examination of policy matters pertaining to the health of the public. The Institute acts under the responsibility given to the National Academy of Sciences by its congressional charter to be an adviser to the federal government and, upon its own initiative, to identify issues of medical care, research, and education. Dr. Kenneth I. Shine is president of the Institute of Medicine.

The National Research Council was organized by the National Academy of Sciences in 1916 to associate the broad community of science and technology with the Academy's purposes of furthering knowledge and advising the federal government. Functioning in accordance with general policies determined by the Academy, the Council has become the principal operating agency of both the National Academy of Sciences and the National Academy of Engineering in providing services to the government, the public, and the scientific and engineering communities. The Council is administered jointly by both Academies and the Institute of Medicine. Dr. Bruce Alberts and Dr. William A. Wulf are chairman and interim vice chairman, respectively, of the National Research Council.

Preface

Limnology, an integrative science of inland aquatic ecosystems, has made immense strides over the past quarter century. Our understanding of the physical, chemical, and biological processes affecting the properties and behavior of lake, flowing water, and wetland ecosystems has increased substantially. In addition, fundamental conceptual advances made by limnologists have contributed to similar advances made in other fields of natural science and to the solution of many practical problems related to human impacts on aquatic resources. The field of limnology has been broadened substantially in scope from a science with a traditional focus on relatively small water bodies (lakes) in temperate latitudes to one that includes the range of inland aquatic ecosystems—lakes of all sizes from the tropics to polar regions, reservoirs, rivers and streams, and a variety of wetland ecosystems. As a multidisciplinary science, limnology also has extended its breadth over the past several decades so that modern limnology in academic institutions involves faculty, courses, and students in a diversity of natural resource, engineering, geoscience, and more traditional bioscience departments. Growth of the field during this period has been reflected in an increasing number of professional societies and journals focused on its several areas of specialization.

In spite of all the above indicators of a vigorous and growing science, limnologists, especially those in academic institutions, increasingly during the past decade have expressed serious concerns about the status of their field. At least three major areas of concern have been cited:

1. inadequate base of funding for fundamental research across the subareas of limnology;
2. inadequate educational programs; and
3. poor linkage between academic programs and academic research in limnology on the one hand and the practice of limnology in the protection, management, and restoration of aquatic ecosystems on the other.

Several recent studies and reports have addressed the first issue; the committee that wrote this report has focused its efforts on the latter two concerns. In brief, the committee concluded that academic training in limnology at North American universities is highly fragmented, lacks visibility, and does not reflect the breadth and vigor of modern limnology. Moreover, existing programs tend to be focused on producing Ph.D.-level researchers with generally narrow interests, rather than professionals interested in applying their knowledge to the management of aquatic ecosystems and the solution of practical problems. This report provides the basis for that conclusion, evidence for the relevance and importance of broad limnological training, and a range of practical recommendations to revitalize education in limnology at both the bachelor's and the graduate degree levels.

The current situation in limnology is analogous to the problems addressed by a predecessor Water Science and Technology Board (WSTB) committee on hydrologic sciences and indeed is a subset of a broad problem in academic programs in aquatic sciences. Courses and areas of specialization within limnology in particular and aquatic science in general are found in many departments and colleges of research universities, but departments or interdepartmental academic programs that encompass the breadth of these subjects hardly exist in U.S. universities. This situation lends itself to a "tragedy of the commons" syndrome: limnology and aquatic science are interests in many disciplines and many departments but are not regarded as the primary responsibility of any. Academic fragmentation may be more pronounced in limnology than in most natural resource sciences; many inherently multidisciplinary natural resource fields, including oceanography, soil science, and forestry, have had departmental and/or collegiate status in some universities for decades. Why limnology has not done so perhaps reflects the unique historical development of the field. Nonetheless, the past need not be a prescription for an indefinite future. The committee hopes that limnologists in universities will consider this report's recommendations carefully and work toward their implementation in revitalizing academic limnology in the United States.

Much of the work in writing this report was done by the committee members at their home institutions and by mail and phone interactions. However, the framework for the report, major recommendations, and initial drafts of Chapter 4 were developed at a week-long workshop held in October 1994 at Wingspread, a conference center of the Johnson Foundation in Racine, Wisconsin. The quiet, picturesque surroundings of the center and the warm hospitality of the Wingspread staff and of Jon Vondracek, then Vice President for Program and Public Communications at Wingspread, contributed greatly to the success and productivity of the workshop. The committee also thanks Art Brooks of the University of Wisconsin–Milwaukee for an informative tour of the Center for Great

Lakes Studies and the Environmental Protection Agency research vessel, the *Guardian*, in Milwaukee harbor.

In preparation for the workshop, during the summer of 1994, various members of the committee developed background papers on a variety of topics related to the committee's charge. The papers formed the basis for many useful discussions at the workshop, and modified versions of the papers are included in this book after the body of the report. I thank Wayne Minshall and Bob Wetzel for serving on the steering committee that developed the topics for these papers and helped to organize the format of the workshop. The committee also owes a special debt of gratitude to Diane McKnight, who served as rapporteur at the workshop and prepared an initial draft of Chapter 4.

The committee completed its discussions on the report and developed a consensus on its recommendations at a second meeting in Washington, D.C., in May 1995. A workshop held in conjunction with this meeting provided the committee with valuable insights on educational needs in limnology from representatives of government agencies and private firms involved in management of aquatic ecosystems.

I am deeply appreciative of the hard work and vital contributions of Jackie MacDonald, senior staff officer for this committee. Her organizational and editorial skills contributed immensely to the task of gathering information and pulling the many pieces of this report together. Anita Hall, administrative assistant, and Greg Nyce, senior project assistant, also contributed significantly to the completion of the committee's work. I would like to thank my secretary at the Water Resources Research Center, Maria Juergens, for help with the survey of universities and for assistance on many other matters.

Several individuals associated with the Water Science and Technology Board (WSTB) were instrumental in the developmental stages of this project. Sheila David, senior staff officer at the WSTB, organized a planning meeting at which the project proposal was written, and Judy Meyer, professor at the University of Georgia and then a member of the WSTB, chaired that meeting. I am grateful to them and to Steve Parker and other members of the WSTB, through whose continued interest and support the project reached active committee status.

Funds for this study were provided by the U.S. Environmental Protection Agency, the Johnson Foundation, Inc., the National Science Foundation, and the National Research Council Basic Science Fund. The committee appreciates the foresight of these institutions in recognizing the need for this study and hopes that their expectations have been realized in this report.

Finally, the committee is grateful to the large number of people who contributed information or text material for this report; without their efforts, this report would be much diminished. A list of these individuals is found in Appendix D.

Patrick L. Brezonik
Chair, Committee on Inland Aquatic Ecosystems

Contents

BACKGROUND PAPERS

Freshwater Ecosystems

REVITALIZING EDUCATIONAL PROGRAMS IN LIMNOLOGY

Executive Summary

From ancient times, civilizations have depended on freshwater bodies—lakes, reservoirs, rivers, and wetlands. Fresh water is essential not only to sustain human life but also to support the activities that form the basis for thriving economies. At the same time that water resources are essential to human societies, activities of these societies can pollute and degrade water resources, limiting their beneficial uses. Since the start of the "environmental movement" in the 1960s and especially since the passage of the Clean Water Act in the United States in 1972, North Americans have realized that manufacturing, agriculture, mining, urban development, and other activities can pose risks to freshwater bodies, and they have taken steps to reduce these risks.

Recently, the U.S. Congress has called for reevaluation of environmental laws, such as the Clean Water Act, based on risk analysis. In theory, risk analysis would indicate the most serious threats to environmental resources, identify the most cost-effective methods to reduce risk, and evaluate whether existing laws provide cost-effective solutions or address secondary threats at high cost. Such risk analysis cannot be carried out in the absence of sound science. For freshwater bodies, risk analysis requires knowledge of how human land use affects the physical, chemical, and biological characteristics of the aquatic ecosystem.

One of the critical sciences required to understand how human actions and natural processes affect lakes, reservoirs, rivers, and wetlands is limnology. Limnology is a multidisciplinary science that integrates the basic sciences (biology, chemistry, physics, and geology) in order to study inland waters as complex ecological systems. Only through such integrated study will it be possible to understand the full range of human impacts on aquatic ecosystems. As the nation has recognized that its resources for addressing environmental problems are limited, the science of limnology has become increasingly vital for devising cost-effective strategies to ensure that freshwater systems not only can serve this generation but also can be preserved for the benefit of generations to come.

1

Because of the increasing importance of interdisciplinary aquatic science in solving environmental problems, in 1994 the National Research Council appointed an expert committee, the Committee on Inland Aquatic Ecosystems, to recommend ways to strengthen limnology programs within U.S. educational institutions. The committee included limnologists and aquatic scientists in related fields from academia, state and federal government, and the private sector. This report presents the committee's findings. It reviews the history of limnology and its role in solving contemporary water problems. Based on this history and society's current needs, it recommends improved ways to educate future limnologists.

Traditionally, limnologists have been perceived as scientists who study primarily the biological properties of lakes. However, modern limnologists define their science broadly; they view it as covering the biology, physics, and chemistry of all inland waters, including rivers and wetlands as well as lakes. Some scientists who study primarily streams or primarily wetlands, including several committee members, identify themselves as limnologists as well as stream or wetland scientists, while others do not. Regardless of where one draws the line in deciding whom to call a limnologist, lakes, rivers, and wetlands are interconnected, and the sciences that study them are closely allied. Understanding of lakes is incomplete without some knowledge of how wetlands and rivers affect them, and vice versa. Whether or not one agrees with the broad definition of limnologist, limnology can serve as a paradigm for interdisciplinary water science in general. Improvements in the teaching and study of limnology will benefit other, closely allied aquatic sciences—as well as society as a whole—by leading to new research breakthroughs and raising the profile of both aquatic ecosystems and the sciences necessary for understanding how to protect them.

CONTEMPORARY WATER MANAGEMENT: ROLE OF LIMNOLOGY

The Clean Water Act and other efforts to control waste discharges to waterways have noticeably improved the condition of some water bodies (well-known examples include Lake Erie and the Potomac River), but many lakes, rivers, and wetlands in the United States remain degraded or at high risk. For example, the Environmental Protection Agency recently reported to Congress that 44 percent of river miles are unsuitable for one or more of the uses (boating, swimming, or fishing) designated by state water managers, and 57 percent of lakes are unsuitable for one or more of these uses. The causes of this degradation extend beyond the pollution sources that have been the focus of pollution control efforts over the past 25 years (primarily direct wastewater discharges from industries and sewage treatment plants); they include various diffuse pollution sources, as well as a variety of physical stresses that directly or indirectly affect aquatic habitats. Limnologists, working with other aquatic scientists, have been involved in developing ways to optimize the preservation of inland waters:

• *Watershed management*: Maintaining and improving the quality of surface waters will require consideration of how the wide variety of human activities on the land surrounding water bodies (their watersheds) affects water quality. Limnologists have been involved in demonstrating how land-use changes alter the yields of chemical constituents and sediments to water bodies, and they should be instrumental in devising scientifically based watershed management plans.

• *Wetland preservation*: Substantial losses of wetlands have occurred in the past. Recognizing the valuable role of wetlands in providing habitats for important species, reducing flood peaks, and filtering runoff before it enters lakes and rivers, the United States has developed a policy intended to prevent further net loss of wetlands. Research by limnologists will be essential for developing a workable way to implement this policy.

• *Control of cultural eutrophication*: Eutrophication—the input of excess nutrients (especially phosphorus and nitrogen) to lakes and rivers from sewage, agricultural fertilizers, and other sources—results in the development of large masses of algae and (often) large aquatic plants. The algae decrease water clarity, deplete deep-water oxygen to the degree that important species of fish cannot survive, and may create taste, odor, and toxicity problems. Work by limnologists has been and will continue to be central to developing plans to control nutrient inputs in order to restore ecosystems as large as Lake Erie and as small as farm ponds.

• *Reservoir management*: More than 80,000 dams exist in North America. Limnologists, working with hydraulic engineers, hydrologists, and fisheries biologists, need to play key roles in developing the scientific understanding necessary to manage the impoundments created by dams and to optimize dam operations to preserve water quality below the dams.

• *Study of the effects of global warming*: Nations worldwide have been discussing strategies to mitigate the impacts of global warming. Limnological studies can shed light on how global warming—which could alter water supply, water temperatures, and related habitat factors—may affect aquatic ecosystems.

• *Evaluation of toxic pollutants*: The release of toxic substances in trace quantities once was thought to cause little harm to the environment, but aquatic scientists have shown repeatedly that such substances can accumulate in fish and fish-eating birds to levels up to 10 million times greater than those in lake-water, resulting in health advisories against fish consumption. Continued limnological research will provide a more complete understanding of the behavior of toxic substances in aquatic environments.

• *Assessment of damage from acid rain and other airborne pollutants*: Nations around the world have signed treaties to control acid rain. Limnological research is increasing the understanding of long-term effects of acid rain on aquatic communities, leading to improved predictive models and mitigation and restoration techniques.

• *Control of exotic species*: Numerous exotic species (both plants and animals) have disrupted aquatic ecosystems in North America, sometimes with high costs. For example, the zebra mussel, accidentally discharged into the Great Lakes with the ballast water of an oceangoing ship in 1986, will cause an estimated $4 billion in damages before the turn of the century, primarily because of clogging of water intake pipes and loss of native mussels and clams. Research by limnologists is a key component of plans to control the undesired proliferation of exotic plant and animal species.

• *Prevention of species extinction*: According to some estimates, at least 30 percent of all North American fish species are rare, at risk of extinction, or extinct. Limnological research can help determine more accurately the distribution of native species, the factors that influence their success in certain habitats, the effects of species loss on aquatic communities, and how best to protect critical species.

In the past, water management decisions have often been made in isolation, without adequately considering the synergistic relationships among a variety of isolated actions. For example, the Clean Water Act focused primarily on reducing water pollution from sewage and industrial facilities and paid much less attention to the range of other human impacts on water bodies. Future water management will demand professionals who are broadly trained to view water bodies as systems that are integrated with the surrounding landscape and in which biological, physical, and chemical processes are interrelated. Education in limnology provides the type of broad perspective needed to understand how water bodies behave in environments without significant human influence and how they are affected by human activities. This broad perspective is essential for any strategy that aims to control environmental problems by considering the full range of risks and benefits.

STATUS OF LIMNOLOGY IN THE UNIVERSITY SYSTEM

Because most of the early limnologists were trained in biology, professors of limnology historically have been housed in departments of biological science (including zoology and botany). However, during the second half of the twentieth century, increased funding for pollution studies of freshwater bodies has drawn scientists from a wide range of fields to limnological research. Today, scientists who study problems in limnology operate from departments as diverse as civil engineering, fisheries and wildlife, botany, zoology, ecology, environmental science, forestry, geology, and geography. For example, in a survey conducted for this report, universities listed 23 types of departments where scientists teach and conduct research in limnology. The net result is increasing fragmentation of the discipline: limnology is an interest within many fields, but it is not the primary focus of any of the traditional departments.

One problem caused by the fragmentation of limnology in the university system is that students who display interest in limnology during their undergraduate years often lack guidance on the mix of courses needed to prepare them for further work in the field. Because U.S. colleges and universities generally do not offer formal undergraduate majors specifically identified as "limnology," students often enter the field accidentally, by taking limnology as an elective within another science major.

A second problem resulting from the fragmentation of limnology is lack of adequate support for laboratory and field programs. Given the tight funding at many universities and the lack of strong departments or interdepartmental programs to serve as advocates for limnology, funding is often insufficient to provide laboratory and field experience for interested students. Typically, laboratory and field courses in limnology must be limited to a small number of students even though laboratory and field studies are critical parts of education in limnology.

A third problem stemming from fragmentation is that graduate students in limnology may not receive adequate training in all of its important subdisciplines. For example, students who obtain their limnology degrees through biology departments might have strong training in fields such as organismal physiology but much weaker knowledge of water chemistry and physical limnology. On the other hand, students specializing in the study of aquatic ecosystems in civil and environmental engineering departments may have strong training in hydrology and water chemistry but relatively little knowledge of the organisms that inhabit aquatic systems. Also, students in these and other types of programs rarely have access to courses that will train them to understand all types of aquatic ecosystems (wetlands, streams, and lakes).

EDUCATION IN LIMNOLOGY: RECOMMENDATIONS FOR CHANGE

Education in limnology, and the strength of the science in general, will depend on providing better coordination of limnological teaching and research. The Committee on Inland Aquatic Ecosystems recommends two possible administrative ways to achieve this coordination:

1. *Establish regional departments of aquatic science with limnology majors.* Creating strong aquatic science departments, with comprehensive limnology majors for undergraduates, in some U.S. universities on a regional basis would establish centers of strength for this science.

2. *Establish strong interdepartmental programs in aquatic science, with an option to specialize in limnology.* Interdepartmental programs may be easier to establish than new departments and can have the advantage of encouraging collaborative work and providing students with access to large numbers of formal courses. On the other hand, faculty loyalties to their primary departments and lack of budgetary authority may be impediments to successful implementation of interdepartmental programs.

The approach appropriate for a given institution will depend on the institution's existing strengths, as well as on the preferences of its faculty and leadership. Strong support from the university administration is essential for the success of interdepartmental programs in order to avoid such practical problems as dividing overhead costs among departments, cross-listing courses, and approving joint appointments.

In addition to instituting well-coordinated programs in limnology, the Committee on Inland Aquatic Ecosystems recommends several other initiatives to improve education in limnology. These actions are needed to (1) inform responsible citizens who will value freshwater ecosystems and understand water resource issues, (2) provide a basic understanding of the science for all who will be involved in decisions and actions that affect freshwater ecosystems, and (3) train future research scientists who will work to advance the understanding of aquatic ecosystems. Following are recommendations for improving limnological education to serve each of these purposes:

- *Educating responsible citizens*: General introductory courses in limnology should be developed and taught at all types of universities and colleges. These courses should include coverage of wetlands and streams as well as lakes, and they should be accessible to all students, with the goal of conveying how freshwater ecosystems function and how they respond to various human activities. Enough faculty support should be provided to allow all interested students to enroll in an introductory limnology course and as many as possible to undertake field and laboratory work.

- *Educating future water managers*: Students should be provided with increased opportunities to gain exposure to practical problems such as the management of freshwater systems in urban areas, in national parks, on grazing lands, and on croplands. Student internships in federal and state agencies and the private sector are one way of providing this exposure and simultaneously furthering the links between universities, agencies, and private firms. Prospective managers of aquatic ecosystems should have the opportunity to gain laboratory and field experience in limnology.

• *Educating future limnologists*: Undergraduates showing special interest in limnology need to be given guidance and opportunities to select a curriculum that will provide the breadth of skills and knowledge required to solve problems of freshwater ecosystems. Opportunities for undergraduate students to attend summer- or semester-long limnology "field camps" should be developed to complement the laboratory portion of conventional limnology courses. At the graduate level, comprehensive programs need to be developed, especially for M.S. degrees, to train limnologists who are knowledgeable across the spectrum of freshwater ecosystem types; who have an integrated understanding of the physical, chemical, and biological processes operating in these ecosystems; who understand the political, economic, and cultural factors that affect aquatic ecosystems and their management; and who have strong problem-solving and communication skills. Graduate programs need to include a research or practical problem-solving component; "course-work only" M.S. programs generally are not adequate to provide limnologists with the above-described skills.

LINKING WATER MANAGEMENT, RESEARCH, AND EDUCATION IN LIMNOLOGY

Beyond strengthening limnology in educational institutions, the Committee on Inland Aquatic Ecosystems recommends additional steps to improve the links between scientific understanding in this field, as produced and disseminated primarily by educational institutions, and the practical management of water resources, as conducted primarily by government agencies and for-profit companies:

• *Support of aquatic science research*: The National Science Foundation should establish a permanent program focused on basic and problem-solving interdisciplinary research on inland aquatic ecosystems.

• *Certification of limnologists*: Leaders of the limnological professional societies should consider establishing professional certification programs in limnology. Such a tactic could raise the value and profile of degrees in limnology and help ensure a minimum level of competence among practicing limnologists. Certification might be based on assessment of an applicant's educational and work experience by a qualified review board. Professional certification programs have pluses and minuses, but a number of environmental science disciplines (for example, hydrology and ecology) and associated professional societies have decided that the former outweigh the latter and have begun certification programs in recent years.

• *Creation of a federal job classification for limnologists*: Establishment of a federal job classification titled "limnologist" would help agencies to identify scientists for water resource management and research positions in which limnological knowledge is essential. Currently, limnologists are classified as "hydrologists" or "microbiologists" by the federal government, making it difficult for agencies to identify job candidates with the training required for many water-related positions.

• *Involvement of water managers in teaching and graduate advising*: Appointing scientists from federal and state agencies and private consulting firms to serve as adjunct professors can benefit practicing professionals by providing them with access to the newest scientific information and can benefit universities by providing students with the opportunity to learn about applied resource problems from practicing professionals. Adjunct professorships also may facilitate student access to internships and other student appointments in government agencies and provide all students with experience on a team studying a practical problem. Although such adjunct faculty appointments need to be handled and monitored with due attention to protecting the interests of all parties and the integrity of academic programs, there is ample precedent for successful academic-government and academic-private sector linkages in other natural resource and engineering fields.

• *Providing university-based continuing education opportunities for water managers*: One of the most direct ways for universities to obtain feedback on the practical relevance of curricula in limnology is by providing continuing education opportunities for water managers. Continuing education courses allow water professionals to improve their comprehensive understanding of ecosystems and to learn newly developed scientific techniques, while providing professors with insights about the questions of most importance to professionals in the field.

• *Increasing public awareness of limnology*: Limnologists can work at several levels to increase public awareness of their field. Examples of possible outreach efforts include participating in elementary and secondary school education, speaking at public hearings, working with university extension services, providing advice to environmental organizations, and participating in citizen-based programs for water quality monitoring and rehabilitation.

In summary, determining the most significant risks to freshwater ecosystems and deciding how to manage those risks effectively will require advances in fundamental science and in the application of this knowledge to practical problems. In turn, achieving these goals will require strengthening limnology within university systems and making better connections among academic limnologists, water resource managers, and the general public.

1

Overview:
Status of Inland Waters

Freshwater systems—lakes, wetlands, rivers, and streams—have been critical to the establishment of civilizations throughout human history. From ancient times, civilizations have been built based on their proximity to water: ancient Mesopotamia thrived because of the ample water provided by the Tigris and Euphrates rivers; ancient Egypt grew up along the Nile; the Romans expended vast resources in building elaborate networks of aqueducts to supply water to their cities. Water bodies are essential to humans not only for drinking but also for transportation, agriculture, energy production, industry, and waste disposal.

Despite the reliance of societies on freshwater systems, only in this century has the importance of protecting the quality of these systems become widely recognized. In the United States, the passage of the Clean Water Act in 1972 reflected widespread awakening to the deteriorating status of the nation's surface waters. The Clean Water Act set a goal of restoring all U.S. lakes and rivers to a "fishable and swimmable" condition by July 1, 1983. A court decision in 1975 ruled that under the act, the federal government also was obliged to protect wetlands (Mitsch and Gosselink, 1993).

The Clean Water Act focused on reducing municipal and industrial wastewater discharges to water bodies, and the results have been significant. In the decade following the act's passage, sewage discharges to U.S. surface waters dropped by 46 percent and industrial discharges by 71 percent, even though the population grew by 11 percent (Frederick, 1991). Two decades later, 85 percent of sewage plants and 87 percent of industrial plants discharging to water bodies were in compliance with the act (Kealy et al., 1993). Nevertheless, according to the Environmental Protection Agency (EPA, 1994), 40 percent of U.S. surface waters remain too degraded for fishing and swimming. Contaminated runoff from expanding urban and agricultural areas, airborne pollutants, and hydrologic modifications such as drainage of wetlands are just a few of the many factors that continue to degrade U.S. surface waters despite reductions in sewage and industrial waste discharges. Determining which of these factors has the most significant influence on the quality of a water body requires knowledge about how the water body interacts with its watershed and airshed and how the various inputs affect its physical, chemical, and biological characteristics.

One of the critical sciences required to understand these freshwater interactions is called limnology (from the Greek *limne,* meaning pool or marshy lake). As defined in this report and other recent reports on the field, limnology includes the study of lakes, reservoirs, rivers, and freshwater wetlands (Edmondson, 1994; Lewis, 1995; Lewis et al., 1995). It is a multidisciplinary science that draws from all the basic sciences relevant to understanding the physical, chemical, and biological behavior of freshwater bodies. There are numerous subspecialties of limnology based on the application of fundamental sciences such as physics, chemistry, geology, and biology; branches of physical science such as optics, fluid mechanics, and heat transfer; and branches of biological science such as microbiology, botany, ichthyology, invertebrate zoology, and ecology. Limnology integrates these other sciences in order to study inland waters as ecological systems (Edmondson, 1994; Lewis, 1995). In a recent essay entitled "What Is Limnology?" limnologist W. T. Edmondson (1994) described limnology as

> the study of inland waters . . . as systems. It is a multidisciplinary field that involves all the sciences that can be brought to bear on the understanding of such waters: the physical, chemical, earth, and biological sciences, and mathematics.

Limnology thus has two distinguishing features: it is an integrative (i.e., interdisciplinary) science, and it consists of many component subspecialties (i.e., it is multidisciplinary). The interdisciplinary, multidisciplinary nature of limnology provides a broad perspective that is critical in identifying the multiple sources of stress that may prevent a water body from serving its essential functions.

Although advances in limnology can play a critical role in improving the quality of fresh surface waters, prominent limnologists have expressed concern that the field is in decline (Naiman et al., 1995). Some have cited lack of a national research budget devoted to limnology (Jumars, 1990), others have identified lack of adequate educational programs (Wetzel, 1991), and still others have suggested inadequate attention by academic limnologists to contemporary environmental problems (Kalff, 1991) as the reasons for this decline. In a recent assessment of the status of limnology, members of the American Society of Limnology and Oceanography warned, "Limnology shows signs of fragmentation, loss of identity, and poor sense of direction, all of which are reducing its potential for solving problems that arise from the escalating demands that society is placing upon inland waters" (Lewis et al., 1995).

Because of the increasing importance of interdisciplinary water science in addressing environmental problems, the National Research Council's Water Science and Technology Board appointed a committee of 13 experts in limnology and related fields to evaluate what changes are needed to strengthen education in limnology. This report presents the committee's findings. The committee included representatives of universities, government agencies, private consulting firms, and research laboratories; members had expertise in ecology, zoology, water chemistry, hydrology, and environmental engineering, as well as limnology. In preparing its recommendations, the committee surveyed major universities with aquatic science programs and sought input from professional societies that include limnologists among their membership. The committee also held a workshop with aquatic resource managers to help develop practical recommendations for improving education in limnology.

Traditionally, many have perceived limnologists as scientists who study primarily the biological properties of lakes. In this report, the Committee on Inland Aquatic Ecosystems has defined limnology broadly to include the biology, physics, and chemistry of all inland waters, including rivers and wetlands as well as lakes. Some scientists who study streams and wetlands, including several committee members, identify themselves as limnologists as well as stream or wetland scientists, while others do not. Regardless of where one draws the line in defining limnologist, lakes, rivers, and wetlands are interconnected, and the sciences that study them are closely allied.

This report is designed to broaden the understanding of limnology and to provide a foundation for developing educational programs in limnology that will better prepare the next generation to address the world's many water quality challenges. Administrators of university and government science programs, students of aquatic science, and aquatic resource managers in government agencies and the private sector can use the report to learn more about the history and current status of human efforts to understand lakes, rivers, and freshwater wetlands. For administrators and teachers of aquatic science, the report recommends ways to strengthen university limnology programs to produce a generation of well-educated citizens, aquatic resource managers, and researchers capable of understanding and alleviating the many sources of water quality degradation.

This chapter highlights the status of inland waters and the role of limnologists in improving these waters. Chapter 2 provides a history of limnology, including profiles of several prominent limnologists and scientists in related fields, and evaluates the strengths and weaknesses of present-day limnology. Chapter 3 describes key contemporary water problems that have been studied by limnologists and whose solution will require further limnological research. Chapter 4 analyzes the current system for teaching limnology in institutes of higher education and recommends ways to restructure the system to strengthen limnological training. Chapter 5 recommends strategies for linking education and research in limnology with aquatic resource management. Also included in this volume are eight background papers prepared by committee members to stimulate the discussions that led to the writing of this report. The papers do not represent the consensus views of the committee as a whole but instead reflect the range of perspectives from which problems in aquatic science education and research can be viewed.

FRESH WATERS AT RISK

The condition of some important water bodies, such as the Potomac River and Lake Erie (see Boxes 1-1 and 1-2) has improved significantly in the two decades since the passage of the Clean Water Act. Nevertheless, many North American fresh waters remain degraded or at high risk. According to EPA reports to Congress on the status of U.S. waters, 43 percent of rivers are too contaminated to provide drinking water if treated with conventional water purification technologies,[1] 40 percent of rivers are too degraded to support aquatic life fully, and 32 percent are too contaminated for swimming, based on state surveys of 18 percent of the nation's total river miles (EPA, 1994). Similarly, 31 percent of lakes are too contaminated for drinking even with conventional treatment, 40 percent are unable to fully support aquatic life, and 36 percent are unsuitable for swimming, based on surveys of 46 percent of total U.S. lake area. Based on these data, it is clear that the United States is still a long way from achieving the goals of the Clean Water Act.

Continued degradation of surface waters causes substantial economic losses, some of the most significant of which are associated with lost fishing revenues, increased costs for treating drinking water, and lost recreational opportunities. The fishing restrictions and fish consumption advisories common in polluted water bodies jeopardize the multibillion-dollar fishing industry. Recreational freshwater anglers spent $19.4 billion on their sport in 1985; the commercial fishing industry generated $3.6

---

[1]For the state surveys that provided the basis for this estimate, the EPA defined conventional drinking water treatment as coagulation and sedimentation followed by disinfection. This type of treatment is designed primarily to remove turbidity and microbiological contaminants.

BOX 1-1   POTOMAC RIVER IMPROVEMENTS

One partial success story of the Clean Water Act is the Potomac River, a 462-km (287-mile) water course that begins in West Virginia and flows into the Chesapeake Bay. Pollution episodes in the Potomac date back as early as the 1840s, when sewers first conveyed human wastes to the river from Washington, D.C. According to Civil War–era reports, President Lincoln frequently left the White House to escape odors emanating from the Potomac (Uman, 1994).

Pollution in the river continued to worsen until, in 1934, the federal government appropriated funds to build the Blue Plains sewage treatment plant to remove settleable solids before the sewage entered the river. The plant's capacity was soon exceeded, however. As a result, discharges of raw sewage became increasingly common, and the river continued to deteriorate. In 1969, participants at a Washington, D.C., conference called the Potomac River "a severe threat to anyone who comes in contact with it" (Adler and Finkelstein, 1993). By 1970, the pollutant loading to the river was higher than it had been in 1932 (Uman, 1994). Algal mats choked a 50-mile stretch of the river downstream from Washington in late summer. Fishing and swimming were prohibited.

Following passage of the Clean Water Act, the government spent $1.6 billion to upgrade the Blue Plains plant and in 1980 spent an additional $500 million to add advanced treatment systems. Recreational boating is now possible on the river. Bottom vegetation and bass have returned after a long absence (Uman, 1994).

Despite these notable improvements, swimming in and consuming fish from portions of the river remain health risks (Interstate Commission on the Potomac River Basin, 1994). Contaminants such as chlordane, polychlorinated biphenyls, and heavy metals remain. In addition, the river is subject to bacterial contamination from discharges of raw sewage when major storms overload Blue Plains and other local sewage treatment plants. Thus, although the quality of the Potomac has improved substantially, it still does not meet the "fishable and swimmable" goal of the Clean Water Act.

---

billion in revenues (including freshwater and saltwater harvests) in 1990 (Adler and Finkelstein, 1993). According to the EPA (1994), 1,279 fish consumption advisories—warning consumers and fishermen to limit intake of certain fish because of contamination—were in effect in 47 states in 1993; contaminant levels in fish tissues can be more than a million times those in surrounding water because of the tendency of contaminants to concentrate in species at higher levels of the aquatic food web.

The need to provide more advanced levels of treatment for degraded sources of drinking water also has significant costs. For example, New York City may eventually be required to filter its drinking water because of degradation of the once-pristine upstate New York reservoirs where the water is stored. The estimated cost for filtration is $1.5 billion to $5 billion for construction plus $300 million per year for operation and maintenance (Adler and Finkelstein, 1993).

Losses of recreational opportunities when water is contaminated have extremely high costs. A 1986 survey by the President's Commission on Americans Outdoors found that 59 percent of Americans over age 12 fish and 27 percent boat each year (Kealy et al., 1993). In 1991, Resources for the Future, an economic study institute in Washington, D.C., surveyed Americans about their willingness to pay for clean surface water. On average, respondents indicated that they would be willing to pay $106 per year for boatable water, an additional $80 per year for fishable water, and an additional $89 per year for swimmable water, for a total of $275 per year (Adler and Finkelstein, 1993). Based on these responses, Resources for the Future estimated the value of restoring nonboatable water to swimmable condition to be $29.2 billion per year in 1990 dollars, with a reasonable range of $24 billion to $43 billion.

Worldwide, contamination of waterways places artificial limits on the supply of potable water. The time required to replenish the water supply of a contaminated lake can be a decade or more (Wetzel, 1983), and the time to cleanse contaminated sediments can be much longer. Consequently, the effects of water quality degradation can be long-lasting and in some cases permanent. Where water resources are scarce, such as in parts of the western United States and other arid areas of the world, contamination of water bodies can threaten the livelihood of surrounding populations.

SOURCES OF STRESS ON INLAND WATERS

Progress in cleaning up U.S. surface waters has been limited in part because the types of problems affecting these waters are more complex than anticipated by the authors of major water pollution control legislation. In particular, the Clean Water Act addressed primarily one form of pollution: it required treatment of "point-source" discharges from municipal sewage treatment plants and industries. Contemporary water problems arise from a multitude of other causes in addition to point sources. Key causes of problems in freshwater ecosystems, described in more detail in Chapter 3, include the following:

• *Runoff of pollutants from agricultural and urban lands:* The EPA (1994) estimates that agricultural runoff impairs 56 percent of the nation's lakes and reservoirs and 72 percent of the nation's rivers and streams. Urban runoff from storm sewers is a cause of impairment of 24 percent of lakes and reservoirs and 11 percent of rivers and streams, according to the EPA. Like point-source sewage discharges, agricultural and urban runoff

BOX 1-2 PARTIAL RECOVERY OF LAKE ERIE

News reports in the 1960s spoke of the demise of Lake Erie, but the lake has recovered in significant ways since 1972—the year the Clean Water Act was passed and the United States and Canada signed the Great Lakes Water Quality Agreement to curtail pollutant discharges to the lake.

In the first part of the twentieth century, the Great Lakes in general and Lake Erie in particular supported one of the world's most valuable commercial fisheries. Lake Erie was more productive than the other lakes, in part because its relative shallowness provided an ideal environment for many species of commercially valuable fish (Egerton, 1987). Beginning in the 1900s, the composition of Lake Erie's aquatic community began to shift from commercially valuable fish such as lake trout and sturgeon toward fish such as catfish and carp that have lower market values. Lake trout, once abundant, had disappeared by 1950 and sturgeon by 1965 (Egerton, 1987). Fisheries biologists attributed loss of these and other valuable species to pollution, loss of habitat, and overfishing (Egerton, 1987).

Several major urban and industrial centers—Buffalo, Cleveland, Toledo, and Detroit—lie along Lake Erie or its major tributaries. Prior to the passage of the Clean Water Act and the signing of the Great Lakes Water Quality Agreement, discharges of untreated or inadequately treated sewage and industrial wastes from these industrial centers into the lake were widespread.

One of the key results of waste discharges to the lake, limnologists discovered, was an overabundance of phosphorus. The excess phosphorus caused excessive growth of algae, which in turn led to the proliferation of bacteria that decompose algae, a decline in the oxygen content of lake bottom waters as the algae decomposed, and loss of the valuable commercial fish that require high oxygen concentrations. For example, the concentration of algae near Cleveland increased 12-fold between 1930 and 1960 (Eos, 1971). By 1953, scientists had discovered that important native organisms retrieved from portions of the bottom sediments were dead as a result of lack of oxygen and that the native species had been replaced by worms and midge larvae, which tolerate low-oxygen environments (Egerton, 1987).

The Great Lakes Water Quality Agreement called for all municipal wastewater treatment plants discharging more than 3,800 m^3 per day (1 million gallons per day) to decrease the phosphorus concentration in their effluents to less than 1 mg per liter. The agreement also called for limits on the phosphate content of detergents—a key source of this element—and for reductions in industrial discharges of phosphorus. Simultaneously, the Clean Water Act provided funding to construct and upgrade sewage treatment plants, allowing most municipalities in the area to achieve or exceed the goals for phosphorus reduction. Between 1968 and 1982, the annual phosphorus load to the lake dropped from 28,000 tonnes to 12,400 tonnes (Great Lakes Water Quality Control Board, 1985). As a result, the massive algal blooms that were prevalent in the 1960s have been eliminated, oxygen concentrations have increased in bottom waters, and the water has become clearer (Great Lakes Water Quality Control Board, 1985).

The partial recovery of Lake Erie is one of the major success stories of the Clean Water Act, but at the same time the lake's problems are not over. The lake continues to receive excess phosphorus from agricultural runoff, and recycling of phosphorus that has accumulated in the bottom sediments continues. Although states surrounding the Great Lakes report that the presence of excess phosphorus and other elemental nutrients is now a major concern along less than 3 percent of Great Lakes shoreline, the presence of toxic compounds such as polychlorinated biphenyls remains a concern and has resulted in advisories against consuming Lake Erie fish (EPA, 1994). Furthermore, invasions of exotic species such as zebra mussels threaten to disrupt the lake's food web and its production of game fish (see Chapter 3).

can convey large quantities of nitrogen, phosphorus, and organic matter that stimulate excess growth of algae and oxygen-consuming bacteria. This process, known as cultural eutrophication, ultimately results in loss of water clarity, loss of oxygen in bottom waters, and a shift in the food web from valuable game fish to less desirable species. High concentrations of pesticides also may be present in agricultural runoff; fish kills are commonly reported to insurance companies or the EPA when major storms follow pesticide applications (National Research Council, 1992). Storm runoff from urban areas also can transport high concentrations of pollutants such as lawn chemicals, metals, automotive oil and grease, and bacteria from animal wastes. Both agricultural and urban runoff contain sediment that can carry adsorbed contaminants and smother fish habitats and spawning areas.

• *Alteration of natural hydrology:* Throughout history, but especially in the twentieth century, humans have manipulated water bodies and their surrounding watersheds to serve purposes such as providing power, supplying water for irrigation, and regulating water flows to allow farming and building on floodplains. As a result, half of the wetlands, which once helped to cleanse the water flowing into rivers and lakes, are now gone from the lower 48 states (see, for example, Box 1-3) (National Research Council, 1995). The more than 80,000 dams in the United States (Frederick, 1991) alter downstream flow patterns in ways that can jeopardize the survival of important fish species. For example, 19 major dams and 100 smaller power projects have been built along the Columbia River in Washington and Oregon since the 1930s; the resulting habitat modifications have contributed to the extinction of 106 stocks of Pacific salmon that once spawned in the river (McGinnis, 1994). The EPA (1994) estimates that hydrologic, habitat, and flow modifications prevent desired uses of 36 percent of U.S. lakes and reservoirs and 7 percent of rivers and streams.

• *Atmospheric transport of pollutants:* Atmospheric circulation can trans-

BOX 1-3   DEGRADATION AND RESTORATION OF
THE FLORIDA EVERGLADES

The Everglades ecosystem is a unique subtropical wetland system located in southern Florida. The Everglades landscape contains a multitude of wetland types, including sawgrass marshes, sloughs, marl- and peat-based wet prairies, tree islands, pinelands, and, at its southernmost extreme, mangroves and the Florida Bay estuary (Davis and Ogden, 1994). Before major human influence, the Everglades landscape was even more diverse and included custard apple swamps, short-hydroperiod wet prairies, and cypress stands. During the last century, the area of the Everglades has decreased by half due to agricultural and urban encroachment (Davis and Ogden, 1994). In addition, natural Everglades hydrology has been altered severely, with coincident disruption of ecological processes, by construction of numerous levees, canals, and pump structures (Davis and Ogden, 1994). Addition of unnatural levels of nutrients from agricultural runoff has exacerbated the decline of Everglades ecological integrity.

The Everglades became a major focus of attention for Florida in 1988, when the federal Department of Justice sued the state, its Department of Environmental Protection, and the South Florida Water Management District for allowing the Everglades to become degraded as a result of nutrient pollution from the 700,000 acres of farms that border it to the north. As a result, in 1994 the state legislature passed a law, the Everglades Forever Act, that requires restoration of the ecosystem. Under the act, agriculture will pay for about half of Everglades water quality cleanup (up to about $300 million), with the other half coming from state and federal sources.

Although there is little disagreement that the Everglades have been altered regrettably during the last century, there is considerable discussion about what a restored system should look like. What vegetation patterns existed before human influence? Were patterns constant or constantly changing? What hydropattern and nutrient levels would be conducive to restoring the landscape? Because half of the Everglades area has been lost to agricultural and urban uses, should the restoration goal be to reproduce the original Everglades in half the space? Alternatively, should restoration efforts return the Everglades that remain to what that particular area was historically, without attempting to restore short-hydroperiod wetlands that existed on the edges of the original Everglades? In addition, given that Florida Bay historically received water from an area twice the size of its current watershed, can sufficient water be sent to the bay through half of the original Everglades without adversely affecting upstream hydropattern restoration efforts? Intensive research and engineering efforts have been initiated by state, federal, and private interests to address these questions.

port pollutants such as heavy metals, pesticides, polychlorinated biphenyls, and sulfuric and nitric acids (the damaging components of acid rain)
to even the most remote environments (Czuczwa et al., 1984). For example,
fish in some subarctic and high-alpine lakes contain toxaphene at levels
above health standards (Kidd et al., 1995). Subsistence fisheries for aborigi-
nal people are threatened in many areas of the far North (Lockhart, 1995).
Pollutants affecting these remote areas are transported by air from the
United States and Eurasia (Barrie, 1986; Lockhart, 1995). The contaminants
concentrate in fatty tissues of organisms high on the food chain, in some
cases rendering top predators 100,000 to 10 million times more contaminated than the rainwater that delivers the contaminants (Schindler et
al., 1995).

• *Introduction of exotic species and loss of native species:* Exotic species
introduced to a water body either on purpose or accidentally can decimate
native species and significantly alter the aquatic food web. For example,
nonnative plants such as water hyacinth, hydrilla, and Eurasian watermil-
foil have spread to thousands of acres of U.S. lakes. The uncontrolled
growth of these plants, some of which were introduced by the aquarium
industry and others because they were regarded as visually attractive,
interferes with swimming, clogs canals and drainage outlets, and alters
the aquatic food web (National Research Council, 1992). Zebra mussels,
introduced to the Great Lakes in 1986 from ship ballast water, are threaten-
ing the survival of important commercial fish and native clams and mussels (Roberts, 1990). In the Boundary Waters Canoe Area wilderness in
northeastern Minnesota, walleyes and smallmouth bass at one time were
stocked in some lakes, where they have taken hold, sometimes to the
detriment of native species (Friends of the Boundary Waters Wilderness,
1992). Scientists have made similar observations in western mountain
lakes: 80 percent of alpine lakes in the western United States have been
stocked with nonnative species (Bahls, 1992). In western Canada, 20 percent of lakes in the mountain national parks have been stocked (Donald,
1987). Research has shown that these exotic species have altered natural
food webs dramatically (Lamontagne and Schindler, 1994; Leavitt et al.,
1994; Paul and Schindler, in press).

Because of these and other problems, eliminating point-source discharges of pollutants to water bodies, while necessary, is insufficient to
prevent further degradation of damaged aquatic ecosystems. Researchers
have attempted to estimate the net effect of reducing point-source discharges, with discouraging results. For example, EPA researchers simu-
lated point-source releases of contaminants along the 630,000 miles of
larger U.S. rivers and streams that receive 85 percent of total point-source
discharges in the United States and modeled the effects of decreases in
discharges of biological oxygen-demanding materials, total suspended

solids, and fecal coliform bacteria as mandated by the Clean Water Act (Kealy et al., 1993). The results were disturbing: just 2.2 percent of the rivers changed their use support status—meaning that nonswimmable waters became swimmable, nonfishable waters became fishable, and so on—as a result of additional point-source controls. These figures may be low because they do not reflect the substantial improvements in water quality that have occurred even where use of the water remains unchanged, nor do they indicate reductions in types of contaminants other than organic matter, fecal coliforms, and suspended solids. Nonetheless, they indicate that controlling point sources of contamination eliminates only a part of the water quality problem. In contrast, the researchers estimated that reducing the nonpoint-source load by half would achieve much more significant reductions—for example, increasing the percentage of swimmable waters from 33 percent to 47 percent. Other researchers have estimated that 99.9 percent of sediment and more than 80 percent of nitrogen and phosphorus enter waterways via nonpoint sources (Shaw and Raucher, 1993).

RESTORING INLAND WATERS: THE ROLE OF LIMNOLOGY

Limnologists have made substantial contributions toward understanding and partially correcting damage to freshwater ecosystems. Notable contributions by limnologists include the following (see Chapter 3 for more details):

• *Understanding the effects of excess nutrients and organic matter:* By the mid-twentieth century, limnologists were conducting research that eventually quantified how human discharges of excess nutrients (primarily nitrogen and phosphorus) and organic matter (such as that contained in sewage) cause water quality to deteriorate rapidly through growth of excess algae and loss of dissolved oxygen (Hasler, 1947; Sawyer, 1947; Vollenweider, 1968). This discovery eventually led to programs and laws to reduce nutrient and organic matter discharges, and as a result the quality of many important water bodies has improved significantly.

• *Identifying damage resulting from acid rain:* Limnologists have shown that acid rain—caused by fossil fuel combustion and metal smelting—can lead to complete loss of important species, such as trout, in affected water bodies (Schindler et al., 1985). Such discoveries have been important catalysts in national and global agreements to control acid rain.

• *Contributing to wetland restoration:* The world's wetlands have been subjected to extensive drainage and destruction, particularly for agriculture and forestry in developed countries (Kivinen, 1980). Limnologists are playing a key role in developing the science needed to restore and protect wetlands.

- *Interpreting past characteristics of damaged water bodies:* By examining the plant and animal remains in and the chemistry of aquatic sediments, limnologists have constructed models of damaged aquatic ecosystems in their pristine state (Bradbury and Megard, 1972; Gorham and Sanger, 1976; Davis and Berge, 1980; Gorham and Janssens, 1992). Through this process, limnologists can help design strategies to restore or partially restore the damaged ecosystems.

As the complexity of freshwater problems increases, the role of limnologists in addressing these problems will become more critical. Environmental engineers can design systems for reducing pollutant inputs to a water body; fisheries biologists can determine water quality changes needed to rescue a threatened species of fish; hydrologists can identify water flow patterns influencing the movement of contaminants. However, the full range of actions required to restore a water body can best be identified by interdisciplinary teams of scientists including limnologists with experience in integrating the many factors that influence aquatic ecosystems into a broad picture of the whole system. The limited gains achieved in water quality to date are a result of focusing too narrowly on reducing inputs to lakes, rivers, and wetlands from point sources at the exclusion of considering the many other factors that influence water quality. In order to ensure that academic institutions and other educational venues are up to the task of training the next generation of limnologists, changes will be needed in the infrastructure underlying limnology education and research, as described in this report.

REFERENCES

Adler, R. W., and J. J. Finkelstein. 1993. The economic value of clean water. Pp. 8-2–8-19 in Clean Water and the American Economy—Proceedings: Surface Water, Vol. 1. EPA 800-R-93-001a. Washington, D.C.: Environmental Protection Agency, Office of Water.

Bahls, P. 1992. The status of fish populations and management of high mountain lakes in the western United States. Northwest Sci. 66:183–193.

Barrie, L. A. 1986. Background pollution in the arctic air mass and its relevance to North American acid rain studies. Water Air Soil Pollut. 30:765–777.

Bradbury, J. P., and R. O. Megard. 1972. Stratigraphic record of pollution in Shagawa Lake, northeastern Minnesota. Geol. Soc. Am. Bull. 83:2639–2648.

Czuczwa, J. M., B. D. McVeety, and R. A. Hites. 1984. Polychlorinated dibenzo-*p*-dioxins and dibenzofurans in sediments from Siskiwit Lake, Isle Royale. Science 226:568–569.

Davis, R. B., and F. Berge. 1980. Atmospheric deposition in Norway during the last 300 years as recorded in SNSF lake sediments. II. Diatom stratigraphy and inferred pH. Pp. 270–271 in Ecological Impact of Acid Precipitation, D. Drablös and A. Tolken, eds. Oslo-Å: SNSF Project.

Davis, S. M., and J. C. Ogden, eds. 1994. Everglades: The Ecosystem and Its Restoration. Delray Beach, Fla.: St. Lucie Press.

Donald, D. B. 1987. Assessment of the outcome of eight decades of trout stocking in the mountain national parks, Canada. N. Am. J. Fish. Manage. 7:545–553.

Edmondson, W. T. 1994. What is limnology? Pp. 547–553 in Limnology Now: A Paradigm of Planetary Problems, R. Margalef, ed. New York: Elsevier.

Egerton, F. N. 1987. Pollution and Aquatic Life in Lake Erie: Early Scientific Studies. Environ. Rev. 11(3):189–205.

Environmental Protection Agency (EPA). 1994. National Water Quality Inventory: 1992 Report to Congress. EPA 841-R-94-001. Washington, D.C.: EPA, Office of Water.

Eos. 1971. U.S. and Canada Agree on Anti-Pollution Measures for Great Lakes. Eos 52(8):581–582.

Frederick, K. D. 1991. Water resources: Increasing demand and scarce supplies. Pp. 23–80 in America's Renewable Resources: Historical Trends and Current Challenges, K. D. Frederick and R. A. Sedjo, eds. Washington, D.C.: Resources for the Future.

Friends of the Boundary Waters Wilderness. 1992. Visitor Use in the Boundary Waters Canoe Area Wilderness. Minneapolis: Boundary Waters Wilderness Foundation.

Gorham, E., and J. A. Janssens. 1992. The paleorecord of geochemistry and hydrology in northern peatlands and its relation to global change. Suo 43:9–19.

Gorham, E., and J. E. Sanger. 1976. Fossilized pigments as stratigraphic indicators of cultural eutrophication in Shagawa Lake, northeastern Minnesota. Geol. Soc. Am. Bull. 87:1638–1642.

Great Lakes Water Quality Control Board. 1985. 1985 Report on Great Lakes Water Quality. Detroit: International Joint Commission, Great Lakes Regional Office.

Hasler, A. D. 1947. Eutrophication of lakes by domestic drainage. Ecology 28:383–395.

Interstate Commission on the Potomac River Basin. 1994. Signs could spur improvements, groups say. Potomac Basin Rep. 50(7):5.

Jumars, P.A. 1990. W(h)ither limnology? Limnol. and Oceanogr. 35(5):1216–1218.

Kalff, J. 1991. On the teaching and funding of limnology. Limnol. Oceanogr. 36(7):1499–1501.

Kealy, M. J., T. Bondelid, and B. Snyder. 1993. Clean water and recreational use support: Has the Clean Water Act made a difference? Pp. 3-2–3-18 in Clean Water and the American Economy—Proceedings: Surface Water, Vol. 1. EPA 800-R-93-001a. Washington, D.C.: Environmental Protection Agency, Office of Water.

Kidd, K. A., D. W. Schindler, D. C. G. Muir, W. L. Lockhart, and R. H. Hesslein. 1995. High toxaphene concentrations in fish from a subarctic lake. Science 269:240–242.

Kivinen, E. 1980. New statistics on the utilization of peatlands in different countries. Pp. 48–51 in Proceedings of the Sixth International Peat Congress, Duluth, Minn. Jyska, Finland: International Peat Society.

Lamontagne, S., and D. W. Schindler. 1994. Historical status of fish populations in Canadian Rocky Mountain lakes inferred from sub-fossil Chaoborus (Diptera: Chasboridae) mandibles. Can. J. Fish. Aquat. Sci. 51:1376–1383.

Leavitt, P. R., D. E. Schindler, A. J. Paul, A. K. Hardie, and D. W. Schindler. 1994. Fossil pigment records of phytoplankton in trout-stocked alpine lakes. Can. J. Fish. Aquat. Sci. 51(11):2411–2423.

Lewis, W. M. 1995. Limnology, as seen by limnologists. Water Resour. Update. (Winter):4–8.

Lewis, W. M., S. Chisholm, C. D'Elia, E. Fee, N. G. Hairston, J. Hobbie, G. E. Likens, S. Threlkeld, and R. Wetzel. 1995. Challenges for limnology in North America: An assessment of the discipline in the 1990s. Am. Soc. Limnol. Oceanogr. Bull. 4(2):1–20.

Lockhart, W. L. 1995. Implications of chemical contaminants from aquatic animals in the Canadian Arctic: Some review comments. Sci. Tot. Environ. 160/161:631–641.

McGinnis, M. V. 1994. The politics of restoring versus restocking salmon in the Columbia River. Restor. Ecol. 2(3):149–155.

Mitsch, W. J., and J. G. Gosselink. 1993. Wetlands, 2nd ed. New York: Van Nostrand Reinhold.

Naiman, R. J., J. M. Magnuson, D. M. McKnight, and J. A. Stanford, eds. 1995. The Freshwater Imperative: A Research Agenda. Washington, D.C.: Island Press.

National Research Council (NRC). 1992. Restoration of Aquatic Ecosystems. Washington, D.C.: National Academy Press.

National Research Council (NRC). 1995. Wetlands: Characteristics and Boundaries. Washington, D.C.: National Academy Press.

Paul, A. J., and D. W. Schindler. In press. Regulation of rotifers by predatory calanoid copepods (subgenus: *Hesperodiaptomus*) in lakes of the Rocky Mountains. Can. J. Fish. Aquat. Sci.

Roberts, L. 1990. Zebra mussel invasion threatens U.S. waters. Science 249:1370–1372.

Sawyer, C. N. 1947. Fertilization of lakes by agricultural and urban drainage. J. N. Engl. Water Works Assoc. 61:109–127.

Schindler, D. W., K. H. Mills, D. F. Malley, D. L. Findlay, J. A. Shearer, I. J. Davies, M. A. Turner, G. A. Linsey, and D. R. Cruikshank. 1985. Long-term ecosystem stress: The effects of years of experimental acidification on a small lake. Science 22:1395–1401.

Schindler, D. W., K. A. Kidd, D. C. G. Muir, and W. L. Lockhart. 1995. The effects of ecosystem characteristics on contaminant distribution in northern freshwater lakes. Sci. Tot. Environ. 160/161:1–17.

Shaw, W. D., and R. S. Raucher. 1993. Recreation and tourism benefits from water quality improvements: An economist's perspective. Pp. 3-19–3-33 in Clean Water and the American Economy—Proceedings: Surface Water, Vol. 1. EPA 800-R-93-001a. Washington, D.C.: Environmental Protection Agency, Office of Water.

Uman, M. F. 1994. Blue Plains: Saga of a treatment plant. EPA Journal 20(1–2):20–21.

Vollenweider, R. A. 1968. Scientific Fundamentals of the Eutrophication of Lakes and Flowing Waters, with Particular Reference to Nitrogen and Phosphorus as Factors in Eutrophication. Das/CSI/68.27. Paris: Organisation for Economic Cooperation and Development.

Wetzel, R. G. 1983. Limnology, 2nd ed. Philadelphia: Saunders College Publishing.

Wetzel, R.G. 1991. On the teaching of limnology: Need for a national initiative. Limnol. Oceanogr. 36(1):213–215.

2

Limnology, the Science of Inland Waters: Evolution and Current Status

The origins of limnology date back many centuries to a time when scientists were called natural philosophers and science was explored by a few, usually wealthy, individuals. For example, Gorham (1953) traced the development of wetland ecology and some of its fundamental premises to studies of British natural philosophers going back to the fifteenth and sixteenth centuries. Similarly, Hutchinson (1967) traced studies on lakes at least as far back as a fifteenth century study on the ponds of a European abbey. However, the evolution of limnology into a modern science depended on the development of concepts and tools in biology, chemistry, and physics that were not available until the late nineteenth century.

Although modern limnology encompasses the study of all inland waters, its development is particularly identified with the study of lakes. Much of the conceptual framework around which the science was built was derived from studies on lakes, and most of the early limnologists were lake scientists. A notable exception was Stephen Forbes, a principal architect of the framework for limnology who was trained as a fish biologist and spent most of his career working on rivers and streams in Illinois, a state with relatively few natural lakes. It is understandable why early limnologists focused on lakes. As systems with easily recognizable boundaries and long residence times for water and substances in it, they are more obvious subjects for systematic scientific analysis than are the open, flowing waters of streams and spatially less defined wetlands. Nonetheless, it must be noted that the historical treatment usually accorded to limnology is colored by the fact that it generally is written by lake scientists, even though scientific studies on flowing waters and wetlands in some cases predate studies on lakes.

Even today, many aquatic scientists in North America associate the word limnology with the study of lakes (and reservoirs). To the extent that there is any general awareness of the word limnology, this perception applies to the public as well. There are historical reasons for this situation, but it has caused difficulties in coalescing the various branches of limnology into a more coordinated and organized science.

This chapter traces the history of the study of lakes, reservoirs, rivers, and wetlands. It includes biographical sketches of some of the individuals (limnologists as well as other scientists) who have contributed to the understanding of inland aquatic ecosystems and significantly influenced the field of limnology. The chapter concludes with an analysis of the current status of limnology with special reference to professional and educational issues in the United States.

EARLY HISTORY

The beginnings of limnology as a modern science usually are traced to the work of a few late nineteenth century biologists who focused on lake studies. The founders of lacustrine limnology defined the scope and nature of the field in a way that survives remarkably intact to the present day; they viewed the subject broadly and integratively. Francois Forel (see Box 2-1) was the first scientist to use the term limnology in a publication. His three-volume treatise on Lake Geneva (bordered by Switzerland and France), published over the period 1892 to 1904, is considered the first book on limnology, and it was encyclopedic in scope. Its 14 chapters define the main supporting fields of modern lake limnology (Edmondson, 1994) and reinforce the idea that lake limnology is the application of all relevant basic sciences to the analysis of lakes as fundamental units of study.

The integrative nature of limnology was stressed even before Forel coined the term limnology. In a prescient article published in 1887, Stephen Forbes (Box 2-2) described lakes as "microcosms," or little worlds. Although the term "ecosystem" was not introduced for another half century (Tansley, 1935), Forbes defined an approach that presaged this concept. He proposed that lake studies should focus on many of the processes that today define the field of ecosystem ecology: mineral cycling, production and decomposition of organic matter, food web interactions and their impacts on the structure of biological communities, and the effects of physical conditions on biological communities. Forbes viewed these topics as essential to understanding lakes as functioning, integrated systems. The notion of lakes as microcosms (or integrated ecosystems) has pervaded their study ever since Forbes' time, even though the concept has

BOX 2-1   FRANCOIS ALPHONSE FOREL (1841–1912)

Francois A. Forel invented the word limnology. The science would have been called "limnography," to match its sister science oceanography, had it not been for the priority of reserving the term "limnograph" for a device used to measure water height in lakes.

Forel was born in Morzes on the shore of Lake Geneva (known to the Swiss as Lac Leman). When he was 13, his father introduced him to ". . . the art of observing and questioning nature," according to a monograph that he later wrote about Lake Geneva (Forel, 1882, 1895, 1904). After graduating from the Academie de Geneve, Forbes completed a medical degree at Würzburg and taught there for three years. In 1870, he joined the faculty of the Academie de Lausanne, where he taught anatomy and physiology and started a lifelong study of Lake Geneva.

One of Forel's earliest observations connected the physical, chemical, and biological properties of Lake Geneva. While looking to see if waves had left ripple marks on the bottom along the shore, he noticed a wriggling nematode in a sample of mud. This "poor worm" piqued his curiosity and led him to invent a bottom dredge with which he discovered that the depths of the lake were not a desert but were occupied by a specialized fauna rich in species and individuals. His studies resulted in a long series of influential papers on benthic fauna, their environmental conditions, and their significance to the fish population; the papers were consistent with modern concepts of ecosystem ecology. He followed with comparative work on other Swiss lakes.

In physical limnology, much of Forel's attention focused on water oscillations that create standing wave patterns, known to limnologists as seiches. In addition to obtaining massive data on Lake Geneva itself, he studied movements of water in small, tilted model lake basins, thus anticipating by many years a kind of experimental limnology. From these studies he developed generalizations about the relations between the dimensions of lakes and the periodicity of their seiches. He also conducted pioneering work in other aspects of physical limnology, devising a color scale and studying light penetration with photographic paper, and considered external influences on the lake, particularly the relation of the water supply to glaciers.

Henri LeBlanc (1912) listed Forel's publications in categories: limnology (126 titles), glaciology (66), seismology (12), meteorology (19), natural history (28), archaeology (12), history (10), and biographies (15). The massive Lake Geneva monograph Le Leman: Monographie Limnologique was published in three volumes in 1882, 1895, and 1904, with a total of 14 chapters. A small textbook appeared in 1901. He continued to work until a few months before his death in 1912, producing about 35 percent of his publications after the appearance of the last volume of the monograph in 1904.

BOX 2-2 STEPHEN ALFRED FORBES (1844–1930)

. . . [A] little world within itself—a microcosm within which all the elemental forces are at work and the play of life goes on in full but on so small a scale as to bring it easily within the mental grasp.

This is how Stephen Forbes described the ecological dynamics of a lake in his often-cited 1887 paper "The Lake as a Microcosm." In this early essay, Forbes described the now familiar concept of the interdependence of living organisms and environmental factors. Owing much to Forbes' influence, the field of limnology developed a strong ecological perspective by the close of the nineteenth century.

Born in 1844 in Silver Creek, Illinois, Forbes was raised on a farm with five siblings. At age 17, he joined the Union cavalry and served four years during the Civil War, including four months as a prisoner of war. After the war, he studied medicine, but within three years he turned to natural history. He attended Illinois State Normal University for a brief time but continued natural history studies on his own.

In 1872, Forbes became curator of the Museum of the State Natural History Society in Normal, Illinois; in 1877, he transformed this institution into the Illinois State Laboratory of Natural History. In 1884, he moved with the laboratory and museum to Urbana, became a professor at the University of Illinois, and completed a Ph.D. from Indiana University. He was chief of the Illinois State Natural History Survey until his death in 1930.

The limnological contributions of Forbes were diverse. He was among the first to study North American inland lakes (other than the Great Lakes). His studies of several lakes in the Rocky Mountains, published in 1893, represented for a number of years the sole biological information on lakes in the western United States. Forbes was an early and notable contributor to limnology of running waters as well; under his direction, the Illinois State Laboratory of Natural History established a floating laboratory on the Illinois River and conducted an extensive, half-century-long study of the river.

In "The Lake as a Microcosm," Forbes fostered the idea that the organisms and dynamics of a water body are isolated from and independent of the landscape. Although today this concept has been supplemented by current understandings about the influence of the catchment basin and airshed, Forbes' cogent view contributed a significant organizing model. The notion of the lake as a microcosm (that is, an isolated, simplified, and understandable system) provided impetus and encouragement for scientists to study lakes from an "ecosystem" standpoint (even though the term ecosystem was not introduced until 1935). The lake as microcosm persists as a vital concept in limnology today. It has inspired studies at all scales, from whole lake to plastic pool and small aquarium, as limnologists have endeavored to understand the components, functions, and interactions of aquatic systems.

BOX 2-3   EDWARD A. BIRGE (1851–1950), CHANCEY JUDAY (1871–1944), AND THE WISCONSIN SCHOOL OF LIMNOLOGY

Birge and Juday are usually included among the founders of limnology. Their research contributed substantially to the basic understanding of a broad range of physical, chemical, and biological characteristics of lakes. They also assisted the development of the field through their roles in initiating a strong educational program and communications networks linking professional limnologists. Several books provide details of their lives (Sellery, 1956; Frey, 1963; Beckel, 1987).

The contributions of Birge and Juday represent a microcosm of the interdisciplinary links that have been essential for progress in limnology. Both began their work with classic zoological studies on the taxonomy and distribution of a major component of lake planktonic communities, the cladocerans. They soon found, however, that little could be understood about the distribution of these animals in lakes without evaluating a range of physical and chemical properties. This led to investigations of water column thermal structure, distribution of dissolved gases, and light penetration, along with the mechanisms controlling these features. Several fundamental aspects of lakes that now comprise a basic component of most modern investigations derive from these efforts (Mortimer, 1956; Frey, 1963). The multidisciplinary effort needed to investigate lake properties led Birge and Juday to involve chemists, physicists, geologists, and other biologists in their research, and they interacted with other scientists in the developing field of limnology around the world. Initially, their work focused on individual lakes in southern Wisconsin. Later, they expanded their efforts at the Trout Lake Limnological Station in northern Wisconsin to compare and evaluate controlling features across a wide range of lake types. Their assessments of the interactions among physical, chemical, and biological processes in lakes helped to develop limnology as an ecosystem science. Substantial portions of the data Birge and Juday collected during their later years never were published, and these archived data remain a useful source of information for present-day limnologists.

Edward A. Birge obtained A.B. and A.M. degrees from Williams College in Massachusetts and a Ph.D. from Harvard. He began his career at the University of Wisconsin in 1875 and remained there for the rest of his life. During his career, he assumed a variety of administrative positions, including president of the university. He also directed the Wisconsin Geological and Natural History Survey, through which he fostered the collection of extensive limnological data. Despite his administrative responsibilities, he maintained his interest in aquatic research, continuing to work at the Trout Lake Station even at the age of 85. Robert Pennak, who completed his graduate work at Wisconsin and now is emeritus professor at the University of Colorado, relates a story of how Birge admonished him, after a Model A car they were using had been turned on its side by slippery road conditions,". . . dammit Pennak, put it back on its wheels, the survey must go on!" (Beckel, 1987).

Chancey Juday arrived at Wisconsin in 1900 as a biologist for the Geological and Natural History Survey. He received A.B. and A.M. degrees from Indiana

University, where he was introduced to aquatic studies during a summer research program, and he was hired to work with Birge at a time when administrative duties were limiting Birge's research efforts. Juday continued at the Geological Survey during his career and served on the faculty of the University of Wisconsin and as director of the Trout Lake Limnological Station. He supervised the graduate training of 13 Ph.D.s, several of whom have made substantial contributions to limnology. Juday was instrumental in establishing the American Society of Limnology and Oceanography and was its first president.

One of Juday's last Ph.D. students, Arthur D. Hasler, was hired by the University of Wisconsin to continue limnological activities. Hasler himself became a major figure in limnology, contributing substantially to the development of limnology as an experimental science (in contrast to its origins as an observational science). During his career at the University of Wisconsin (1940–1975), Hasler supervised the training of numerous M.S. and Ph.D. limnologists, including several who have attained international status in limnology and ecology.

been broadened and refined as twentieth century science has become more sophisticated (see the background paper "Organizing Paradigms for the Study of Inland Aquatic Ecosystems" at the end of this report). Today, limnological studies focus on lakes as "mirror images of the landscape around them" (A.D. Hasler, quoted in Beckel, 1987)—in other words, as open systems that receive inputs of water, solar energy, and chemical substances from terrestrial and atmospheric sources.

Limnology began to take its place as a recognized field for research and scholarly activities near the turn of the century. The first limnological research institute in Germany was founded at Plön in 1891; it still is one of the major centers for limnological research (Overbeck, 1989). Edward Birge and his colleague Chancey Juday (see Box 2-3), usually regarded as the founders of academic limnology in North America, began their limnological studies at about the same time. Both spent their careers at the University of Wisconsin in Madison, and they began a rich limnological tradition that continues at that university to the present (Mortimer, 1956; Frey, 1963; Beckel, 1987; Kitchell, 1992). Birge was a zoologist and was attracted to lake studies during his student days in the 1870s in the context of the life cycles of microscopic animals (zooplankton). Juday also was trained as a biologist and was hired by Birge in 1897 to help conduct lake surveys. Birge and Juday soon branched into the physics and chemistry of lakes as they realized that the dynamics of plankton could not be understood without knowledge of these subjects. Their studies on temperature stratification and dissolved gases provided limnologists with information needed to understand virtually all biological cycles in lakes. Birge and Juday sought collaboration with physicists and chemists to study lake phenomena beyond their own field of expertise. Together, these scientists developed many new techniques to measure physical properties and processes and many chemical characteristics of lakes.

REGIONAL AND DESCRIPTIVE ERA

Limnology continued to develop as a field of study and expand its geographic base during the first half of the twentieth century. Limnologists of the 1920s and 1930s founded many field stations, used them to collect a wealth of information on individual lakes, and synthesized this information at the regional scale. As practiced during these decades, limnology was essentially an observational science: knowledge gained was largely from sample collection and analysis of the resulting data rather than from

Edward Birge and Chancey Juday with plankton trap on Lake Mendota in Madison, Wisconsin, circa 1917. SOURCE: State Historical Society of Wisconsin, Visual and Sound Archives.

controlled experiments. This regional/descriptive approach reflected the pervading notion of lakes as microcosms in that studies on individual lakes usually were multidisciplinary: physical, chemical, and biological measurements were included in most studies, reflecting at least implicitly the idea that lakes are complex organized systems. Efforts at the regional scale during this period also focused on classifying lakes into major types based on a multidimensional set of descriptors. For example, the scheme that classifies lakes according to trophic state (meaning general nutritional status) was developed by August Thienemann and Einar Naumann (see Box 2-4) in the 1920s. According to this scheme, an array of indicators—including a physical measure (transparency), chemical concentrations (of nutrients), and biological characteristics (species types and abundance and primary production)—was used to classify lakes according to their overall nutritional status and productivity. These and other classification efforts provided an impetus for integration and synthesis, leading to generalizations about lakes as ecosystems.

In 1922, the international limnology society, Societas Internationalis Limnologiae (SIL), known in English as the International Association for Theoretical and Applied Limnology, was founded in Germany under the aegis of Thienemann and Naumann. Limnologists in the United States were organized as the Committee on Aquaculture in 1925 and as the Limnological Society of America in 1936. From a starting base of 221 members in 1936, the American society grew to include 4,000 scientists today. It joined with oceanographers to become the American Society of Limnology and Oceanography in 1948; its journal, *Limnology and Oceanography*, one of the premier research periodicals on lake limnology in the world, was launched in 1955.

MIDCENTURY EXPANSION

Most major universities in North America and Europe had hired limnology professors by the middle of the twentieth century. In almost all cases, these faculty were in departments of biological science (including zoology and botany as well as biology), and the field developed a distinct biological focus. With few exceptions, limnology programs in universities were staffed by one faculty member, and the success of the program rose or fell with the intellectual ability and initiative of that individual. In contrast, natural sciences that are related more directly to resource utilization and economic production (such as forestry, soil science, and fisheries and wildlife) typically developed academic programs with larger and more diverse faculties. Thus, their long-term success was less dependent on that of a single individual.

G. Evelyn Hutchinson, who spent most of his career at Yale University, was a dominant figure in North American limnology during the middle

**BOX 2-4   AUGUST THIENEMANN (1882–1960) AND
EINAR NAUMANN (1891–1934)**

August Thienemann dominated the development of comparative limnology in Europe for much of the first half of this century. With strong zoological interests, Thienemann conducted detailed analyses of numerous lakes of different geomorphological, chemical, and biotic characteristics. By induction from these analyses, he synthesized common functional relationships in lake typology that were essential to the young discipline and stimulated extensive further studies throughout the world.

Born in 1882 in Thüringen, Germany, Thienemann began his studies in 1901, primarily in botany and later in zoology and philosophy, at the Universities of Greifswald, Innsbruck, and Heidelberg. He initiated extensive research programs while holding positions in zoology at the Universities of Greifswald and Münster. Between 1910 and 1914, he conducted studies on the volcanic Eifel Maar lakes, which provided the basis for his organization of lakes in terms of bottom-dwelling invertebrate communities and their relationships to chemical conditions, in particular the oxygen content, of bottom waters of lakes.

In 1917, Thienemann, then associate professor of hydrobiology at the University of Kiel, became director of the Hydrobiologische Anstalt der Kaiser-Wilhelm-Gesellschaft in Plön, which up to that time had been operated as a private biological station since its founding by another pioneering limnologist, Otto Zacharias, in 1891. Under Thienemann's leadership, the hydrobiological station became one of the foremost limnological research and advanced educational institutions of Europe. That foundation of limnological excellence has continued to the present as the Max-Planck-Institut für Limnologie—among the leading experimental limnological research facilities of the world.

Thienemann conducted pioneering studies in many places and on many topics. For example, he led limnological expeditions to remote tropical areas such as Java. He developed ecosystem concepts in the 1920s that influenced subsequent conceptual developments by Hutchinson and Lindeman (see Boxes 2-5 and 2-6). A tireless student of limnology, he authored nearly 500 publications and 25 books.

Thienemann collaborated in the early 1920s on lake typology and regional limnology with the Swedish limnologist Einar Naumann, who was an assistant professor of botany and later the first professor of limnology at the University of Lund, Sweden. Naumann's research on phytoplankton distribution and sediment formation in relation to nutrient conditions in lakes complemented the zoological interests of Thienemann. Despite highly disparate viewpoints, the two men developed a general system of classifying lakes that persists to this day. In 1921, Naumann and Thienemann founded the International Association of Theoretical and Applied Limnology (Societas Internationalis Limnologiae), drafted its statutes, and organized the first international congress of limnology in 1922. Their leadership guided this organization during its early development; it subsequently has evolved into a global association that provides its more than 3,000 members in 80 countries opportunities to exchange limnological information.

third of the century and a leader in the development of ecology in general (see Box 2-5). A man of wide-ranging interests and enormous intellect and insight, he brought a theoretical approach to aquatic ecology to complement its empirical underpinnings (Lewis et al., 1995). He attracted outstanding students to his program, many of whom developed prominent academic programs and had influential careers of their own. Some of Hutchinson's students eventually developed entirely new subdisciplines within ecology (see Box 2-6).

BOX 2-5   G. EVELYN HUTCHINSON (1903–1991)

In 1979, G. Evelyn Hutchinson joined the eminent select, such as Einstein, Edison, and Max Planck, by being awarded the Franklin Medal "for developing the scientific basis of ecology." The most voluminous scientific contributions of Hutchinson were in the biogeochemistry of lake ecosystems and included a monumental treatise on limnology (in four volumes) that demonstrated his remarkable abilities to interpret and synthesize disparate information into meaningful concepts. These scientific foundations in biogeochemistry (and population dynamics) led to major contributions in evolutionary ecology. His development of the ecological concept of multidimensional niches is a most fundamental scientific contribution. Several of his former students have led the subsequent development of ecology as a discipline. He was generous in sharing his conceptual advances with colleagues, such as in his work with Raymond Lindeman on trophic food web relationships (see Box 2-6).

Hutchinson's propensity for natural history was nurtured in Cambridge, England, in a stimulating intellectual environment. After undergraduate studies at Cambridge University and brief research positions at the Stazione Zoologica in Naples and the University of Witwatersrand in South Africa, Hutchinson accepted an instructorship in zoology at Yale University in 1928. He spent the remainder of his career at Yale, continuing years of high productivity after his official retirement in 1971. In addition to teaching in natural history, ecology, limnology, and biogeochemistry, he developed a research program of enormous breadth. He made seminal contributions to knowledge of processes in lake bottom waters and sediments, oxygen deficits, benthic invertebrates, paleolimnology, and biogeochemical cycling, especially of phosphorus. He was a pioneer in the development of innovative experimental techniques, using radioisotopes of phosphorus in lakes as early as the 1940s and bioassays of nutrient effects on phytoplankton population dynamics as early as 1941.

Hutchinson had penetrating understanding of many fields of science. He contributed significantly to geochemistry, oceanography, anthropology, paleontology, sociology, and behavioral sciences, as well as to his primary research areas in biogeochemistry and limnology. He received numerous national and international awards in science, and as the foremost ecologist and limnologist of the twentieth century, he left a substantial legacy in his scientific writings and the students he trained.

BOX 2-6   RAYMOND L. LINDEMAN (1915–1942) AND
H. T. ODUM (1924–): EXAMPLES OF THE
INTELLECTUAL LEGACY OF G. E. HUTCHINSON

Raymond Lindeman was a young aquatic ecologist who developed an important concept for synthesizing ecological principles based on energy flow through food chains. His trophic-dynamic concept, published posthumously in 1942, emphasized the importance of short-term nutritional functioning to an understanding of long-term changes in the dynamics of lake communities. Drawing from conceptual works of the plant ecologist Tansley and the limnologists Thienemann and Hutchinson, Lindeman showed how organic and inorganic cycles of nutrients are integrated. His theoretical model of nutrient cycling, expressed in terms of energy flow, allowed evaluations of biological and ecological efficiencies of energy transfer over long periods.

Lindeman did his graduate studies in zoology at the University of Minnesota under Samuel Eddy and W. S. Cooper. His doctoral research involved a detailed evaluation of trophic (food web) structure in Cedar Bog Lake and provided support for the general tenets of trophic-dynamic concepts.

In 1941, Lindeman began postdoctoral studies at Yale University with G. E. Hutchinson. Many of Lindeman's trophic-dynamic ideas were melded into conceptual and mathematical treatments from Hutchinson's then-unpublished writings. The combined efforts of these two scientists led to many major conceptual breakthroughs. Hutchinson also assisted with the publication of Lindeman's synthesis paper (Lindeman, 1942), which was rejected at first because of its theoretical nature. Trophic dynamics and ecosystem concepts are so embedded in modern ecology that it is difficult to comprehend how revolutionary his theoretical model was at the time. The paper provided much of the intellectual framework on which subsequent development of ecosystem ecology was based. Lindeman died prematurely in 1942 at age 27.

Howard Thomas Odum, known as H. T. or Tom, is a major figure of modern aquatic ecology whose influence extends beyond the confines of traditional ecology. His innovations spurred the development of several new disciplines—in particular, systems ecology, ecological economics, and ecological engineering—that relate ecology to other sciences in analyzing major environmental problems.

Odum was born in Durham, North Carolina, the son of Howard W. Odum, a renowned sociologist at the University of North Carolina. He received an A.B. in zoology in 1947 from that institution and a Ph.D. in 1951 under G. E. Hutchinson at Yale University. His career was spent at several academic institutions, including the University of Florida, where he is now professor emeritus. His work on the energetics of Silver Springs, Florida (Odum, 1957), is a landmark whose impact on flowing water ecosystems is analogous to the impacts of Lindeman's trophic-dynamic work on lakes. He advanced experimental ecology through work on mesocosms and by refining the diurnal oxygen method for measuring primary production. He directed several large-scale experiments in a tropical rain forest in Puerto Rico (Odum and Pigeon, 1970) that were classic examples in forest ecology and the assessment of how radionuclides affect ecological processes. During the past 25 years, his work in aquatic ecology has focused on experimental wetland ecology, including the use of wetlands as natural treatment systems for wastewaters.

Odum has authored or coauthored several books that have significantly influenced ecology and related fields. He contributed to the classic ecology text written by his brother, Eugene Odum, also a major figure in ecology in the second half of this century. H. T. Odum's first book, *Environment Power and Society* (1970), presented a computer language and modeling technique to describe energy flow through ecosystems. The language and modeling approach became the tool of a group of followers who modeled energy flows associated with the movement of commodities in both natural ecosystems and human-dominated systems. This work led to the concept of "embodied energy," since termed "emergy," which accounts for the direct and indirect energy flows (those from "free environmental services" and those supplied by the economy) required to produce a substance. In turn, this led to efforts to conduct economic analyses in terms of energy units.

Odum coined the term ecological engineering in 1962, and he has contributed much to its development as a field distinct from but related to environmental engineering (Mitsch, 1994). He continues to promote the development of university curricula to produce ecological engineers (Odum, 1994). Odum has received many awards, including the Mercer Award of the Ecological Society of America; the AIBS (American Institute of Biological Science) Distinguished Service Award; the Prize of the Institut de la Vie, Paris; and the Crafoord Prize of the Swedish Academy of Sciences. The last two prizes were shared by the two distinguished brothers, H. T. and Eugene.

Experimental Limnology

Experimental lake limnology has involved at least three types of manipulations: (1) stress-response experiments, in which a lake (or a basin in a lake) is treated with a chemical or biological stressor (such as excess nutrients, acid, or a top predator) and the responses of the lake system are studied; (2) hydrologic, physical, chemical, and/or biological manipulations aimed at lake remediation or rehabilitation; and (3) tracer additions to measure rates of physical processes, such as use of radiotracers to follow water movement and noble gases to monitor air-water-gas exchange. Most lake manipulations of the first type employ the modern, expanded concept of the lake as a microcosm in that their aim is to apply a stress to an ecosystem and observe the changes that it causes in various properties of the ecosystem. (Preferably, these properties are measured for a given number of years before the stress is applied, and the experiment continues for several years and includes a recovery phase following the removal of the stress.) Limnologists conducting these experiments typically have studied a wide range of responses—from changes in chemical concentra-

Raymond Lindeman, year and photographer unknown.
SOURCE: Eville Gorham, University of Minnesota.

tions to changes in individual organisms, populations, communities, and ecosystem-level processes (Schindler et al., 1992; Brezonik et al., 1993). The idea that whole lakes can serve as subjects for experimental manipulation developed slowly, beginning in the late 1930s and 1940s. Juday was the first to conduct an experiment on a whole lake. During the mid-1930s, he added various fertilizers to a small pond in northern Wisconsin to study their effects on plankton production and fish populations (Juday and Schloemer, 1938). Einsele (1941) performed a similar experiment on a small lake in northern Germany a few years later. Neither of these manipulations had much immediate impact on the development of experimental limnology, perhaps because of the disruptive influence of World War II on natural science. Whole-lake experiments by Arthur Hasler and his group at the University of Wisconsin in the 1950s and 1960s were more influential in establishing the usefulness of this approach (see the background paper "Organizing Paradigms for the Study of Inland Aquatic Ecosystems" at the end of this report).

Experimental limnology did not play a prominent role in lake science until the late 1960s and 1970s, probably because of the lack of funding to support such complicated and expensive initiatives. Widespread concern about excessive nutrient enrichment (eutrophication) of lakes led to government research programs in industrialized countries during the 1960s, and these programs facilitated wider use of experimental approaches in lake limnology. Even so, large-scale experiments (such as whole-lake manipulations) have been relatively few in number because of their comparatively high cost, the long time (at least several years) required to complete them, and the limited availability of lakes that can be dedicated to such purposes. Consequently, experimental approaches at smaller scales using enclosures of one to a few meters in diameter—often called mesocosms, limnocorrals, limnoenclosures, or limnotubes—that are installed in the lake have become popular in Europe and North America. This intermediate scale has enabled limnologists to complete a great variety of experiments, under conditions that can be controlled and replicated, on systems more similar in complexity to whole lakes than one can achieve in laboratory-scale systems. Nonetheless, mesocosms cannot duplicate the complicated ecosystems of whole lakes and are especially inadequate to study populations of large fish over long periods.

Because manipulations of whole aquatic ecosystems generally cannot be duplicated, limnologists have focused considerable effort over the past decade on developing sophisticated statistical methods and other techniques to evaluate data from such unreplicated experiments (Carpenter et al., 1989; Rasmussen et al., 1993). Of special importance is the gathering of adequate baseline data prior to manipulation. Paleolimnological techniques (described later in this chapter) also can help to carry such baseline information backward in time.

Despite the difficulties involved in conducting and interpreting whole-lake experiments, a strong consensus has developed among limnologists that observing responses to manipulations made at the whole-system level is a highly useful technique. Whole-lake experiments conducted over the past 30 years have been important both in advancing the understanding of fundamental limnological processes and in providing critical evidence for the management or solution of major pollution issues such as eutrophication and acidification. Their strengths for both purposes lie in their ability to test hypotheses and to provide a "platform" for related laboratory or field experiments at a range of scales.

Paleolimnology

During the middle of this century, the field of paleolimnology (see Box 2-7) developed into one of the key subdisciplines of limnology. Paleolimnology, closely related to paleoecology and paleoclimatology, has its origin in early nineteenth century botanical and chemical studies on peat cores and late nineteenth century geological studies on lithified sediments of ancient lake beds. By the early 1920s, limnologists had begun to collect sediment cores from lakes and to interpret stratigraphic data on plant and animal fossils as a record of the lake's history. Nipkow (1920) was

A whole-lake experiment to investigate the causes of eutrophication. SOURCE: David Schindler, University of Alberta.

the first to observe and explain the existence of thinly laminated sediments. He showed that laminae result, in some lakes, from an annual depositional cycle in which photosynthesis during warm months causes precipitation of calcium carbonate, which settles and forms a thin light layer on the sediments. Deposition of organic matter during the rest of the year forms a dark layer on top of the calcium carbonate; other mechanisms also

BOX 2-7   PALEOLIMNOLOGY: THE LIMNOLOGIST'S ARCHAEOLOGY

Lake and wetland sediments contain detailed archaeological records of how natural events and human activities have affected the overlying ecosystems— records revealed through "paleolimnology." Paleolimnology is in many ways analogous to the reconstruction of past civilizations by examining their stratified remains, a well-known technique in archaeology. Paleolimnological data are important in providing baselines against which to assess the damage done by activities such as land clearance, drainage, water pollution, and air pollution. They are also useful in assessing the rate of recovery from such damage when the cause has been mitigated or brought to an end.

Lake sediments and wetland peat deposits form a selective trap for a variety of plant and animal remains and elements, such as carbon, phosphorus, sulfur, iron, and manganese that are stored at varying concentrations depending on the activities that were occurring at the time the sediment layer was formed. Changes in the sedimentary profile inevitably record a good deal of lake and peatland history as well as the history of the surrounding catchment.

Following are three examples of paleolimnological studies:

1. At Shagawa Lake in northeastern Minnesota, paleolimnology has been used to trace the history of human development along the lakeshore (Bradbury and Megard, 1972; Bradbury and Waddington, 1973; Gorham and Sanger, 1976). Analyses of a 1-meter sediment core taken in 6.5 meters of water reveal striking changes in the flora, fauna, and chemistry of the lake following human settlement in its catchment, which began in the late 1880s. Settlement is clearly marked in the sediments by rapidly increasing concentrations of hematite grains from the dewatering of iron mines, which led to the founding of the town of Ely on the southern shore of the lake. The town's population grew to about 6,000 people in 1930, after which it decreased somewhat as mining declined; decreases in hematite grains in the sediments follow the decrease in mining activity. Settlement is also marked by a rise in the concentration of ragweed pollen that is typical of land clearance and the replacement of forests by agricultural fields. Also reflecting human activities in the catchment are marked shifts in the depth profiles of different types of siliceous diatom shells (frustules) in the sediments. The cause of these changes was severe nutrient enrichment due to discharge of the town's sewage directly into the lake. Most notably, *Stephanodiscus minutus*, *Fragilaria crotenensis*, and *F. capucina*, all characteristic of lakes in more fertile regions, increased greatly in abundance after settlement, while *Cyclotella comta* and several other species characteristic of the absence of pollution and lower nutrient concentrations declined and disappeared.

2. Paleolimnologists have used the Red Lake Peatland in northwestern Minnesota to develop tools for documenting the occurrence of acid rain and global warming (Gorham and Janssens, 1992; Janssens et al., 1992). The peat deposit contains a diverse array of fossil mosses whose environmental tolerances can be inferred from those of modern mosses at other sites in order to reconstruct pH and water table changes in the bog over time. Figure 2-1 shows profiles for two cores from a particular bog "island" in the Red Lake Peatland,

one (RLP8112) from its center and the other (RLP8104) from its periphery. The site at the center of the island (RLP8112) began to accumulate peat with a pH close to 7 about 3,300 years ago. About 1,100 years later the fen began to be invaded by *Sphagnum* mosses, which in less than three centuries transformed it into an acid bog with a pH close to 4. This change was accompanied by a lowering of the water table relative to the peat surface. The site at the margin of the present bog (RLP8104) exhibits very different depth profiles: drier, acid bog conditions alternated with wetter, near-neutral conditions, presumably as a result of island expansion and contraction in response to differing degrees of ground water upwelling. The utility of such depth-time profiles as baselines for assessing the effects of human disturbance lies in the boundaries they set. For example, if acid deposition were to lower the pH of the bog surface substantially, we should expect to see the development of moss assemblages characteristic of much more acid conditions and distinctly different from any of the assemblages observed in these peat cores over the past three millennia. Similarly, if global warming were to lower the bog water table significantly, we should expect to see moss assemblages characteristic of drier conditions and quite different from those observed at any time since peat began to accumulate.

3. By analyzing the pH preferences of individual species of diatoms and chrysophytes, paleolimnologists can calculate the past pH of a lake from the composition of diatom and/or chrysophyte remains at various dated strata in the lake's sediments (e.g., Smol et al., 1984a,b; Charles et al., 1989; Cumming and Smol, 1993; Cumming et al., 1994). The pH preferences for these species are evaluated by examining their remains in surface sediments from numerous lakes spanning a broad range of pH, and transfer functions are derived from modern data by multivariate statistical methods for extrapolating the data to past periods. Using such techniques, Cumming et al. (1994) showed that of 20 acid-sensitive lakes examined in Adirondack Park, approximately 80 percent have acidified since preindustrial times. This information refutes the contention of some that lakes in the Adirondacks are naturally acidic. Lakes that became acidic around 1900 generally were smaller, higher-elevation lakes with lower preindustrial pH values than the lakes that did not acidify or acidified more recently. Post-1970 pH trends in the lakes have been small and variable, suggesting that the lakes have been unresponsive to post-1970 declines in sulfate deposition.

can produce annual laminae in some lakes. These annual laminae allow limnologists to count back in time and to date individual strata of a sediment core. By studying plant and animal microfossil remains (such as pollen, diatom shells, and remains of zooplankton bodies) in laminae and by knowing the environmental tolerances of modern assemblages of the organisms being fossilized, paleolimnologists can reconstruct historical conditions in a lake and/or its drainage basin (see Box 2-7).

Relatively few lakes deposit clearly laminated sediments, however,

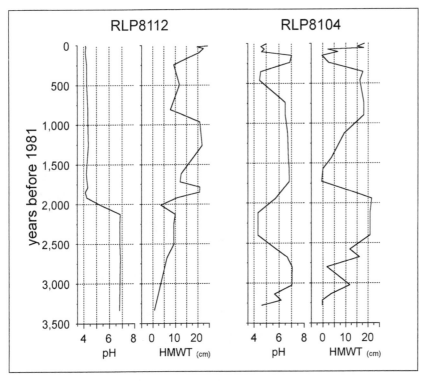

FIGURE 2-1   Depth-time profiles from the Red Lake Peatland. The profiles show the changes in pH and the height of the peat surface above the mean water table (HMWT) over the past 3,300 years in two peat cores, designated RLP8112 and RLP8104. SOURCE: Reprinted, with permission, from Gorham and Janssens (1992). ©1992 by *Suo.*

and consequently the application of paleolimnology was limited until radioisotope dating methods were developed in the 1950s and 1960s. The tools of paleolimnology include "radiochronometers," which use radioisotopes such as carbon-14 and lead-210 to date the time a sediment stratum was deposited; pollen, which indicates what types of terrestrial vegetation were present; various plant and animal fossil remains, including cell fragments and molecules such as plant pigments, which provide further clues about vegetation and aquatic life; and organic pollutants and trace elements, whose biogeochemical cycles have been influenced by human activity. Over the past 30 years, analyses of the layers in long sediment cores from lakes and wetlands have provided information about regional variations in past climatic conditions and watershed vegetation patterns, from which paleoecologists have sought to answer questions about the causes of environmental change. Paleolimnological studies on more recently deposited lake sediments have provided evidence for the timing and causes of lake pollution, including information about the

Sediment core being taken through a frozen lake for a paleolimnological analysis. SOURCE: Thomas M. Frost, University of Wisconsin, Trout Lake Station.

effects of excess nutrient inputs to lakes and about atmospheric transport of various pollutants.

RECENT HISTORY

Since the early 1960s, limnology in North America has been characterized by four related trends:

1. increasing emphasis on research related to the effects of pollution on aquatic resources and on ways to restore and manage these resources;
2. increasing diversity in the disciplinary backgrounds of limnologists;
3. increasing variety in the types of lakes studied; and
4. increasing focus on other types of inland aquatic ecosystems (streams, wetlands, and reservoirs), broadening the field of limnology from its traditional focus on natural lakes.

Research Driven by Pollution Concerns

Public concern about declining water quality and impaired ecological conditions in many aquatic resources caused by various human activities has resulted in greatly increased public expenditures for limnological studies in North America since the 1960s. Much of this funding has been associated with two major pollution problems: eutrophication and acid deposition, both of which generated large government-sponsored research programs for approximately 15-year periods (early 1960s to mid-1970s for eutrophication; late 1970s to early 1990s for acid deposition). A substantial portion of this research support was supplied by mission-oriented agencies such as the Environmental Protection Agency rather than by basic science agencies such as the National Science Foundation.

These large programs, although criticized by some scientists as inefficient (Roberts, 1987), resulted in significant practical advances. They defined the nature and extent of the problems, quantified sources of pollution, developed relationships for determining the responses of water bodies to various levels of pollution, and identified a variety of control and restoration measures. Beyond these practical results, the research supported or otherwise stimulated by these initiatives produced many conceptual advances and much new fundamental information about limnological processes, as well as advances or refinements in field and laboratory techniques. For example, eutrophication-related research led to improved understanding of aquatic food web interactions and to conceptual advances regarding the factors that control material and energy flows through aquatic food chains and webs. Similarly, research stimulated by concerns about lake acidification led to greatly improved understanding of the chemical and microbial processes affecting alkalinity and acid-base balances in dilute lake waters, new information about the biogeochemical cycling of sulfur in such systems, and advances in the understanding of mineral weathering rates. Acidification research also helped foster the development of techniques to understand the natural variability of ecosystems as a benchmark against which to measure the effects of human-caused stress. Both problems led to substantial advances in the ability to describe lake ecosystem processes mathematically and to develop predictive models of these systems. Finally, both led to improved understanding of the role of sediment-water interactions in the cycling of elements in lakes and in the control of water column conditions.

Diversification of Disciplines Participating in Limnological Research

The focus of limnology on chemical pollutants over the past quarter century helped to attract greater numbers of scientists from disciplines other than biology into the field. For example, environmental aquatic chemistry, a field that has developed its own identity primarily since the early 1960s, has close ties with limnology; many water chemists work essentially as chemical limnologists and focus their expertise on developing a more complete understanding of the behavior of natural chemicals, such as nutrients and acids, and synthetic chemicals, such as pesticides and various chlorinated compounds, in inland aquatic ecosystems. Similarly, other physical scientists and environmental and hydraulic engineers brought their analytical skills and mathematical modeling techniques into limnology as a result of the opportunities provided by government research programs on eutrophication and acid rain.

In most academic situations, individuals from these other disciplines have maintained their disciplinary identity and are not directly associated with traditional limnology programs in departments of biological science. Scientists who carry out studies related to limnology today operate from departments as diverse as civil engineering, fisheries and wildlife, botany, zoology, ecology, environmental science, forestry, geology, and geography. For example, in a survey conducted for this report, 42 of 69 universities housed professors who teach limnology-related courses in biology departments and 32 housed them in civil and environmental engineering departments, but these schools also listed 21 other types of departments in which limnologists and related aquatic scientists teach and do research (see Table 2-1). Some scientists in civil engineering, fisheries, environmental science, and other departments listed in Table 2-1 call themselves limnologists, whereas others study components of aquatic ecosystems but do not necessarily identify with the field of limnology.

Although the interdisciplinary nature of limnology is one of its important assets, it has created difficulties in organizing and sustaining the infrastructure needed to conduct the science. Most universities have no single "home" for limnologists, and limnological studies are scattered among several departments at major research universities. Typically, the individual programs are small and focus on only one or a few aspects of limnology. Levels of departmental support vary depending on the perceived importance of those aspects to the parent discipline of the department.

TABLE 2-1 Departments Housing Faculty Working in Limnology and Related Aquatic Sciences

Department	Number of Schools
Agriculture/soil science	6
Biology	42
Botany	12
Chemistry	3
Civil/environmental engineering	32
Ecology	6
Entomology	9
Environmental science/studies	10
Fisheries	17
Forestry	10
Geography	3
Geology/geophysics/geoscience	18
Hydrology	4
Landscape architecture	1
Life science	2
Marine science and limnology	1
Natural resources	9
Oceanography	3
Public health	2
Rangeland management	3
Urban and regional planning	1
Zoology	15
Zoology and limnology	1

SOURCE: This table was compiled from surveys of 69 universities belonging to the Universities Council on Water Resources and/or housing a U.S. Geological Survey Water Resources Research Institute (see Appendix A).

Limnology Beyond Small Temperate Lakes

The third major limnological trend of the past quarter century is a diversification of the types of aquatic systems to which limnologists have directed their attention. Most lake limnologists through the first part of the century focused on relatively small lakes in temperate latitudes—reflecting the latitudes of the industrialized countries where limnology developed. Limnologists have always liked to travel and study lakes in exotic places. As long ago as 1910, Juday traveled to Central America and sampled lakes in Guatemala and El Salvador (Juday, 1915). Thienemann conducted limnological studies in Indonesia in 1928 (Thienemann, 1931), and European limnologists have had long-term research programs in various tropical regions for many decades. Nevertheless, the number and intensity of such tropical studies were quite limited until the development of high-speed, low-cost air travel made the world seem smaller and more interconnected. North American limnologists now have significant research programs in tropical latitudes (such as the Rift Valley lakes in equatorial Africa), the high Arctic, and even in Antarctica.

In recent decades, lake research also has expanded to include the large, multipurpose reservoirs that have been constructed in regions of the United States (such as the mid-Atlantic states and arid West) where natural lakes are rare. The differences between reservoirs and natural lakes generally are a matter of degree rather than kind, and the same organizing principles and research techniques apply to both. Nonetheless, reservoirs often present special management problems. For example, the relationships between nutrient concentrations and chlorophyll levels are significantly different in lakes and reservoirs because the latter systems generally have much higher levels of inorganic turbidity (which limits light penetration and primary production) than do natural lakes. Consequently, nutrient loading criteria developed for lakes (to prevent or limit eutrophication problems) may not be directly applicable to reservoirs.

Lake research in North America during the past few decades has also expanded to the Great Lakes. In spite of their vast economic and ecological importance, limnological studies on these lakes were extremely limited before about 1960, but concern about pollution and declining water quality in the lakes stimulated Canadian and U.S. monitoring and research programs. Several government agencies and universities in both countries have significant research programs on the limnology of the Great Lakes. Relative to their size and importance, these lakes are still understudied, at least in part because of the high costs and large research infrastructure required for studying them. In some ways, Great Lakes limnological research has more in common with oceanographic research, and it is no accident that the major U.S. funding agency for university research on the Great Lakes, the Sea Grant program, is primarily a marine research program.

Stream Limnology

Over the past two to three decades, stream science has developed the integrative approach characteristic of limnological studies on lakes. Until relatively recently, the physical, chemical, and biological components of stream limnology were independent subdisciplines that were associated more with parent basic science disciplines than with a discipline called stream limnology (see Box 2-8). For example, physical studies on water flow in streams were associated with the fields of hydrology and hydraulics, typically taught in engineering and geoscience departments. Similarly, the origin and development of stream channels and drainage networks is a topic in the field of geomorphology, which is a subdiscipline of geography and geology. Until recently, chemical studies on flowing

BOX 2-8 RUTH PATRICK (1907–)

Ruth Patrick was a pioneer in predicting ecosystem risk well before these words were commonly used. She began her innovative studies of stream pollution in 1948 as curator of limnology at the Academy of Natural Sciences in Philadelphia. At that time, most pollution assessments were carried out by sanitary engineers (now called environmental engineers). The prevailing view was that if river water met certain chemical and physical conditions, particularly related to dissolved oxygen and pH, there was little need to examine the biota. Patrick was one of the first to consider how pollutants might affect the organisms that inhabit streams.

Patrick spent much time persuading water pollution professionals in engineering that biological information was useful in determining ecosystem conditions. Most biologists were not especially interested in environmental pollution in those days—a fact now hard to comprehend—and the assessment and prediction of ecological risk went by default to those involved with water supply and wastewater treatment (predominantly sanitary engineers). In a pioneering survey of the Conestoga River basin in Pennsylvania, Patrick demonstrated that aquatic community structure was changed dramatically not only by pollution from human sewage but also by industrial pollution. Moreover, she demonstrated that there was a pattern in how aquatic communities respond to pollution that transcends the particular organisms present in any given stream. She tested her theories about the structure of aquatic communities in pristine environments during an expedition to the Amazon, which she led in 1955.

Patrick stands out not only for her important scientific findings but also for her contributions to the way science is carried out. She was a scientist when women scientists were exceedingly uncommon, a leader in developing the scientific team approach to problem solving when "lone-wolf" scientific specialists were dominant, a pioneer in the systems approach of looking at entire drainage basins, and perhaps most important, living proof that theoretical and applied science not only can coexist but are commonly synergistic.

Acknowledgment of her many contributions was a long time in coming, but at age 63, Patrick began to receive wide recognition for her work. She was elected to the National Academy of Sciences in 1970 and received the prestigious Tyler Ecology Award in 1975. She also has received awards from the Botanical Society of America, Ecological Society of America, American Water Resources Association, American Academy of Arts and Sciences, and Society of Environmental Toxicology and Chemistry. She has received honorary degrees from two dozen schools and in 1972 was appointed to the board of directors of E. I. du Pont de Nemours and Company—the first woman board member.

Patrick was born in Topeka, Kansas, in 1907. She developed an interest in microbiology at a young age when her father, an attorney, let her look through his microscope. She received a B.S. from Coker College in 1929 and a Ph.D. in botany in 1934 from the University of Virginia. In 1939, after a brief term at Temple University, she joined the Academy of Natural Sciences, where today she is Francis Boyer Chair of Limnology and senior curator.

SOURCE: Adapted with permission from the dedication in Cairns et al. (1992).

waters were done primarily by geochemists, environmental engineers, and lake scientists interested in quantifying fluxes of minerals and pollutants from watersheds or into standing bodies of water (lakes and the oceans), and there was little interest in studying the chemical processes of streams and rivers themselves.

Historically, stream science has been identified most strongly with stream biology and stream ecology. Stream biologists often are identified primarily with other parent disciplines such as public health microbiology, fisheries biology, and aquatic entomology. Stream biology was mostly descriptive through the first half of the twentieth century and focused on the distribution and taxonomy of stream organisms. The development of stream ecology or stream limnology as a discipline analogous to lake limnology grew out of initiatives that began in the 1950s and 1960s. Hynes' 1970 book *The Ecology of Running Waters* is usually regarded as the first book on stream ecology (see Box 2-9).

The use of benthic invertebrates as indicator organisms for organic pollution and the division of streams into zones of pollution and recovery based on the presence of indicator species or groups of organisms have been major driving forces in stream biology since the early twentieth century; the first stream classification system based on the benthic organism species composition was the European "Saprobien" system of Kolkwitz and Marsson (1908, 1909). This paradigm stimulated much research on the structure of stream communities through the middle of this century. It can be considered a precursor of broader and more recent classification schemes and indices of biological integrity and biodiversity, which are currently popular subjects for research in stream ecology (see the background paper "Bringing Biology Back into Water Quality Assessments" at the end of this report).

Stream scientists have developed several organizing principles in recent decades to integrate the separate physical, chemical, and biological disciplines that contribute to studies of streams as ecosystems. The River Continuum Concept (RCC) (Vannote et al., 1980) is the most important of these. The RCC describes river systems as a continuously integrated series of physical changes that cause adjustments in the associated biota (Cummins et al., 1995). Geomorphological and hydrological characteristics of rivers provide a fundamental physical template that changes in a predictable fashion from the headwaters to the river mouth. Biological communities develop in adaptation to the fundamental physical template. The concept thus has a watershed orientation and focuses on terrestrial and aquatic interactions.

Since its development, the RCC has been modified in many ways to accommodate a broad range of factors, such as climate, geology, tributary effects, and local geomorphology, that influence streams. In addition, several organizing concepts developed as alternatives to the RCC have

---

BOX 2-9   H. B. NOEL HYNES (1917–)

Noel Hynes has been a major force in the development of the subdiscipline of limnology dealing with the study of flowing water ecosystems. His treatise *The Ecology of Running Waters*, published in 1970, summarized the knowledge of the field to that time on an international level and served as a foundation for modern stream ecology. An earlier book, *The Biology of Polluted Waters*, published in 1960, strongly influenced the development of biological assessments of water quality in streams.

Hynes was born in Devizes, England. His formal training was in zoology and entomology; he earned B.Sc., Ph.D., and D.Sc. degrees from the University of London. Assignments as an entomologist during World War II took him to various locations in Africa, where he remained until 1946. Returning to England, he taught at the University of Liverpool until 1964. In 1964, he was awarded a professorship in biology at the University of Waterloo, Canada, where he remained until his retirement.

Hynes specialized in the taxonomy and biology of freshwater Plecoptera and Amphipoda (invertebrate organisms that live in bottom sediments); their wide occurrence in flowing waters led him early to study stream ecology. His broad interests led to influential contributions in community ecology, trophic dynamics, and secondary production, and he was among the first to identify the importance of imported organic matter (allochthonous detritus) and the hyporheic zone (the interface between surface and ground water) in streams. His essay "The Stream and Its Valley," originally delivered to the International Association of Theoretical and Applied Limnology upon receipt of the association's prestigious Baldi Award, presaged the modern focus on watershed science in stream ecology. In 1984, Hynes received the Hilary Jolly Award from the Australian Limnological Society and, in 1988, was the recipient of the first Award of Excellence in Benthic Science from the North American Benthological Society. In all, Hynes has published more than 150 research articles, mainly on stream ecology.

Hynes' career spanned the transition period between the eras of descriptive and experimental science in limnology, and his work reflected the best of both. His skill as a seasoned natural historian was superbly balanced against his brilliance and intuition as a scientist, and he often pointed out that much good science was still to be done with only a few simple devices, a keen eye, and a sharp mind. An astute critic and articulate spokesman, his painstaking reviews, polished presentations, and enthusiastic audience participation improved numerous journal articles and enlivened many scientific meetings.

---

been subsumed into a broader RCC. Examples include the nutrient spiraling concept (Newbold et al., 1981), in which the down-gradient flow of streams causes nutrient cycles to be open rather than closed, and the patch dynamics concept (Pringle et al., 1988), which is based on the idea that disturbance and variations in time are primary determinants of community organization in streams. An outgrowth of the patch dynamics concept is the flood-pulse model (Junk et al., 1989), developed to describe biological communities in rivers that regularly overtop their banks and inundate the floodplain (Cummins et al., 1995) (see the background paper "Organizing Paradigms for the Study of Inland Aquatic Ecosystems" at the end of this report for further information on the RRC and its variants).

The RCC is a very broad framework for describing flowing water ecosystems. It is sufficiently robust that it can accommodate many other organizing concepts to account for various physical factors or processes. Consequently, the RCC, like the microcosm concept for lakes, is likely to continue to be modified and expanded rather than abandoned or supplanted. Of course, there are many other organizing concepts for stream limnology. Many of them, such as the idea that streams can be used as experimental units, are similar to the organizing concepts for lake limnology (see the background paper "Organizing Paradigms for the Study of Inland Aquatic Ecosystems" for further examples).

The concept of the watershed or catchment as the basic unit in stream hydrology dates back at least to the 1920s, and the watershed perspective has been used both in organizing hydrologic concepts and in data collection. More recently, the watershed concept has been used in an ecosystem context as a unifying theme to link both aquatic and terrestrial scientists (e.g., Likens and Bormann, 1974), and it was an organizing paradigm for the U.S. effort in the International Biological Program of the 1970s. Limnologists increasingly recognize that streams, lakes, and wetlands must be considered as interconnected systems in the context of their watersheds and airsheds.

Wetland Ecology

Integrated studies using the techniques of modern science to examine wetlands as ecosystems are primarily a phenomenon of the past 25 years (Mitsch and Gosselink, 1993), but the basic concepts on which wetland ecology is based can be traced back at least as far as the mid-sixteenth century (Gorham, 1953). Natural philosophers in the sixteenth and seventeenth centuries described and classified wetlands with names similar to those used today, and they related these classes to basic hydrologic conditions. The idea of wetland development and succession dates back to an account on Irish bogs published in 1685, and some fundamental ideas about the chemistry of bogs were developed in the late eighteenth and early nineteenth centuries. Nonetheless, most of these pioneering studies were overlooked by twentieth century botanists and ecologists as they developed similar concepts (Gorham, 1953).

Throughout most of the nineteenth and twentieth centuries, wetlands were commonly regarded as wastelands or nuisances to be "reclaimed" by draining or dredging, and this attitude hindered the development of wetland science (if for no other reason than financial—"wastelands" are unlikely to attract funds for research studies). The usefulness of wetlands in regulating hydrologic processes (see Box 2-10), such as floods, and in providing ecological benefits, such as buffering adjacent lakes and streams from impacts of upland human activity and serving as habitat for wildlife, has been widely recognized only in the past 25 years or so. With this recognition and the simultaneous understanding that human activities are causing wetlands to disappear at alarming rates came an impetus to study them, and the field of wetland ecology received a major stimulus.

As scientific and public appreciation for wetland ecosystems has grown, there ironically also has been a trend to view them as useful in serving society's needs as natural analogues or extensions of engineered systems. Thus, a common view has developed that wetlands can be used as natural treatment systems. Research programs have been developed in association with academic institutions and water management agencies to study the effectiveness of wetlands in removing nutrients and other contaminants from domestic waste effluent or in purifying stormwater runoff before it reaches lakes and streams. Both natural and constructed wetlands now are being used for such purposes on a wide scale within the United States. For some proponents of this approach, the quality and ecological integrity of the wetland itself appear to be less important than its ability to perform the desired function. Nonetheless, studies done to support this approach have advanced the previously meager understanding of the ways in which wetland ecosystems function and have provided some basis for preserving wetlands that otherwise might be destroyed by drainage and urban or agricultural development.

Many of the organizing and integrating principles on which wetland limnology is based are similar to those for lake and stream limnology, and a few scientists have combined interests in lakes and wetlands (see Box 2-11). However, some organizing concepts are linked primarily to wetlands. For example, wetlands can be considered as seres and ecotones—gradient ecosystems linking terrestrial and open-water aquatic systems and products of delicate, evolving interactions of hydrology and vegetation that produce unique, patterned landforms. Wetlands also are unique repositories of organic carbon (peat) and play an important role in the cycling of trace gases, including those implicated in global climate change (carbon dioxide and methane). Consequently, wetlands have become important ecosystems to study with regard to the global carbon cycle and global warming processes.

CURRENT STATUS

During the past few decades, limnological research has led to impressive conceptual and practical advances on all types of inland aquatic

---

BOX 2-10   KONSTANTIN E. IVANOV (1912–1994)

Konstantin Ivanov led the investigation of peatland (mire) hydrology in the Soviet Union for several decades after World War II. This was a time when the field was little studied outside that country despite the primary importance of hydrology for peatland development. His pioneering studies received due attention outside the Soviet Union only after translation of his book *Water Movement in Mirelands* (*Vodoobmen v bolotnyk landshaftakh*) in 1981 by Arthur Thompson and H.A.P. Ingram.

Ivanov was born in St. Petersburg and educated in hydraulic engineering. During World War II, he was called on to investigate the physical properties of ice and peat because they were important for military operations in the northern part of Russia. In 1944, Ivanov organized the first swamp station to study peatland hydrology. In 1948, Ivanov returned to Leningrad to head the State Hydrological Institute's Department of Peatland Hydrology, from which he directed permanent stations in peatland areas and led expeditions to other such areas. In 1949, he began to lecture at the University of Leningrad on peatland hydrology and other topics and was made professor in 1957. In 1963, he was appointed deputy scientific director at the State Hydrological Institute. In 1969, he joined the faculty of geography in the University of Leningrad and became full-time head of the Department of Land Hydrology. Ivanov was elected vice-president of the International Association of Hydrological Sciences and headed the Hydrological Commission of the Russian Geographic Society.

Ivanov's early investigations of water storage in and runoff from peatlands led him to collaborate with Vladimir V. Romanov in developing an earlier concept of V. D. Lopatin that peatlands can be divided fundamentally into two layers: an upper *acrotelm* that is aerated at least part of the time and in which most biological activity takes place lying above a deeper *catotelm* that is permanently waterlogged, anaerobic, and relatively inactive biologically. He went on to unite studies of vegetation, climate, topography, water supply, water chemistry, and pattern of development into a scheme of peatland classification that spanned the complete range of peatland scales from small mossy hummocks and hollows on individual peatlands to large-scale geographical units such as the water divides between major river systems. He also used aerial photography, in conjunction with detailed examination of representative peatlands that provided "ground-truth," to predict the hydrologic properties of peatlands from regional climatic data. His work provided the basis for investigating the stability of peatland landscapes, which depends strongly on the balance between water supply and water loss.

SOURCE: This sketch is derived from a more detailed biography in Ivanov (1981).

BOX 2-11 WILLIAM HAROLD PEARSALL (1891–1964)[1]

William Harold Pearsall was a botanist of many talents who carried out important studies in limnology, wetland ecology, and algal physiology. He also was a pioneer landscape ecologist long before the term was invented. He is fondly remembered by colleagues and students as a veritable fountain of ideas that he shared generously—along with endless stories—with all who came into contact with him, particularly on the hills of northern Britain, where he was most at home.

Pearsall was educated at the University of Manchester. After serving in World War I he joined the faculty of the University of Leeds, where he became a reader in botany. In 1938, he was appointed professor of botany at the University of Sheffield, and from 1944 until retirement, he was Quain Professor of Botany at University College, London. He was elected a fellow of the Royal Society in 1940 and received the Gold Medal of the Linnaean Society of London in 1963. Among his many service activities, he was a founder of the British Freshwater Biological Association, its honorary director from 1931 to 1937, and chairman of its council from 1954 onward. His influence on the staff was immense, and his frequent visits to the laboratories in the Lake District were greatly appreciated. He also was a charter member of the British Nature Conservancy and served as editor of the *Journal of Ecology* and the *Annals of Botany*.

Pearsall's research began with his father, a well-known amateur botanist. Together they began, in 1913, to study the depth distribution of aquatic macrophytes of the English lakes in relation to light penetration. After the war, they extended their studies to phytoplankton. At the same time, Pearsall mapped the vegetation of Esthwaite North Fen, which he repeated 40 years later to show considerable changes. This research led to a series of papers outlining the development of the English Lakes, their sediments, and their planktonic and macrophytic vegetation. A seminal study of redox potentials in soils, sediments, and peats in relation to waterlogging, organic content, and the forms of nitrogen and iron provided the basis for a landmark presidential address that he delivered to the British Ecological Society in 1947. Pearsall's years of studying the hill country of northern Britain culminated in his 1947 book *Mountains and Moorlands*, a masterly account of soils and vegetation that shows his insight into the broad patterns of landscape development. At the same time, his aesthetic appreciation of that landscape was manifested in a delightful series of water colors.

As John Lund has written, "One of his most wonderful characteristics was that it was out of disagreements that some of the most fruitful researches of his colleagues were likely to come, much to his delight. Their regard for him grew all the time, irrespective of whether they agreed with his ideas or not. Moreover, his hypotheses were by no means always incorrect; they might equally be too novel for people to appreciate . . . ."

[1] This biographical sketch owes much to an obituary by J. W. G. Lund (1965) and an account by A. R. Clapham (1971).

ecosystems. Chapter 3 describes many of these advances in more detail and explains the roles of limnologists, along with aquatic scientists in related discipline, in assessing and developing solutions for contemporary problems related to the degradation of inland waters. Many advances in limnology were made by academic limnologists and other scientists working in departments not traditionally focusing on limnology (such as civil and environmental engineering, environmental science, and earth sciences); others were made by interdisciplinary research teams associated with government agencies and with contract research and consulting firms. Thus, limnological research has spread beyond its traditional base of operations in academic departments of biological science.

Activity in limnology in recent decades is reflected by the vitality of its professional societies and scholarly journals. Within the past 15 years, three new North American societies have formed, each resulting from the expanding activities in a particular aspect of limnology and its related aquatic sciences:

1. The North American Benthological Society (NABS) expanded from an older regional organization (the Midwest Benthological Society) in 1974, emphasizing stream ecology and processes occurring at the interface between water and land.

2. The North American Lake Management Society (NALMS) was established in 1980 as an outgrowth of expanding interest in restoring and rehabilitating lakes and reservoirs degraded by human activity.

3. The Society of Wetland Scientists (SWS) was founded in 1980 to promote research for understanding and managing wetlands.

Memberships in the above three societies plus the two older limnological societies, the American Society of Limnology and Oceanography (ASLO) and the International Association for Great Lakes Research (IAGLR), total more than 12,000 (see Appendix B). Many limnologists also belong to SIL and to discipline-based societies. The aquatic section of the Ecological Society of America, for example, has more than 1,000 members (although many of these individuals also belong to one or more of the five primary limnological societies listed above).

The scholarly and technical journals published by the limnological societies continue to grow in circulation and pages published annually (see Table 2-2), and several new journals that focus on different aspects of limnology (for example, *Lake and Reservoir Management, Ecological Engineering*, and *Wetlands*) have appeared within the past two decades. Moreover, annual meetings of the societies attract growing numbers of presentations and attendees. For example, the number of presentations at the annual meeting of NABS increased from 234 to 409 between 1984 and 1994, while the number of papers at the annual SWS meeting increased

TABLE 2-2 Recent Publication Trends for Limnological Journals

	Year	Journal[a] L&O	CJFAS	JGLR	JNABS[b]	LRM[c]	W
Circulation	1984	5,215	2,000	na	943	na	460
	1989	4,763	1,900	na	1,179	na	1,930
	1994	5,171	1,675	na	1,441	na	3,848
Pages published	1984	1,358	1,862	466	325	390	220
	1989	1,766	2,437	728	375	242	327
	1994	2,025	3,187	800	617	352	320[d]
Publication frequency	1984	6	12	4	4	1	1
(issues per year)	1989	8	13	4	4	2	3
	1994	8	13	4	4	3	4
Number of papers	1984	na	416	na	65	na	19
submitted	1989	na	421	na	73	na	37
	1994	316[e]	474	na	96	na	80

NOTE: na = information unavailable.

[a]Journal abbreviation and responsible society: L&O: *Limnology and Oceanography*, American Society of Limnology and Oceanography; CJFAS: *Canadian Journal of Fisheries and Aquatic Science*, National Research Council of Canada; JGLR: *Journal of Great Lakes Research*, International Association for Great Lakes Research; JNABS: *Journal of the North American Benthological Society*, North American Benthological Society; LRM: *Lake and Reservoir Management*, North American Lake Management Society; W: *Wetlands*, Society of Wetland Scientists.
[b]First set of numbers for this journal represents 1986, the journal's first year of publication.
[c]First set of numbers for this journal represents 1985, the journal's first year of publication.
[d]Page size increased by approximately 70 percent.
[e]This number is for 1993; no earlier statistics are available.

from 51 to 306. It is apparent from this growth in presentations and publications that activity in limnology is continuing to increase.

Limnology courses continue to be offered at most major research universities. For example, 59 of 69 universities surveyed for this report indicated that they offer an introductory limnology course for undergraduates (see Appendix A). Furthermore, introductory limnology courses at major universities generally have stable or increasing enrollments; 55 of 57 universities responding to a survey question about student interest in limnology reported that interest is increasing or holding steady (see Appendix A). Graduate training in limnology is available at many of these institutions, even though only a few universities have distinct degrees or programs called limnology. In addition, during the past decade or so an increasing number of colleges and non-Ph.D.-granting universities (such as the University of Wisconsin, Stevens Point, described in Chapter 4) have developed M.S.-level programs related to limnology or aquatic science. Several new textbooks on stream and wetland limnology have been published, as have new editions of popular limnology books that focus on lakes (e.g., Mitsch and Gosselink, 1993; Horne and Goldman, 1994; Allan, 1995).

However, not all is well in limnology. Indeed, some of the positive characteristics of limnology at the close of the twentieth century can also be interpreted as indicators of underlying problems. For example, the formation of new societies is symptomatic of increased fragmentation, as well as an unwillingness on the part of the original society (ASLO) to embrace fully some of the newer aspects of the field, in particular applied limnology, resource management-oriented activities, and wetland ecology. In general, problems in the conduct of modern limnology can be grouped into six major areas:

1. inadequacy or instability of research support, especially in certain areas (such as physical and chemical limnology and wetland ecology);

2. loss of some prominent academic positions, especially in biological limnology;

3. growing fragmentation in academic programs (an ironic situation for an inherently interdisciplinary field);

4. inadequate educational programs, both at the general education level and at the professional level;

5. growing professional separation among various kinds of limnologists; and

6. poor public understanding of limnology and failure to identify it as a field that can contribute to the solution of aquatic problems important to human society.

Limnologists have not been reluctant to express concerns about the viability of their field. Discussions on these issues have appeared in limnological journals over the past decade, most notably in *Limnology and Oceanography* (e.g., Jumars, 1990; Kalff, 1991; Wetzel, 1991). These discussions have led to several studies dedicated to critical self-examination and to the development of recommendations to overcome perceived deficiencies and problems. Major self-analyses include (1) the Freshwater Imperative (Naiman et al., 1995), a broad initiative of a diverse group of aquatic scientists to address research needs in limnology, develop plans for government-supported interdisciplinary research programs on freshwater ecosystems, and otherwise promote the professional development of the field; and (2) ASLO's self-assessment of the field (Lewis et al., 1995), which contains a broad range of recommendations to reinvigorate the field and reverse the trend toward fragmentation among its component disciplines and subject areas.

As discussed in this chapter, much of the recent research support for limnology has been tied to targeted research programs in mission-oriented agencies focused on practical pollution problems. Although there is much to be gained from the focus that such programs provide, their funding levels rise and decline over a relatively small number of years as public and congressional interest waxes and wanes or as scientists develop solutions to problems. This approach simply is not adequate to support the basic scientific research and consistent training of new generations of scientists needed if limnology is to play a role in supporting wise and sustainable management of inland aquatic ecosystems. Some contend that the lack of a central federal program for limnology has led to a lack of research funding opportunities in limnology and has contributed to a decline of this field in academia (Jumars, 1990). Others disagree with this pessimistic attitude and point to the many advances in understanding and practical accomplishments achieved by limnologists in recent decades. To a certain extent, these diverging opinions reflect basic philosophical differences (is the "limnological cup" half full or half empty?) that can never be resolved completely. Nonetheless, it is likely that few regard current mechanisms for funding limnological research as optimal.

Within the National Science Foundation (NSF), there is no program that deals specifically with limnology in its broad definition. Instead, support for limnological research is subsumed in a variety of programs, including NSF's Divisions of Atmospheric Sciences, Earth Sciences, Environmental Biology, Integrative Biology and Neuroscience, International Programs, Ocean Sciences, and Polar Programs (Firth and Wyngaard, 1993). In contrast, NSF has a specific division (Ocean Sciences) devoted to the study of oceanography, which is bolstered by a similar funding program administered by the Office of Naval Research (Jumars, 1990). Most of the support for limnological research in NSF comes from the Division of Environmental Biology. As a result, support for research in chemical and physical limnology is not emphasized, and opportunities for support of interdisciplinary and long-term research have been limited. For example, 90 percent of NSF's grants on subjects directly or indirectly related to limnology in 1991 included work with a biological component, but only 25 percent supported studies with a chemical component and only 2 percent supported studies with a physical focus (Firth and Wyngaard, 1993).

One encouraging sign of change to promote interdisciplinary research on aquatic ecosystems within NSF is its new "Water and Watersheds" initiative. A joint effort between the NSF and the Environmental Protection Agency (EPA), this program targeted $10 million in competitively awarded funds in fiscal year 1995 to university-based aquatic research. In terms of generating proposals, the first competition announced in February 1995 was overwhelmingly successful: more than 650 proposals were submitted for review. Financial limitations meant that only about 30 proposals (roughly 5 percent) received funding. Nonetheless, the large response to the call for proposals suggests that the program addresses a major, unmet research need. The Water and Watersheds research initiative was funded primarily by the EPA (roughly 70 percent) in 1995, and the program does not yet have a permanent home in the NSF administrative structure. Consequently, it is premature to conclude that this program will provide a long-term source of funds for interdisciplinary research in limnology, but developing a permanent program to fund such research should be a priority.

Over the past decade or two, limnological programs have been eliminated or significantly downsized at some leading research universities as prominent faculty limnologists retired. Notable examples include G. E. Hutchinson of Yale University (Box 2-5); D. G. Frey of Indiana University, highly regarded for his examination of biological remains in sediments to chronicle past histories of lake conditions; and W. T. Edmondson of the University of Washington, noted for his pioneering studies of eutrophication in Lake Washington (see Chapter 4). Although Frey retired and was not replaced, Indiana University still offers a limnology course through its Department of Public and Environmental Affairs, but Yale no longer offers courses or employs faculty in limnology. The University of Washington offered a course in limnology for more than 30 years through the Department of Zoology, but when Edmondson's retirement was followed by cuts in state funding, the frequency of the course was reduced, and it was taught by visiting professors for several years (in 1996, the university again hired a limnologist to serve on its faculty). On a national basis, it is fair to say that some faculty positions have been lost because of declining financial resources in some universities, but in other cases, positions vacated by limnologists have been converted to other subject areas, usually some aspect of subcellular biology, which reflects a trend in many academic biology programs away from organismal and higher-level biology and toward subcellular and molecular scales.

The loss of highly visible academic positions in biological limnology probably is the single most important factor contributing to the perception among academic limnologists that all is not well within their profession, but in some respects the concern about lost positions may not be well founded. As limnological positions have been lost in traditional biology departments, others have been added in departments and colleges of environmental science and engineering, fisheries science, natural resources, and other resource-oriented programs. It is possible that larger numbers of faculty are involved in teaching and research across the broad field of limnology at research universities in the 1990s than ever before. However, they are dispersed more widely across departments and colleges than they were in earlier decades, when limnology was a narrower and simpler field that focused on temperate lakes. In summary, the elimination or deemphasis of limnology in the biology departments of major research universities has left a leadership vacuum in limnology, at least at many institutions, as well as a vacuum in the training of biologists with skills in basic systematics and the organismal biology of aquatic ecosystems. Overall, the dispersion of limnology into so many different academic programs in universities has led to severe fragmentation of the field. Few universities are producing limnologists with truly interdisciplinary backgrounds, an ecosystem perspective, and an ability to integrate across the sciences and major categories of aquatic ecosystems. In addition, most universities do not provide adequate course offerings in limnology at the general education level.

The fragmentation of limnologists during their education continues at the professional level. There is no limnological society or organization that represents the field and its practitioners as a whole. Although most, if not all, of the traditional journals publish occasional articles on streams and wetlands, their focus generally is on lake science. There is no journal that covers both the fundamental and applied aspects of limnology and all major categories of water bodies within the domain of limnology. Lake, stream, and wetland limnologists and fisheries scientists largely go their separate ways when joining professional societies, attending conferences, and publishing scientific papers. Although the Ecological Society of America includes theoretical and applied ecologists in its membership and has a special applied ecology section, fundamental and more practical, management-oriented aspects of limnological science are covered by separate societies (ASLO and NALMS). Both aspects of limnology have much to gain from closer interactions. Limnologists involved in research on the Great Lakes also have their own society and journal, a situation that is particularly ironic given that ASLO combines limnologists and oceanographers. As noted earlier, Great Lakes research combines elements of both limnology and oceanography (the latter particularly in terms of the scale of research vessels and equipment needed to conduct the research). The American Fisheries Society combines interests in fundamental science (fish physiology and genetics) with fisheries management, and it formulates and publicizes positions on the application of science to resource management issues. Nonetheless, fisheries science is not well integrated into limnology at the professional level (or for that matter in academic programs), in spite of the fact that fish are obviously integral components of aquatic food webs.

Although limnology is a diverse field, it is no more so than many other fields that have managed to bring their varied elements under the umbrella of one professional society that provides a sense of identity and public visibility to the field. Civil engineering, for example, sometimes is referred to as a "holding company" rather than a discipline because of the breadth and diversity of activities in which civil engineers are engaged. Nonetheless, one society, the American Society of Civil Engineers, represents the entire field. Similar situations prevail in chemistry, where the American Chemical Society includes theoretical and applied chemists working in fields as diverse as quantum mechanics and polymer science, and in microbiology, where the American Society for Microbiology publishes six subdisciplinary journals dealing with a broad range of fundamental and applied studies.

The fragmentation of limnology at the educational and professional levels has left the field with an identity crisis. Limnology is poorly understood by scientists and the general public. Even many aquatic scientists do not realize their ties to the general field of limnology or (in some cases) their effective involvement in limnological research. This problem of visibility tends to relegate limnologists and their science to secondary positions in public policy debates and decisions about aquatic resource management—situations in which their expertise is highly relevant.

Root causes of the problems described in preceding paragraphs are numerous and include factors that are internal to the field of limnology and external factors over which limnologists have had little control. Professional fragmentation—exemplified by the lack of a single society and single journal that represents all major areas of the field—would seem to be a problem of limnologists' own making. In part, it reflects an unwillingness by the older limnological societies and journals to diversify themselves and fully embrace some of the newer trends, such as the emphasis on restoration- and management-oriented activities that spawned the formation of organizations such as NALMS and SWS. Clearly, limnologists have it within their own power to overcome this professional fragmentation, and indeed only they can do so.

Causes of educational fragmentation are complicated and more difficult to assign. In large part, they reflect the long developmental history of academic disciplines and department structures within universities that have been defined for many decades by a primary orientation toward the basic science disciplines. Limnology, defined by the objects it studies (inland aquatic ecosystems), is inherently multidisciplinary and interdisciplinary, and components of the field have developed in many departments. The problems of academic limnology are a mirror of the problems of water science as a whole in higher education: water science is an interest within many fields, but it is not the primary focus of any of the traditional departments or disciplines. Moreover, "turf" problems inhibit any existing department from assuming too strong a leadership role over the entire field. At the same time, there probably is little enthusiasm among university administrators (deans and department heads) in this time of declining financial resources for reorganizations that would remove limnologically oriented faculty from existing programs and place them in new administrative units.

It is interesting to note that the fragmentation typically found in water resource science and limnological programs in academic institutions does not hold for natural resource subjects that traditionally have been linked directly to production-oriented economic activities. For example, soil science is an inherently multidisciplinary and interdisciplinary subject, but it has achieved departmental status in many land-grant universities. A similar situation exists in the field of forestry. No one considers it unusual or improper to have microbiologists, chemists, and physicists in the same department of soil science; it should not be so difficult to convince university faculty and administrators that broad-based departments or schools of aquatic science are both feasible and desirable.

Other recent critical examinations of limnology have dealt with the broad array of deficiencies and problems that face the field (e.g., Lewis et al., 1995) and with the development of a research prospectus and recommendations for better funding support (e.g., Naiman et al., 1995). Consequently, this report focuses on educational issues in limnology. In this context, the committee that wrote the report examined both the training of professional limnologists at all levels (B.S. to Ph.D.) and the provision of limnological information in the general education of college students and the public. Although the major emphasis of this report is education, the problems of limnology cannot be solved in academia alone. Also needed are improved links between those who conduct research in limnology and those who manage water resources. The professional societies in limnology have a critical role to play in helping to develop these links and in serving as advocates for the discipline of limnology as a whole.

REFERENCES

Allan, J. D. 1995. Stream Ecology: Structure and Function of Running Waters. New York: Chapman Hall.

Beckel, A. L. 1987. Breaking New Waters. Transactions of the Wisconsin Academy of Sciences, Arts, and Letters, special issue. Madison, Wisc.: Wisconsin Academy of Sciences, Arts, and Letters.

Bradbury, J. P., and R. O. Megard. 1972. Stratigraphic record of pollution in Shagawa lake, northeastern Minnesota. Geol. Soc. Am. Bull. 83:2639–2648.

Bradbury, J. P., and J. C. B. Waddington. 1973. The impact of European settlement on Shagawa Lake, northeastern Minnesota. Pp. 289–307 in Quaternary Plant Ecology, H. J. B. Birks and R. G. West, eds. Oxford, England: Blackwell.

Brezonik, P. L., J. G. Eaton, T. M. Frost, P. J. Garrison, T. K. Kratz, C. E. Mach, J. H. McCormick, J. A. Perry, W. A. Rose, C. J. Sampson, B. C. L. Shelley, W. A. Swenson, and K. E. Webster. 1993. Experimental acidification of Little Rock Lake, Wisconsin: Chemical and biological changes over the pH range 6.1 to 4.7. Can. J. Fish. Aquat. Sci. 50:1101–1121.

Cairns, J., Jr., B. R. Niederlehner, and D. R. Orvos, eds. 1992. Predicting Ecosystem Risk. Princeton, N.J.: Princeton Scientific Publishing.

Carpenter, S. R., T. M. Frost, D. Heisey, and T. K. Kratz. 1989. Randomized intervention analysis and the interpretation of whole-ecosystem experiments. Ecology 70:1142–1152.

Charles, D. F., R. W. Battarbee, I. Renberg, H. Van Dam, and J. P. Smol. 1989. Paleoecological

62   FRESHWATER ECOSYSTEMS

analysis of lake acidification trends in North America and Europe using diatoms and chrysophytes. Pp. 207–276 in Soils, Aquatic Processes and Lake Acidification, S.A. Norton, S.E. Lindberg, and A.L. Page, eds. New York: Springer-Verlag.

Clapham, A. R. 1971. William Harold Pearsall, 1891–1961. Biographical Memoirs of Fellows of the Royal Society of London 17:511–540.

Cumming, B. F., and J. P. Smol. 1993. Scaled chrysophytes and pH inference models: The effects of converting scale counts to cell counts and other species data transformations. J. Paleolimnol. 9: 147–153.

Cumming, B. F., K. A. Davey, J. P. Smol., and H. J. B. Birks. 1994. When did acid-sensitive Adirondack lakes (New York, USA) begin to acidify and are they still acidifying? Can. J. Fish. Aquat. Sci. 51: 1550–1568.

Cummins, K. W., C. E. Cushing, and G. W. Minshall. 1995. Introduction: An overview of stream ecosystems. Pp. 1–10 in River and Stream Ecosystems, C. E. Cushing, K. W. Cummins, and G. W. Minshall, eds. Ecosystems of the World, vol. 22. New York: Elsevier.

Edmondson, W. T. 1994. What is limnology? Pp. 547–553 in Limnology Now: A Paradigm of Planetary Problems, R. Margalef, ed. New York: Elsevier.

Einsele, W. 1941. Die Umsetzung von zugeführtem anorganischen Phosphat in eutrophen See and ihre Rückwirkung auf seinen Gesamthaushalt. Z. Fisch. 39:407–488.

Firth, P., and G. Wyngaard. 1993. Limnology support at the National Science Foundation. Bull. Ecol. Soc. Am. 72:170–175.

Forbes, S. A. 1887. The lake as a microcosm. Bull. Peoria, Illinois. Sci. Assoc. Reprinted in Bull. Ill. Nat. Hist. Surv. 15(1925):537–550.

Forel, F. A. 1882, 1895, 1904. Le Leman: Monographie Limnologique, Vols. 1–3. Lausanne: F. Rouge.

Frey, D .G. 1963. Wisconsin: The Birge-Juday era. Chapter 1 in Limnology in North America, D. G. Frey, ed. Madison: University of Wisconsin.

Gorham, E. 1953. Some early ideas concerning the nature, origin and development of peat lands. J. Ecol. 41:257–274.

Gorham, E., and J. A. Janssens. 1992. The paleorecord of geochemistry and hydrology in northern peatlands and its relation to global change. Suo 43:9–19.

Gorham, E., and J. E. Sanger. 1976. Fossilized pigments as stratigraphic indicators of cultural eutrophication in Shagawa Lake, northeastern Minnesota. Geol. Soc. Am. Bull. 87:1638–1642.

Horne, A. J., and C. R. Goldman. 1994. Limnology. New York: McGraw-Hill.

Hutchinson, G. E. 1967. A Treatise on Limnology, Vol. II: Introduction to Lake Biology and the Limnoplankton. New York: Wiley-Interscience.

Hynes, H. B. N. 1960. The Biology of Polluted Waters. Liverpool, England: Liverpool University Press.

Hynes, H. B. N. 1970. The Ecology of Running Waters. Toronto: University of Toronto Press.

Ivanov, K. E. 1981. Water Movement in Mirelands, translated by A. Thompson and H. A. P. Ingram. London: Academic Press.

Janssens, J. A. , B. C. S. Hansen, P. H. Glaser, and C. Whitlock. 1992. Development of a raised bog complex in northern Minnesota. Pp. 189–221 in Patterned Peatlands of Northern Minnesota, H. E. Wright, Jr., B. Coffin, and N. Aaseng, eds. Minneapolis: University of Minnesota Press.

Juday, C. 1915. Limnological studies on some lakes in Central America. Trans. Wis. Acad. Sci. Arts Lett. 18:214–250.

Juday, C., and C. L. Schloemer. 1938. Effects of fertilizers on plankton production and on fish growth in a Wisconsin lake. Progr. Fish-Cult. 40:24–27.

Jumars, P. A. 1990. W(h)ither limnology? Limnol. Oceanogr. 35:1216–1218.

Junk, W. J., P. B. Bayley, and R. E. Sparks. 1989. The flood pulse concept in river floodplain systems. Pp. 110–127 in Proceedings of the International Large River Symposium. D. P. Dodge, ed. Can. Spec. Publ. Fish. Aquat. Sci. 106.

Kalff, J. 1991. On the teaching and funding of limnology. Limnol. Oceanogr. 36:1499–1501.

Kitchell, J. F., ed. 1992. Food Web Management: A Case Study of Lake Mendota. New York: Springer-Verlag.

Kolkwitz, R., and M. Marsson. 1908. Ökologie der pflanzlichen Saprobien. Ber. Deut. Botan. Ges. 26a:505–519.

Kolkwitz, R., and M. Marsson. 1909. Ökologie der tierischen Saprobien. Int. Rev. Ges. Hydrobiol. Hydrol. 2:126–152.

LeBlanc, H. 1912. LeProfesseur Dr. Francois Alphonse Forel, 1841–1912. Actes Soc. Helv. Sci. Nat. 95:109–148.

Lewis, W. M., S. Chisholm, C. D'Elia, E. Fee, N. G. Hairston, J. Hobbie, G. E. Likens, S. Threlkeld, and R. G. Wetzel. 1995. Challenges for limnology in North America: An assessment of the discipline in the 1990s. ASLO Bull. 4(2):1–20.

Likens, G. E., and F. H. Bormann. 1974. Linkages between terrestrial and aquatic ecosystems. BioScience 24:447–456.

Lindeman, R. L. 1942. The trophic-dynamic aspect of ecology. Ecology 23:399–418.

Lund, J. W. G. 1965. Prof. W. H. Pearsall, F.R.S. Nature 205:21.

Mitsch, W. J. 1994. Energy flow in a pulsing system: Howard T. Odum. Ecol. Eng. 3:77–105.

Mitsch, W., and J. G. Gosselin. 1993. Wetlands, 2nd ed. New York: Van Nostrand Reinhold.

Mortimer, C. H. 1956. E. A. Birge, an explorer of lakes. Pp. 165–206 in E. A. Birge, a Memoir, G. C. Sellery, ed. Madison: University of Wisconsin Press.

Naiman, R. J., J. M. Magnuson, D. M. McKnight, and J. A. Stanford, eds. 1995. The Freshwater Imperative: A Research Agenda. Washington, D.C.: Island Press.

Newbold, J .D., J. W. Elwood, R. V. O'Neill, and W. Van Winkle. 1981. Measuring nutrient spiraling in streams. Can. J. Fish. Aquat. Sci. 38:860–863.

Nipkow, F. 1920. Vorläufige Mitteilungen über Untersuchungen des Schlammabsatzes im Zür. Rev. Hydrol. 1:100–122.

Odum, H. T. 1957. Trophic structure and productivity of Silver Springs, Florida. Ecol. Monogr. 27:55–112.

Odum, H. T. 1970. Environment Power and Society. New York: Wiley-Interscience.

Odum, H. T. 1994. Ecological engineering: The necessary use of ecological self-design. Ecol. Engr. 3:115–118.

Odum, H. T., and R. F. Pigeon, eds. 1970. A Tropical Rain Forest. Oak Ridge, Tenn.: U.S. Atomic Energy Commission.

Overbeck, J. 1989. Plön—History of limnology, foundation of SIL and development of a limnological institute. Pp. 61–65 in Limnology in the Federal Republic of Germany, W. Lampert and K. O. Rothhaupt, eds. Kiel, Germany: International Association for Theoretical and Applied Limnology.

Pringle, C. M., R. J. Naiman, G. Bretschko, J. R. Karr, M. W. Oswood, J. R. Webster, R. L. Welcomme, and M. J. Winterbourn. 1988. Patch dynamics in lotic systems: The stream as a mosaic. J. N. Am. Benthol. Soc. 7:503–524.

Rasmussen, P. W., D. M. Heisey, E. V. Nordheim, and T. M. Frost. 1993. Time-series intervention analysis: Unreplicated large-scale experiments. Pp. 138–158 in Design and Analysis of Ecological Experiments, S. M. Scheiner and J. Gurevitch, eds. New York: Chapman and Hall.

Roberts, L. 1987. Federal report on acid rain draws criticism. Science 237:1404–1406.

Schindler, D. W., T. M. Frost, K. H. Mills, P. S. Chang. I. J. Davies, L. Findlay, D. F. Malley, J. A. Shearer, M. A. Turner, P. J. Garrison, C. J. Watras, K. E. Webster, J. M. Gunn, P. L. Brezonik, and W. A. Swenson. 1992. Comparisons between experimentally- and atmospherically-acidified lakes during stress and recovery. Proc. R. Soc. Edinburgh 97B:193–226.

Sellery, G. C. 1956. E.A. Birge, a Memoir. Madison: University of Wisconsin Press.

Smol, J. P., D. F. Charles, and D. R. Whitehead. 1984a. Mallomanadacean (Chrysophyceae) assemblages and their relationships with limnological characteristics in 38 Adirondack (New York) lakes. Can. J. Bot. 62: 911–923.

Smol, J. P., D. F. Charles, and D. R. Whitehead. 1984b. Mallomonadacean microfossils provide evidence of recent acidification. Nature 307: 628–630.

Tansley, A. G. 1935. The use and abuse of vegetational concepts and terms. Ecology 16:284–307.

Thienemann, A. 1931. Tropische Seen und Seetypen-lehre. Arch. Hydrobiol. Suppl. 9:205–231.

Vannote, R. L., G. W. Minshall, K. W. Cummins, J. R. Sedell, and C. E. Cushing. 1980. The river continuum concept. Can. J. Fish. Aquat. Sci. 37:130–137.

Wetzel, R. G. 1991. On the teaching of limnology: Need for a national initiative. Limnol. Oceanogr. 36:213–215.

3

Contemporary Water Management: Role of Limnology

Understanding how aquatic systems function is complex because of the interdependencies among chemicals in the water and sediments, populations of aquatic organisms, water temperature, the shape of the water body, and the nature of the surrounding landscape. When one considers humans as a part of aquatic ecosystems, the dynamics of these systems become even more difficult to comprehend. Nearly every human activity—from farming and gardening to road building, shipping, fishing, and fuel combustion—affects rivers, lakes, and wetlands in some way. Limnology provides the tools necessary for understanding how water bodies behave in environments without significant human influence and how they are affected by the full range of human activities.

This chapter highlights risks to North American surface waters and describes the role of limnologists and scientists in closely allied disciplines in improving understanding and stewardship of these waters. The contributions of limnologists range from establishing a detailed understanding of the extent and causes of an environmental problem to developing techniques to solve the problem or minimize its impact. Although limnologists, often drawing on the work of water scientists in fields such as environmental engineering and hydrology, have made major contributions toward understanding and solving the major problems of freshwater ecosystems during the past few decades, much remains to be learned. Consequently, the chapter also describes how additional limnological research would be helpful in defining and solving problems.

The chapter divides problems in aquatic ecosystems according to whether they originate from modifications in the watershed or physical characteristics of the water body; from changes in the water's chemical composition; or from alterations in the ecosystem's biological communities. It is critical to realize, however, that the physics, chemistry, and biology of a water body are interrelated. Changes in the physical landscape surrounding water bodies can affect the chemical inputs to them, which in turn can affect aquatic biota. Similarly, changes in the chemical composition and biota of an aquatic ecosystem can affect the physical landscape.

Much of the research described in this chapter has been conducted in response to problems caused by human activities. In a general sense, this research could be considered goal oriented (or directed research) rather than basic research conducted for the pursuit of knowledge itself. Nonetheless, much of this work has contributed to the understanding and solution of aquatic ecosystem problems because it advanced understanding of the fundamental behavior of these ecosystems. Similarly, many of the research needs identified in this chapter address basic limnological questions even though the results could be applied toward the solution of practical problems. (For more detailed information about research needs in limnology, see Naiman et al., 1995, and background papers at the end of this report.)

PHYSICAL CHANGES IN WATERSHEDS AND WATER BODIES

In many locations, the most serious causes of water quality decline are not direct inputs of pollutants but indirect effects resulting from changes in the landscape and atmosphere surrounding the water body and alteration of the water's natural flow path. Countless freshwater systems also have been affected by direct physical alterations to the shoreline or shape of the water body. For example, vegetation along lake and stream banks often is cleared to allow recreational or commercial access. Outlets to lakes often are dammed to provide downstream flow controls and allow water-level regulation in the lake. Channels are constructed between lakes and rivers, and littoral areas of lakes are dredged to allow ship and boat traffic. In addition, wetlands often are drained for agriculture and forestry. These physical changes can have subtle or dramatic impacts on the structure and functions of aquatic ecosystems, depending on the severity of the change. In many cases, the impacts are caused by excessive diversion of water from a stream for crop irrigation or other water supply purposes to the extent that so-called in-stream uses of the water (for example, maintenance of fish populations) may be impaired. Limnologists have made and continue to make critical contributions toward understanding how water bodies are disrupted by physical changes to the water bodies themselves or to their watersheds.

Dam and Impoundment Building

More than 80,000 dams exist in the United States (Frederick, 1991), creating impoundments that range in size from small millponds to large multipurpose reservoirs. Until recently, dams were viewed as good, clean ways to generate hydroelectric power (Abelson, 1985; Bourassa, 1985), control floods, and provide storage for water supplies. Hence, their environmental impacts were not understood, evaluated, or monitored. Recently, society has become more aware of these impacts and the economic value of resources, such as fisheries, that have been damaged by dam construction.

Limnologists, along with hydrologists and fisheries biologists, have been involved in documenting how dam building affects river ecosystems in several ways (National Research Council, 1987):

• *Temperature alterations:* Because the water passing through reservoirs often originates from points near the middle or bottom of the water column, it may be much colder during the summer months than natural flows would have been. Sustained low temperatures during the warm months may support cold-water fisheries for species such as trout in streams and rivers that otherwise would not provide appropriate temperature regimes for these fish. At the same time, however, temperature alteration may suppress important native fish (Minckley, 1991) and other aquatic animals (Ward and Stanford, 1979).

• *Changes in dissolved oxygen, nutrient, and suspended solids concentrations:* Dams may affect the amounts of oxygen, suspended solids, and nutrients in water flowing downstream (Gordon et al., 1992). The concentrations of dissolved oxygen in the lower water column of reservoirs may be low or zero in some instances, hampering the development of fisheries or altering the native fauna below dams (Petts, 1984). Further, water released from dams is likely to have a lower sediment content than water entering a reservoir (Andrews, 1991), causing substantial biotic changes such as enhanced growth of algae (Blinn and Cole, 1991) as well as physical changes in the downstream sediment balance (Simons, 1979).

• *Hydraulic modifications:* Dams may stabilize the natural variation in the flow of rivers, alter seasonal extremes, or induce entirely new patterns. In addition, hydropower production facilities associated with some dams may establish a regular daily pulse in stream discharge and mean depth. These hydraulic peculiarities in turn can have significant biological effects. For example, decreased variability in streamflow below dams may cause habitat losses for fish and other aquatic organisms (Kellerhals and Church, 1989). In addition, wetland areas can suffer massive losses of important habitats (Baumann et al., 1984). The Atchafalaya Delta of the Mississippi River and the Peace-Athabasca Delta in northern Alberta are important examples. Aquatic scientists have estimated that the latter delta, which supports many unique species of wildlife and several hundred indigenous people, will disappear in fewer than 50 years unless Bennett Dam is decommissioned (see Box 3-1).

BOX 3-1 BENNETT DAM AND THE DISAPPEARANCE OF THE PEACE-ATHABASCA DELTA

In the past, most dams and reservoirs were constructed without adequate study of their consequences for aquatic ecosystems, which has lead to irreparable or costly damage. One example is the installation of Bennett Dam on the Peace River. The Peace River flows from headwaters in northern British Columbia across Alberta to Lake Athabasca. Historically, the spring melt flood of the Peace backed up water into the lake and delta of the Athabasca River, flooding small, perched lakes and wetlands along the dendritic channels in the delta. The area was rich in wildlife and was home to more than 1,500 indigenous people.

Except in 1974, when an ice jam caused flooding, there has been no flooding of the Peace-Athabasca Delta since 1969, when Bennett Dam was constructed on the Peace River near the British Columbia-Alberta border. Muskrat, the staple of a thriving trapping industry, disappeared within a few years. Rich fisheries declined. Waterfowl numbers decreased dramatically as marshlands were invaded by willows and other trees. Many of the dendritic channels filled in, making boat travel impossible. Grazing lands for wood bison declined as range quality deteriorated after the annual deposition of rich sediments ceased (Carbyn et al., 1993).

Few indigenous people now live on the land. Most remain in the community of Fort Chipeweyan, where their use of natural foods is being replaced by less-nutritious alternatives (Wein et al., 1991). Damming of the river has resulted in the end of a traditional way of life for people in the area. Studies done to document the deleterious downstream impacts of the Bennett Dam provide resource managers and water resource planners with knowledge about the consequences of dam building so that these problems can be avoided in the future.

* *Increase in mercury levels:* In some cases, the construction of reservoirs has caused the mercury content of fish to increase rapidly to values that exceed guidelines for human consumption, as shown in the example in Figure 3-1 (Bodaly et al., 1984; Rosenberg et al., 1995). Increased mercury levels have led to losses of commercial, recreational, and subsistence fisheries. The increase appears to be largely the result of low oxygen concentrations caused by the decay of flooded vegetation. Such conditions promote the increased activity of bacterial species that transform inorganic mercury into the methylated form, which is greatly biomagnified (Rudd, 1995).

* *Release of greenhouse gases:* The release of the greenhouse gases carbon dioxide (CO_2) and methane (CH_4) following the flooding of forests and peatlands is another major concern identified by limnologists and other environmental scientists. The total area of reservoir surface in North

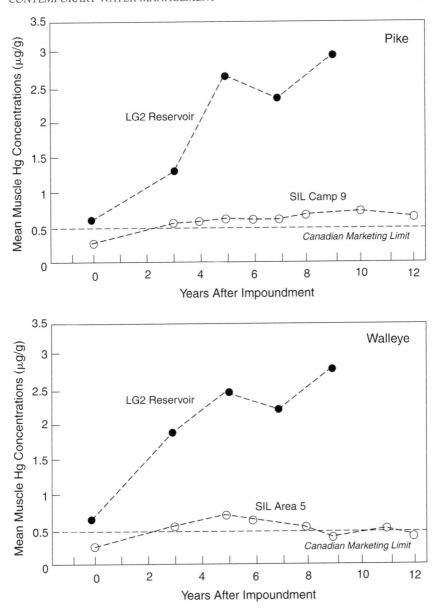

FIGURE 3-1 Mercury concentrations in pike and walleye flesh from two Canadian reservoirs after impoundment. LG2 is part of the James Bay Phase 1 development in Quebec. SIL is Southern Indian Lake on the Churchill River in northern Manitoba. SOURCE: Reprinted, with permission, from Rosenberg et al. (1995). © 1995 by *Global Environmental Change*.

America is substantial, roughly equivalent to that of Lake Ontario (Rudd et al., 1993). Under natural conditions, wetlands are significant sinks for CO_2 and major sources of CH_4, while forests are minor sinks for CH_4 and about in equilibrium with atmospheric CO_2 (Rudd et al., 1993; Gorham, 1995). Following flooding, these areas become strong net sources of greenhouse gases for several years. The rate of greenhouse gas emission per unit of hydropower production may be comparable to emissions from a fossil fuel plant producing an equivalent amount of power, depending on the area of land that is flooded to create the hydroelectric facility (Rudd et al., 1993).

The broad range of effects of dams on water quality, physical habitat, and biotic communities presents numerous research opportunities relevant to resource management that are now being actively explored with the involvement of limnologists (National Research Council, 1987; Cheslak and Carpenter, 1990). For example, multiple-level outlet structures can be used to manipulate water temperature (Larson et al., 1980) or to mix waters of various oxygen or nutrient contents at the dam. Input from limnologists and other aquatic scientists can help to ensure that management of existing dams is optimized.

As dams age and fill with sediment, their utility decreases and so does the economic incentive to maintain them. However, removing dams can create severe problems from remobilization of upstream sediment, which is likely to be rich in nutrients and oxygen-consuming organic matter deposits and may contain hazardous chemicals. Limnologists can help assess and avoid such problems when dams are removed.

Wetlands Destruction

Wetlands were long regarded by many as wastelands fit for nothing until "reclaimed" by human manipulation. In recent times, however, human societies have begun to recognize the broad array of values that wetlands can provide (Greeson et al., 1979; Richardson, 1989, 1994; National Research Council, 1992, 1995). Among the physical values of wetlands are such properties as shoreline stabilization, flood-peak reduction, and ground water recharge. Wetlands also function as filters, transformers, and sinks for materials delivered to them by human activities, thus improving water quality. For instance, they can filter 60 to 90 percent of suspended solids from wastewater and as much as 80 percent of sediment in runoff from agricultural fields (Richardson, 1989). In addition to these important physical and chemical functions, wetlands provide a great variety of biological benefits. Riparian wetlands serve as protective nurseries and food sources for young fish before they move out into open water. The microorganisms and vegetation in wetlands are essential on both local and global scales in cycling carbon, nitrogen, and sulfur between the earth and the atmosphere. In addition, wetlands provide commercial products such as lumber, cranberries, marsh hay, wild rice, waterfowl, muskrat, beaver, and mink.

Despite their benefits, wetlands in many parts of the world are disappearing at an alarming rate due to human activity. In the contiguous United States, more than 50 percent of the original wetlands had been destroyed by the 1970s, and an additional 2.6 percent were lost through the 1980s (Frayer, 1991). In some states—such as California, Ohio, and Iowa—less than 10 percent of the original wetlands remain, with consequent losses of waterfowl, furbearing animals, and fish.

The largest human uses of wetlands are for forestry and agriculture (Kivinen, 1980). Across large sections of the Midwest, for example, drainage tiles have been installed beneath wetlands to allow use of their fertile soils for crop production. Urban development also has led to the dredging and filling of many wetlands. Others have been flooded or drained by

Example of a wetland: a patterned peatland in the Hudson Bay Lowlands in Canada. Because of upwelling ground water, large peatlands such as this often develop intricate landscape patterns, which represent perhaps the most delicate mutual interaction between hydrology and vegetation on the earth's surface (Sjörs, 1961; Heinselman, 1963; Wright et al., 1992). SOURCE: Paul H. Glaser, Limnological Research Center, University of Minnesota.

highway construction. Stormwater and wastewater diversion to wetlands has added both nutrients and toxic materials (Environmental Protection Agency, 1994), leading to major alterations of flora and fauna. The United States currently has a policy of "no net loss" of wetlands, which has spawned work in restoring damaged wetlands and attempting to construct artificial wetlands to replace natural ones that have been destroyed. As presently practiced, however, wetland restoration is a very imperfect science (Mitsch and Gosselink, 1993). The best candidates for restoration are early successional wetlands such as cattail marshes. Restoration of natural prairie wetlands also may be possible in some cases. Restoration of forested wetlands is difficult, given the long lives of trees, and restoration of the large patterned peatlands of the boreal zone—developed over millennia by interactions of local and regional hydrology—must be regarded as essentially impossible once they are seriously degraded. Replacing mature wetlands that have taken centuries or millennia to develop with cattail marshes cannot be considered in compliance with the no-net-loss policy because it results in a major shift in biodiversity and ecological function.

Construction of new wetlands, allowed in the United States under Section 404 of the Clean Water Act (Kusler and Kentula, 1989), to replace ones that have been destroyed for development is even more difficult than restoration of damaged wetlands. There often is no record of what the destroyed wetland was like. Moreover, follow-up investigations of created wetlands are rare, so that the success of mitigation efforts is seldom evaluated (Erwin, 1991). In a rare follow-up study, Erwin reported very limited success for wetland mitigation in South Florida. According to Bedford (in press), adequate attention has been given neither to reproducing the original wetland hydrology nor to establishing specific types of plant communities.

Further fundamental limnological research is essential for managing, restoring, and creating wetlands and for developing a workable no-net-loss policy. Additional scientific understanding is needed to address issues such as the following:

• What have been the spatial patterns of wetland development through the centuries?

• To what degree has wetland development been controlled by environmental factors (hydrology, in particular), and to what degree has it resulted from autogenic processes such as peat accumulation?

• Do environmental changes generally cause slow and steady changes in wetlands, or do threshold phenomena dictate relatively sudden responses to environmental stresses (natural or anthropogenic) that build up over time?

• How will global warming affect wetlands, particularly peatlands, and vice versa?

• How are wetlands linked ecologically and biogeochemically to other aquatic and terrestrial ecosystems?

Limnological research to answer such questions is critical in the quest to protect and manage the remnants of the valuable wetland resource in North America.

Other Modifications to Watersheds

For millennia, humans have been modifying watersheds in ways other than by building dams and draining wetlands, often with adverse consequences for aquatic biota and water quality (Solbe, 1986; Harriman et al., 1994). Forests and grassland have been transformed to agricultural fields or urban pavements as societies have established themselves around water bodies; forest ecosystems have been replaced with tree plantations designed to meet human needs for timber. Studies by limnologists and their fellow water scientists have provided valuable insights about impacts of human development on water bodies. Examples of limnological work in this area include the following:

• *Effects of early agricultural and urban activities:* Studies by limnologists have shown that even the earliest stages of agricultural and urban development caused changes in water quality. Fires built by native people significantly altered the landscape of North America (Lewis and Ferguson, 1988). Work by several limnologists has indicated that fire changes water quality, causing increases in runoff of water, nutrients, and mineral ions (Schindler et al., 1980; Bayley et al., 1992 a,b; MacDonald et al., 1993). Paleolimnological studies by Hutchinson et al. (1970) showed that construction of the Appian Way (Via Appia) by the Romans in the second century A.D. changed drainage patterns for Lago di Monterosi in Italy in such a way that the lake became eutrophic. Similarly, Frey (1955) showed that early agrarian societies in the catchment of Längsee, Austria, caused the lake to become meromictic (meaning the bottom waters no longer mix with the remainder of the lake) through land clearing and associated activities.

• *Changes in nutrient loads caused by land use:* Limnological studies have shown that land-use changes alter the yield of nutrients from watersheds to lakes. For example, Dillon and Kirchner (1975) showed that the transformation of forested land into pasture causes a considerable increase in nutrient yields from catchments to lakes. Transformation to agricultural land causes still greater increases. Sorrano et al. (in press) showed that urbanization increases nutrient inputs and that in addition to land use, the proximity of modified land to stream edges is an important factor controlling nutrient inputs to Lake Mendota, Wisconsin.

• *Effects of forestry:* Clearcut logging is well known to increase water yield and the transport of nutrients, sediments, and other substances to streams (Bosch and Hewlett, 1982; Murphy et al., 1986; Harr and Fredriksen, 1988; Harriman et al., 1994). Likens et al. (1970) demonstrated that when catchments are not revegetated, the yield of nutrients, sediments, and other substances can increase manyfold. Even managed forestland can affect aquatic ecosystems. For example, when deciduous forests in catchments are transformed into conifer plantations, soils, ground water, and runoff typically become more acidic because of the increased trapping of strong acids from the atmosphere by conifers and the acidifying effect on soils of decomposing conifer needles (Feger, 1994); the result is modification in the acid-base balance of lakes whose watersheds encompass the forestland. Even when no changes are made in species of vegetation, different stages of the growth and harvest cycle cause differences in the chemistry of streams (Harriman et al., 1994; Kreuzer, 1994).

Global Warming

Most atmospheric scientists agree that average global temperatures are likely to rise as a result of increasing levels of greenhouse gases, particularly CO_2 and CH_4, in the atmosphere (Houghton et al., 1995). Although scientists who believe that global warming is occurring or will soon occur outnumber the dissenters, there is disagreement about how much the temperature might increase and how fast (American Society of Limnology and Oceanography and North American Benthological Society, 1994; Houghton et al., 1995). Despite this disagreement, it is widely accepted that substantial changes in the temperature of the earth are likely to have significant effects on inland aquatic ecosystems (see Box 3-2).

Atmospheric CO_2 concentrations have been rising since the industrial revolution as the result of fossil fuel burning. Deliberate burning of forests as land is cleared is an important second source (Houghton et al., 1995). Levels of CH_4 in the atmosphere also have been increasing, but for reasons that are not well understood. Although the abundance of CH_4 in the atmosphere is much lower than that of CO_2, it is 7.5 to 62-times more effective as a greenhouse gas, depending on the time scale under consideration (Houghton et al., 1995). Important CH_4 sources include wetlands, rice paddies, termites, and domestic animals such as cows. Wetland limnologists have contributed greatly to understanding the role of wetlands in CH_4 production (Bartlett and Harriss, 1993; Harriss et al., 1993), but a mechanism that fully explains the increase in the atmosphere still needs to be found (Houghton et al., 1995). Research has scarcely begun on an alternative explanation that CH_4 is increasing because human activities have slowed the mechanisms removing it from the atmosphere.

Refined estimates of the magnitudes and rates of global warming are problematic because of uncertainties in emissions from human activities and variabilities in different storage mechanisms for carbon. For example,

Unless properly designed with appropriate vegetative buffers, clearcuts can be a source of sediments and nutrients to aquatic ecosystems. SOURCE: Elizabeth Rogers, White Water Associates, Inc.

seawater, organic detritus, and vegetation all tie up large amounts of carbon. Furthermore, cloud cover, aerosol haze, and oceanic circulation all affect the rate and degree of change, thereby complicating predictions. Recently, Mitchell et al. (1995) modeled the global interactions of greenhouse gases, aerosols, and ocean circulation and concluded that, overall, aerosols will reduce the warming expected from greenhouse gases by about one-third. These many uncertain factors will have to be better understood before the timing and extent of global warming can be predicted. Limnological research can help to reduce these uncertainties.

Limnologists have been centrally involved in evaluating the effects of climate warming on fresh waters and have described a wide range of effects that climatic warming is likely to have on lakes and streams (Firth and Fisher, 1992; Schindler et al., in press, a,b). Increased temperature will increase evapotranspiration, so that lower soil moisture, ground water flows, and streamflows are likely to result (except where climatic warming is accompanied by large increases in rainfall). Stream temperatures track air temperatures reasonably closely; therefore, some streams may become too warm for some organisms. Periods when small streams are dry may increase dramatically. The early melting of snow will cause less pronounced spring flow pulses, which are vital to the maintenance of wetland and riparian habitats. The reduction in water flows, coupled with the

---

BOX 3-2 HOW CLIMATE CHANGE AFFECTS LAKES

Freshwater lake temperatures respond to changed atmospheric conditions, and changes in lake water temperatures and temperature stratification dynamics may have a profound effect on fish and other aquatic organisms (Meisner et al., 1987; Coutant, 1990; Magnuson et al., 1990; Stefan et al., 1995).

The response of lake water temperatures to climate changes can be investigated by several methods. One approach is to examine long-term records. In a few lakes, such as those in the Experimental Lakes Area in Ontario, Canada, and Lake Mendota in Wisconsin, weekly and biweekly vertical profiles of water quality and biological parameters have been collected for many decades (Robertson, 1989; Schindler et al., 1990). Observations from such lakes indicate rising average surface water temperatures, increased evaporation, and increases in transparency and diversity of phytoplankton. Such trends can be determined, however, only if lake temperature records are long enough. Where records are short but detailed, a second approach is to compare individual warm and cold or wet and dry years (Hondzo and Stefan, 1991). A third approach, useful for extrapolating to lakes of different geometries and latitudes and to possible future climates, involves numerical simulation models. Such models calculate heat transfer from the atmosphere to the water and within the lake water column (Blumberg and DiToro, 1990; Croley, 1990; McCormick, 1990; Schertzer and Sawchuk, 1990).

In one investigation using a numerical modeling method to simulate daily water temperatures in 27 classes of lakes characteristic of the north-central United States, Hondzo and Stefan (1993) determined that a doubling of atmospheric CO_2 would increase surface water temperatures by less than the corresponding air temperature increase but that bottom water temperatures in seasonally stratified lakes would be largely unchanged. The effect on fish would be a decrease in habitat for cold-water species and an increase in habitat for species that prefer warmer water, with the greatest effects on lakes in southern latitudes (Stefan et al., 1995). Atmospheric warming also would increase evaporative loss of lake water by as much as 300 mm for the season (Hondzo and Stefan, 1993).

---

decrease in weathering of rocks and soils, may cause decreased export of chemicals—including nutrients, base cations, strong acid anions, silica, and dissolved organic carbon—to lakes and larger rivers (Schindler et al., in press a,b). In lakes, longer ice-free seasons, warmer water temperatures, and deeper thermoclines may cause dramatic increases in habitat warmth (measured as degree-days) and decreases in habitat suitable for cold-water species such as trout, thus affecting the composition of fish communities (Magnuson et al., 1990; Stefan et al., 1995).

Limnologists have shown that in addition to affecting lakes and streams, global change will markedly affect the balance of productivity and decomposition in wetland ecosystems (Billings, 1987; Gorham, 1995). Peatland ecosystems sequester an estimated 4×10^{14} kg of carbon, about one-quarter of the total soil pool (Woodwell et al., 1995) and thus play a major role in the global carbon cycle. A general warming would lead to longer growing seasons and might simultaneously increase net primary productivity in wetlands. Under wetter conditions, production and emission of CH_4 in wetlands would increase relative to production and emission of CO_2. In contrast, if climatic warming leads to a drawdown of wetland water tables, CO_2 emissions are likely to increase and CH_4 emissions to decline. The possibility of melting of the extensive permafrost beneath northern peatlands complicates this scenario (Gorham, 1995). Fire is another significant variable in this context because if droughts become increasingly frequent and severe, surface peats may catch fire and smolder for years, releasing both CO_2 and CH_4 to the atmosphere (Gorham, 1995). Continued limnological research is needed to reduce the large uncertainties—whether water tables will rise or fall, whether CH_4 levels will increase or decrease, how much fire frequency and intensity may increase—about how global warming will affect wetlands and how wetlands, in turn, will affect global warming.

Long-term records of climatic, hydrological, and ecological variables are invaluable for analyzing and evaluating the effects of climate change. For example, long-term studies of ice records (Assel and Robertson, 1994, in press) and the changes in lakes and streams (Schindler et al., 1990, 1992, in press a) under climatic warming have greatly increased understanding of the effects of climate change on aquatic ecosystems and on land-water linkages, as have paleoecological studies of wetlands. Experimental studies in limnology are particularly needed to assess the secondary effects of climate alteration (mediated through heating, erosion, nutrient alterations, etc.) on organisms and biogeochemical processes in aquatic communities.

CHEMICAL CHANGES IN AQUATIC ECOSYSTEMS

Humans have long used the water bodies beside which they have settled as receptacles for their wastes. For example, archaeological investigations in the Indus River Basin have uncovered brick sewer systems dating back as early as 2500 B.C. (McHenry, 1992). As the human population has grown and concentrated in urban areas, the quantity of wastes discharged to water bodies has increased, changing the chemistry of the receiving waters in ways that, in turn, affect aquatic organisms. Emissions of contaminants to the air through combustion also have caused significant changes in the chemistry and, in turn, the biology of aquatic systems. A wide variety of limnological research addresses the problems created by chemical changes in water bodies.

Municipal, Industrial, and Agricultural Waste Discharges

Municipal wastewater treatment plants currently discharge more than 110 million cubic meters (30 billion gallons) of sewage per day into U.S. water bodies (van der Leeden et al., 1990). Manufacturing industries discharge an additional 91 million cubic meters (24 billion gallons) per day (van der Leeden et al., 1990). More difficult to manage than these point sources of pollution, however, are diffuse sources of pollution such as agricultural and urban runoff. For example, state water resource managers identified agricultural runoff as a cause of impairment of 49 percent of the damaged lakes, rivers, and streams they assessed; they identified urban runoff as a factor in the decline of 24 percent of damaged lakes and 10 percent of damaged rivers and streams (Environmental Protection Agency, 1994).

Three of the major impacts of excessive waste loadings to surface waters are loss of dissolved oxygen, cultural eutrophication, and buildup of toxic compounds.

Loss of Dissolved Oxygen

Excess discharges of organic wastes cause the depletion of the receiving water's oxygen supply. Low oxygen levels threaten the survival of desirable sport fish species such as trout, salmon, and bass, which may be replaced by populations of catfish and carp. If the oxygen level drops low enough, even catfish and carp cannot survive. In the worst cases, when the oxygen concentration reaches zero for extended periods, no higher organisms can survive, and the only life remaining in the water body consists of anaerobic bacteria that produce gases such as CH_4 and hydrogen sulfide.

Much of the earliest aquatic science research focused on understanding the effects of sewage and other organic waste discharges on dissolved oxygen concentrations. As a result of these early efforts, the effects of adding excess organic matter to a water body are well understood. For example, in 1884, Dupré recognized that oxygen depletion in a stored bottle of water occurred because of the activity of microscopic organisms, which he called "microphytes" (Phelps, 1944). By the early decades of this century, scientists had developed standardized procedures to determine the amount of oxygen that organisms will use when degrading a given waste; this amount is known as "biochemical oxygen demand" (BOD) (Streeter and Phelps, 1925). H. W. Streeter, a Public Health Service researcher, and E. B. Phelps, a professor of stream sanitation at Columbia University, developed an equation that predicts how much the oxygen concentration of a river will decline at given points downstream of a waste discharge (Streeter and Phelps, 1925). This equation is historically important as the first mathematical model used to predict water quality. It predicts oxygen levels based on rates of two processes: microbial oxygen

---

**BOX 3-3 RECOVERY OF THE MISSISSIPPI RIVER
NEAR MINNEAPOLIS-ST. PAUL**

Early in the twentieth century, the Mississippi River near Minneapolis-St. Paul, Minnesota, was severely polluted. The discharge of large amounts of untreated sewage from the Twin Cities, combined with construction of a dam that reduced the naturally cleansing river currents, had resulted in the near elimination of dissolved oxygen from the portion of the river flowing through the Twin Cities region (Johnson and Aasen, 1989). As shown in the lower curve on Figure 3-2, mean August dissolved oxygen concentrations were near zero between St. Paul and Lock and Dam 2 in 1926; the oxygen concentration increased below the dam due to biodegradation of the pollutants upriver and natural reaeration of the water. The low dissolved oxygen levels in the reach of river between the Twin Cities and the dam encouraged the proliferation of pollutant-degrading bacteria that produce hydrogen sulfide and methane gases, creating an unbearable stench and lifting mats of sludge to the water surface (Johnson and Aasen, 1989). In addition, native fish and other pollution-sensitive organisms died off.

Research by A. H. Wiebe in the late 1920s documenting the poor condition of the Mississippi encouraged the Twin Cities to plan their first sewage treatment plant (Johnson and Aasen, 1989). Following the opening of the plant (the first on the Mississippi River) in 1938, water quality improved markedly. As shown in Figure 3-2, mean August dissolved oxygen concentrations increased to more than 4.0 mg per liter from 1942 to 1955. However, during the 1950s the population of the Twin Cities expanded dramatically, and river flows decreased at the same time. By 1960, water quality had declined again between the cities and the dam, as shown in Figure 3-2. Because of the drop in dissolved oxygen levels and the increase in pollution levels, the only life present in the portion of the river near the sewage treatment plant was tubified worms (Johnson and Aasen, 1989).

Following a significant increase in the capacity of the wastewater treatment plant and upgrades in the type of treatment provided in 1966, water quality improved again. Dissolved oxygen levels rose to more than 4.0 mg per liter (Johnson and Aasen, 1989). With the addition of more advanced treatment technology in the mid-1980s, water quality improved still further. Mean August dissolved oxygen levels now greatly exceed 5.0 mg per liter (the current water quality standard) along the full stretch of river through the Twin Cities (see Figure 3-2). Pollution-sensitive species such as mayflies have returned after a 50-year absence (Fremling, 1989).

---

demand (that is, BOD), which is a sink or loss term, and atmospheric reaeration, which is a source term. The output of this equation is known as the "oxygen sag curve," because the oxygen concentration decreases downstream of a waste discharge and, further downstream, increases again once the waste is biodegraded (see Box 3-3 and Figure 3-2). Although

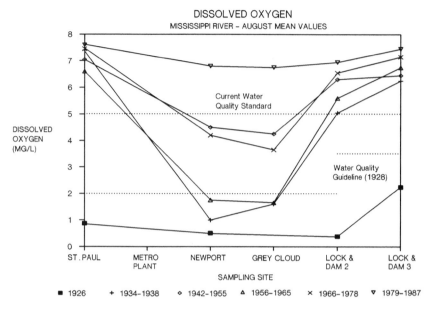

FIGURE 3-2 Mean August dissolved oxygen concentrations in the Mississippi River below St. Paul, Minnesota. Note the "sag" in the curves: the oxygen level decreases immediately downstream of the city due to oxygen consumption by microorganisms that degrade wastes; it increases again further downstream due to atmospheric reaeration once the wastes are biodegraded. SOURCE: Reprinted, with permission, from Johnson and Aasen (1989). © 1989 by the *Journal of the Minnesota Academy of Science.*

Phelps did not call himself a limnologist, he identified himself as an expert in "the science of rivers"—with knowledge of the biology, chemistry, and geology of river systems (Phelps, 1944). His work helped to advance the field of stream limnology.

Early research on how sewage affects streams was an important source of the scientific basis for federal legislation to control water pollution. For example, the opening paragraphs of the 1972 Clean Water Act state that its objective is "to restore and maintain the chemical, physical, and biological integrity of the nation's waters." The act requires states to issue permits— known as National Pollutant Discharge Elimination System (NPDES) permits—to municipalities and industries that discharge wastes to waterways. For municipal sewage treatment plants, two primary requirements specified in the permits are reductions in BOD (the basis for the Streeter-Phelps equation) and suspended solids in the waste to levels that are based on the capabilities of standard sewage treatment technologies. Thus, the early work by Streeter, Phelps, and others examining the effects of pollution on streams helped to provide the technical basis for this permit system.

Although the goals of the Clean Water Act have not been fully achieved (see Chapter 1), the act has been effective in reducing discharges of untreated municipal sewage and, as a consequence, in combating oxygen depletion in waterways. In 1960, for example, 110 million of 180 million U.S. residents were served by sewage systems, but only 36 percent of the sewage was treated before discharge to a water body. For the remaining 64 percent, waterways were used as conduits for raw sewage. By 1984, 170 million of 236 million U.S. residents were served by sewers, and the sewage was treated for 169 million (more than 99 percent) of those served (van der Leeden et al., 1990). By 1992, the 50 states reported to the Environmental Protection Agency (EPA) that municipal sewage posed a "major" problem in just 2 percent of rivers they assessed and 3 percent of assessed lake acres (EPA, 1994). Similarly, the states reported that dissolved oxygen depletion is a major source of impairment in 2 percent of assessed rivers and 3 percent of assessed lakes (EPA, 1994). The work by river specialists and other aquatic scientists has been a major factor underlying the success of the Clean Water Act in addressing the dissolved oxygen depletion problem.

Cultural Eutrophication

Cultural eutrophication, the detrimental increase of nutrient inputs to lakes by humans, has been one of the major freshwater problems of the twentieth century. Typical symptoms include the development of nuisance blooms of algae and aquatic macrophytes, with associated problems of taste, odor, and even toxicity; the depletion of deep-water oxygen due to increased decomposition; and the loss of species of fish and invertebrates that require high oxygen concentrations (see Box 3-4). Limnologists began documenting these problems as early as the mid-nineteenth century, but paleolimnological studies have revealed that the problem existed at least as early as the second century (Hutchinson, 1969). By the early twentieth century, limnologists had developed a system for classifying lakes relative to their degree of eutrophication (Thienemann, 1925). By the mid-twentieth century, cultural eutrophication was recognized as a pervasive problem in much of western Europe and North America.

Despite early recognition that nutrients were responsible for eutrophication, many questions were unresolved in the early 1960s, which prevented the adoption of effective prevention, management, and restoration strategies. Public concern about the problem led governments in many countries to establish large-scale research programs to address it. Important questions that needed answers were (1) the role of various nutrients in contributing to eutrophication problems; (2) the quantitative importance of various natural and cultural sources of nutrients; (3) the extent to which natural factors such as lake shape and water renewal rates influence the ability of a lake to assimilate nutrient inputs without devel-

---

BOX 3-4 THE CULTURAL EUTROPHICATION OF LAKE GEORGE

Lake George, located in the eastern Adirondack Mountains of New York State, is sometimes called the "queen of American lakes." Its water clarity is among the highest of natural lakes in the United States. Most of the lake's watershed is forested, but the southwestern shore has residential developments, and a major commercial area is located on the lake's southern tip. The year-round population is about 5,000, but in summer the population increases to about 50,000. Residents and visitors use the lake for recreation and potable water.

As the population surrounding Lake George has increased, the lake's characteristics have changed. Projected future development of the basin has raised concerns about increasing eutrophication, because the recreation-based economy of the area depends heavily on maintaining the lake's water quality.

Limnological studies of Lake George identified a logarithmic increase in algal production between 1974 and 1978. Distinct spatial water quality patterns were observed, with lower transparencies, lower dissolved oxygen concentrations in deep waters, higher phosphorus and chlorophyll-a concentrations, and a greater occurrence of blue-green algal blooms in the southern portion of the lake than in the undeveloped northern portion. In addition, fecal coliform counts in the southern end of the lake were five times above permissible levels for contact recreation. Of the 80 streams draining into the lake, those in the southern basin, nearest to human developments, had significant problems with sedimentation, affecting lake water quality and the fish spawning areas in the streams and the lake. The studies attributed the reduction in water quality to runoff from urban developments. As an example, the Lake George National Urban Runoff Program project determined that the increased algal production was due to phosphorus (EPA, 1982). Because there are no point-source discharges, the phosphorus loading was attributed to urban runoff. The study estimated that urban runoff accounted for 20 percent of the annual phosphorus load in the southern basin.

---

oping undesirable symptoms of eutrophication; (4) the relative importance of internal nutrient cycling versus external nutrient loading in causing or maintaining symptoms of eutrophication; and (5) the degree to which controlling external nutrient sources would result in rapid improvements in water quality conditions or whether additional in-lake restorative measures would be needed for most lakes. There was a lack of consensus among limnologists even on a question as basic as "How should eutrophication be defined?" Limnologists responded rapidly to the challenge of these questions, stimulated by the widespread occurrence of eutrophication problems and the availability of government research funds. Proceedings of several important symposia during this period (e.g., National Research Council, 1969; Likens, 1972) document the breadth of research activities aimed at solving the problem. Such symposia played key roles in exchange of information and ideas among limnologists and in the development of recommendations for how eutrophication might be controlled.

Within a decade, many of the most critical issues had been resolved, at least to the extent that effective management programs could be established. For example, limnologists developed ways to express the consequences of eutrophication quantitatively in terms of changes in lake transparency and chlorophyll content of phytoplankton (e.g., Carlson, 1977), and they developed quantitative predictive models in the late 1960s and early 1970s (e.g., Vollenweider, 1969; Dillon and Rigler, 1975) that facilitated the determination of loading criteria for phosphorus. A key limnological study of an early lake restoration effort on Lake Washington in Seattle resolved questions regarding the reversibility of eutrophication. W. T. Edmondson, a University of Washington limnologist, demonstrated that this lake recovered rapidly following diversion of nutrient-rich sewage effluent from the lake (Edmondson, 1970) (see Box 3-5). However, it should be noted that control of external nutrient sources is not sufficient to alleviate the symptoms of eutrophication in all lakes. In many cases, in-lake treatment to decrease the rate of phosphorus recycling from botton sediments or to restore damaged habitat also is necessary to reverse long-term and severe impacts of excess nutrient loadings.

A hotly debated controversy developed over the role of phosphorus in eutrophication (see Box 3-6). Members of the detergent industry (a major source of phosphorus in sewage) proposed that carbon dioxide was the primary nutrient controlling phytoplankton blooms in eutrophied lakes and that phosphorus was of little importance (Edmondson, 1991). Confusion about which nutrients to control ultimately was resolved through a variety of field experiments, laboratory studies, and modeling exercises that identified phosphorus as the critical element in most lakes.[1] Limnologists at Canada's Experimental Lakes Area in the late 1960s and early 1970s designed whole-lake experiments to resolve the controversy. An early experiment in one lake showed that controlling carbon sources would have little effect on eutrophication, because even in the most severely carbon-limited lakes, sufficient carbon could be drawn from the atmosphere by photosynthesis to allow the development of large algal blooms (Schindler et al., 1972). A second experiment in another lake showed that phosphorus control was effective at preventing eutrophication and that other nutrients would not cause eutrophication in the absence

---

[1]Lakes in glaciated regions of the northeastern and north-central United States are phosphorus limited in the absence of anthropogenic nutrient loadings. Coastal waters and lakes in some other geographic regions may be nitrogen limited. A well-known example of a nitrogen-limited inland lake is Lake Tahoe in the Sierra Nevada.

BOX 3-5 RECOVERY OF LAKE WASHINGTON

Lake Washington serves as a remarkable example of the use of basic scientific research in a public action to protect environmental quality. The lake began to show a response to eutrophication in 1955. Although its condition was not bad at that time, the lake was protected by public action before serious deterioration took place. The action involved passage of a special law by the state legislature, creation of a new kind of municipal government organization, a public education campaign, and a public vote in which the community agreed in effect to tax itself to protect the lake.

The lake had been receiving increasing amounts of nutrient-rich secondary sewage effluent since 1941. By 1955, the daily input was about 75,000 cubic meters (20 million gallons). In that year there was a modest bloom of *Oscillatoria rubescens*, a species of cyanobacteria that had appeared early in the eutrophication of several European lakes. The bloom served as an early warning signal. W. T. Edmondson, at the University of Washington, started a research program designed as a basic scientific, experimental study of lake fertilization. At the same time, public awareness of a number of pollution problems in the area was growing. The mayor of Seattle appointed a committee of public-spirited citizens to study the problems and propose solutions. Meanwhile, Edmondson's studies led him to predict that Lake Washington would develop nuisance conditions in a few years and that diversion of sewage effluent would not only protect the lake against further deterioration but permit it to recover from the damage already shown. The committee proposed diversion of the effluent in such a way as not to cause the same problem elsewhere.

Two difficulties had to be overcome. First, since sewage was coming from several cities around the lake, financing would have to be shared. State law

---

of phosphorus (Schindler, 1974). Overall, these experiments demonstrated that atmosphere-water exchanges of carbon and nitrogen could correct algal deficiencies of these elements. They also supported newly developed quantitative models relating phosphorus input and water renewal rates to eutrophication; these models form the basis for modern eutrophication control strategies (Vollenweider, 1969, 1975, 1976; Dillon and Rigler, 1975). Following these limnological discoveries, in North America and most European countries, phosphorus inputs to lakes from sewage and detergents were controlled in an effort to solve eutrophication problems.

In many cases, the control of point sources of nutrients alone has not been enough to prevent or reverse eutrophication. As nutrient control measures took effect on municipal effluents, the relative importance of contributions from nonpoint sources began to rise. By 1980, nonpoint sources contributed approximately 95 percent of the total nitrogen load and 98 percent of the total phosphorus load to surface waters (van der Leeden et al., 1990). Also, as population and industry continue to grow, at that time did not provide for such financial arrangements among municipalities, so the first hurdle was to get a new law passed by the state legislature. A persuasive presentation of the case was made by a state senator who later became governor and was well aware of the limnological situation. The enabling legislation permitted the formation of a governmental entity, the Municipality of Metropolitan Seattle (Metro), to deal with several kinds of problems associated with urbanization. The next step was to persuade the citizens to work for Metro and to commit $125 million for the diversion project, essentially a tax on themselves; there would also be a monthly fee for each household for treatment.

Opposition to the Metro project developed, based on financial and political considerations. In 1958, federal funding was not available for such a project. An educational campaign was launched with lectures and with debates presenting both sides of the issue. It made full use of the limnological information and predictions coming from Edmondson's laboratory, as well as the results of an engineering study.

After a positive vote in favor of Metro and the sewage diversion, five years passed while plans were drawn up and construction began. During that time cyanobacterial populations increased and transparency decreased, a condition widely noticed by the public. Diversion took five more years to complete. With the first partial diversion, the lake stopped deteriorating and then improved to a condition better than it had been in 1950, five years before the *Oscillatoria* bloom.

A full account of the limnological situation and the public action is given by Edmondson (1991), and recent developments are described in Edmondson (1994).

---

atmospheric precipitation and windblown dust have become sources large enough to aggravate the problem (e.g., Brezonik, 1976; Hendry et al., 1981). In some cases, atmospheric inputs of nitrogen have exceeded "critical loads" to natural catchments, causing saturated ecosystems to release excess nitrogen to lakes and streams, where both acidification (via nitric acid) (Nilsson and Grennfelt, 1988) and eutrophication (in nitrogen-limited waters such as Lake Tahoe and the Baltic Sea) result (Rosenberg et al., 1990). Nitrate concentrations in the Great Lakes and in lakes of Scandinavia have increased dramatically in the past few decades (Bennett, 1986; Henriksen and Brakke, 1988).

Even with increasing land-use controls to prevent excess nutrient discharges, the eutrophication problem will not be eliminated entirely. Limnologists have shown that in some cases; internal recycling of nutrients from bottom sediments can help to maintain eutrophic conditions in lakes for long time periods after external loading rates are controlled (e.g., Chapra and Canale, 1991).

Limnologists have experimented with altering the structure of food

BOX 3-6 THE GREAT CARBON CONTROVERSY

During the 1960s, considerable public concern developed over the rapidly increasing incidence and intensity of algal blooms in lakes of Europe and North America, including the Great Lakes. Dr. R. A. Vollenweider, a limnologist, spent several years reviewing possible causes of this eutrophication. In a widely circulated report, he concluded that increased inputs of phosphorus and nitrogen were the most probable causes of the problem (Vollenweider, 1968). His report convinced the International Joint Commission (a joint U.S.-Canadian body that recommends Great Lakes management strategies) to recommend that phosphorus be removed from sewage effluents and eliminated from laundry detergents, the major sources of the element to the lower Great Lakes.

The soap and detergent industry resisted this recommendation. Industry representatives argued that laboratory experiments with Great Lakes water had shown carbon, not phosphorus, to be limiting to algal growth. They also presented other calculations, based on the amount of carbon in lake water, suggesting that carbon would limit algal production and that the carbon in sewage effluents, rather than the increasing phosphorus, was causing the eutrophication problem.

A whole-lake experiment was conducted to resolve the carbon-phosphorus controversy. Limnologists at Canada's Experimental Lakes Area (ELA) selected a small oligotrophic lake containing a very low concentration of inorganic carbon (much lower than concentrations in the Great Lakes and typical eutrophic lakes). They added phosphorus and nitrogen to the lake in order to determine whether carbon limitation would prevent the lake from becoming eutrophic. Within weeks after fertilization began, an enormous algal bloom developed in the lake, proving that low carbon would not limit eutrophication (Schindler et al., 1973). Further studies in the lake showed that the carbon necessary to produce and maintain the algal bloom was entering the lake as carbon dioxide from the atmosphere (Schindler et al., 1972). The laboratory experiments used as evidence by proponents of the carbon limitation theory had provided erroneous results. The experiments had been carbon limited because carbon dioxide was prevented from entering the test vessels. The atmosphere had been ignored as a carbon source. This important limnological experiment provided strong evidence to refute the carbon limitation theory. In subsequent whole-lake experiments using combinations of phosphorus, nitrogen, and carbon, limnologists at ELA were able to show that phosphorus control would indeed control eutrophication in most lakes (Schindler, 1974, 1977). In most parts of the world, control of phosphorus input is now used as the basis for controlling eutrophication.

webs, termed "biomanipulation," to reduce populations of nuisance algae in situations where nutrient and water inputs cannot be altered sufficiently (Shapiro et al., 1975; Shapiro and Wright, 1984; Shapiro, 1995). Biomanipulation controls algal blooms by maximizing the population of grazing zooplankton. The zooplankton population can be increased by minimizing the population of zooplanktivorous fish by establishing large populations of piscivorous predators. Carpenter et al. (1985) developed a set of specific predictions on the cascading impacts of shifting populations of large piscivores on the primary producers of lakes and tested the predictions through a series of whole-lake experiments (Carpenter and Kitchell, 1993). These experiments supported the basic biomanipulation concept, although some limnologists have criticized it (see Shapiro, 1995). The efficacy of biomanipulation as a management tool is currently being explored in Lake Mendota, Wisconsin, through a collaborative effort involving the University of Wisconsin–Madison and the Wisconsin Department of Natural Resources (Kitchell, 1992).

The response of limnologists to improve understanding of the eutrophication problem is one of the important success stories in ecological science. Key investigations involved both fundamental and applied science. Practical solutions were implemented that reversed the trend of increasing eutrophication in many important lakes, including Lakes Erie, Michigan, and Tahoe. However, many lakes continue to suffer from eutrophication problems in part because of the economic difficulties involved in implementing effective controls on nonpoint nutrient sources. Development of cost-effective strategies for controlling nonpoint sources of nutrients, minimizing their effects, and restoring degraded lakes remains a key research need. Moreover, many fundamental and practical limnological questions remain unanswered or incompletely resolved. For example, limnologists cannot predict which specific assemblage of blue-green algae will result from specific nutrient loadings, nor can they predict which other algae will replace them when nutrient levels are decreased or biomanipulation is used.

Accumulation of Toxic Pollutants

Municipal and industrial waste discharges, urban runoff, and agricultural runoff all contain trace concentrations of toxic compounds (Novotny and Chesters, 1981; Lazaro, 1990). Often present in these discharges are organochlorine compounds, some of which are among the most toxic substances known to humans. Examples include dioxins and PCBs (polychlorinated biphenyls), which enter waterways from industrial sources, as well as DDT, toxaphene, and many other pesticides, which enter lakes and streams via runoff from agricultural lands and managed forests.

Originally, people believed that release of toxic compounds in trace quantities caused little harm to the environment because concentrations in water were very low—usually in the part per billion, trillion, or even quadrillion range. However, such views did not account for biomagnification in aquatic food chains (Gilbertson, 1988; Peakall, 1993). Many limnological studies have shown that in the passage up food chains, organochlorine compounds can accumulate in predatory fish or fish-eating birds and mammals at concentrations up to 10 million times greater than those in lake water. Even in very remote areas, the concentrations of organochlorine compounds in fish and marine mammals can be high enough to require that consumption by humans be limited.

Much of the runoff of toxic pollutants is episodic in nature, yet past research has focused mostly on long-term ambient conditions. Limnologists are developing better approaches for capturing episodic pulses, identifying their consequences, and determining the synergistic effects of multiple contaminants and sources (Davies, 1991; Herricks et al., 1994). Limnologists also continue to study long-term, sublethal effects of toxic compounds on organisms and populations; many management decisions are based on lethal doses, whereas the true long-term consequences may result from more subtle effects. Finally, limnologists are advancing knowledge of the effects of toxic compounds that have accumulated in the sediments of rivers and lakes, but much more needs to be learned about the long-term effects of these compounds. It is known, however, that some pesticides have subtle, even intergenerational effects on endocrine systems (Colborn and Clement, 1992; Colborn et al., 1993).

In summary, limnologists, along with other water scientists, have a long history of identifying the sources, routes, and consequences of pollutants from urban, industrial, and agricultural areas. Limnologists have been and will continue to be instrumental in identifying the linkages among waste discharges, runoff, water quality, and ecosystem functioning (see Box 3-7) Their involvement in developing solutions to the problems of nonpoint-source pollution from runoff will be especially critical as these problems increase.

Acid Rain and Toxic Air Pollutants

Acid rain has been known as a localized problem resulting from industrial development since the middle of the nineteenth century; in this century it has become recognized as a broad regional to global environmental problem, with major occurrences in Europe, eastern North America, the former Soviet Union, and more recently in East Asia. It is a consequence of two major human activities. One is the use of fossil fuel. For example, many sources of coal contain substantial amounts of sulfur, which upon combustion is released to the atmosphere as sulfur dioxide (SO_2). In addition, the high-temperature combustion of all fossil fuels,

BOX 3-7 INVOLVEMENT OF LIMNOLOGISTS IN IMPROVING LAKE WATER QUALITY: TWO EXAMPLES

Two places where limnologists have helped develop plans to improve lake water quality are Lake Tahoe, in the Sierra Nevada of northern California, and Lake Mendota, in Madison, Wisconsin.

Lake Tahoe, the third-deepest lake in North America, is known for its high water clarity. Because of its unique scenic beauty, the basin surrounding the lake became an attraction for naturalists and developers. Since 1960, when Lake Tahoe hosted the winter Olympics, the resident population has grown from 20,000 to more than 50,000. Each year, more than 12 million tourists visit the lake.

The development of Lake Tahoe during the 1960s brought with it a reduction in water clarity and a doubling of phytoplankton production. By the late 1960s, water clarity had decreased from 40 meters (130 feet) to 30 meters (100 feet); today, water clarity is close to 24 meters (80 feet). Extensive algal mats developed along the shoreline in late spring, creating a decaying scum. The visible signs of eutrophication concerned citizens and conservationists.

Charles Goldman, a limnologist at the University of California, Davis, determined that the nitrogen content of the lake had increased with urbanization, and this had promoted the growth of excess algae (Goldman, 1981, 1985, 1988, 1989). With the increase in urbanization had come an increase in nitrogen from septic systems, lawn fertilizers, stormwater runoff from impervious surfaces, and erosion of topsoil associated with land clearing and development (Goldman, 1981, 1985, 1988, 1989). Goldman's work led to the development of a new sewage disposal plan for the basin. Working with a group of environmental engineers, he facilitated conversion of the septic system to a municipal sewage treatment network that incorporated special facilities for removing nitrogen from the sewage. Goldman's work continues to provide a basis for educating the public and is the foundation of conservation efforts to control the effects of future development on Lake Tahoe.

Lake Mendota, a prominent feature of the landscape of Madison, Wisconsin, has a long history of limnological research (Brock, 1985; Kitchell, 1992). Despite some major successes in mitigating the effects of pollutants from point sources, the lake's water quality continues to have undesirable attributes that are direct consequences of nutrients entering the lake from nonpoint sources of pollution. Two Wisconsin limnologists, Patricia Soranno and James Kitchell, are involved in efforts to reduce the impact of these excess nutrients. One program involves manipulating the food web to increase water clarity by increasing the number of predators of zooplankton-consuming organisms, thus lowering the populations of these secondary consumers and increasing the populations of algae-eating zooplankton (Kitchell, 1992). A second effort involves an assessment of the effects of land use on nonpoint-source pollutants. Because agricultural land use dominates the Lake Mendota watershed (presently accounting for 86 percent of the surface area), it contributes most of the basin's nonpoint-source nutrient loading. Current land-use trends are to urbanize existing agricultural land, and these changes are expected to degrade water quality further (Soranno et al., in press). These shifts are being evaluated in a major project sponsored by the Wisconsin Department of Natural Resources.

especially in internal combustion engines of automobiles and trucks, partially oxidizes atmospheric nitrogen to the compounds nitric oxide (NO), and nitrogen dioxide (NO_2) (collectively referred to as NO_x). Once released to the atmosphere, sulfur and nitrogen oxides are oxidized further to sulfuric acid and nitric acid. The other major human activity leading to acid rain is the smelting of sulfide ores of various metals—including iron, copper, and lead—which releases SO_2 to the atmosphere.

Sulfur dioxide and NO_x have residence times in the atmosphere of up to a few days, by which time they can spread hundreds of kilometers from their sources. In this way, they are able to create broad patches of air pollution over, for instance, much of Europe and most of eastern North America (National Research Council, 1981). The arctic haze phenomenon also is caused by sulfate aerosols, which are carried north in winter from industrial sites in Eurasia (Barrie, 1986; Welch et al., 1991). The polar ice caps show evidence of increasing deposition of sulfate and nitrate beginning with the industrial revolution (Boutron and Delmas, 1980; Wolf and Peel, 1985).

Acid rain is neutralized readily if it falls on fertile soils rich in bases, the more so if the soils contain particles of calcium carbonate weathered from soft limestone rocks. If, however, it falls on infertile soils, poor in bases and derived from hard, slowly weathering rocks such as granite and quartzite, the acid is neutralized only partly by soil bases and by microbial and plant uptake of nitrate. The remainder, chiefly sulfuric acid, is washed into streams and lakes, where further neutralization may take place owing to sulfate reduction in lake sediments and loss of nitrate by algal assimilation or denitrification in the sediments (Cook et al., 1986; Brezonik et al., 1987; Rudd et al., 1988) (see Box 3-8).

If neither soils nor aquatic sediments can neutralize acid rain completely, the pH of the lake or stream water declines, causing a variety of deleterious effects. Marked changes occur in the species abundance and composition of all types of aquatic communities, including open-water and shoreline algae, larger aquatic plants, open-water microscopic zooplankton, bottom-living invertebrates, and fish. For instance, increasing acidity can interfere with spawning, so that the population fails to reproduce (Dillon et al., 1984). Alternatively, embryonic development or the development of species in juvenile life stages may not take place normally (Rosseland and Staurnes, 1994). Acidification of lake and stream water also releases toxic forms of aluminum (Al^{3+} and $AlOH^{2+}$) that clot the mucus on fish gills, interfering with their function. Finally, acidification can alter food webs. For example, acidification may lead to a reduction in prey species of minnows and invertebrates so that predators such as lake trout starve to death (Schindler et al., 1985). Minns et al. (1990, 1992) estimate that acid precipitation has eliminated many species of organisms in thousands of sensitive lakes in eastern Canada alone. They further

---

BOX 3-8 INTERNAL ALKALINITY GENERATION

In the early 1970s it was widely believed that lakes acidified by acid precipitation would not recover. Geological materials in the catchments of lakes were believed to be the primary source of alkalinity for neutralizing acid in rain and snow. It was thought that these would become exhausted, after which neutralization of incoming acids could not occur. Early acidification models were constructed on this belief.

This view was peculiarly at variance with limnological studies. More than 30 years earlier, G. E. Hutchinson (1941) and C. H. Mortimer (1941-1942) had published observations showing that alkalinity was produced by anoxic lake sediments, although the mechanisms by which this occurred were not elucidated.

A whole-lake experiment at Canada's Experimental Lakes Area quantified the extent of in-lake alkalinity production and revealed that when sulfuric acid was the primary strong acid added, microbial reduction of sulfate to sulfide was the most important process (Cook and Schindler, 1983; Cook et al., 1986). Subsequent whole-lake experiments with nitric acid showed that algal uptake of nitrate and microbial denitrification to N_2 similarly neutralized incoming nitric acid (Rudd et al., 1990). Investigations of lakes in other regions showed that the acid-neutralizing processes are widespread in lakes (Baker et al., 1986; Rudd et al., 1986; Schindler, 1986; Brezonik et al., 1987).

Although the microbial in-lake acid neutralizing mechanisms are not 100 percent efficient, they greatly reduce the effect of acid precipitation and allow lakes to recover when acid precipitation is reduced. Limnologists have developed successful models to predict the rate of internal alkalinity generation in different lakes with different inputs of sulfuric and nitric acids (Kelly et al., 1987; Baker and Brezonik, 1988).

---

predict that recently imposed controls on sulfur oxide emissions will reduce acidification damage only to about half of that caused by emissions at early 1980s levels.

Limnologists have shown that the degree of biological change occurring in aquatic communities as a result of acid rain can be related clearly to the degree of acidification (Brezonik et al., 1993), but in only a few cases can the chain of cause and effect be specified in detail; these cases relate chiefly to fish of recreational and economic importance, in particular salmonids (Baker et al., 1994). Simple direct responses by individual species to changes in chemical conditions can be ruled out as the mechanisms underlying many organismal responses to acidification (Webster et al., 1992). Likewise, it has been shown that standard laboratory bioassays are of limited use in predicting organismal responses to acidification in actual ecosystems (Gonzalez and Frost, 1994). Despite expenditures of many millions of dollars on research related to acidic deposition, especially

Vinyl curtain being placed in a lake for an experiment on acidification. The curtain divides the lake into two basins. One will be treated to lower the pH, and the other will be used as an experimental control. SOURCE: Dave Cornelius, University of Wisconsin, Center for Limnology.

during the 1980s, information about the full range of ecosystem effects of acid rain is lacking for nearly all groups of aquatic organisms other than commercially important fish. Therefore, it still is not possible to predict with great accuracy the long-term consequences for the functioning of aquatic ecosystems of the alterations brought about by acid stress, and further research is needed.

Studies by paleolimnologists show that many ecosystems have not come to equilibrium with the acid loading now affecting them (Charles et al., 1990). Limnologists have also shown that climatic warming and acid deposition may have synergistic effects (Bayley et al., 1992a; Lazerte, 1993; Schindler et al., in press,b). Therefore, ecological and biogeochemical responses induced by acid rain can be revealed only by whole-ecosystem experiments and long-term studies over several decades. Too few such studies have been carried out, and more are needed in different terrains with differing biotic communities.

Now that many, but by no means all, countries have passed legislation to partially control emissions of sulfur, the continuation of studies of ecological effects of acid rain has been curtailed severely, as have studies of the recovery of ecosystems damaged by acid deposition. For example, research funds to study acidification problems have plummeted in the

United States since 1990, when Congress passed the Clean Air Act Amendments, which call for a decrease in sulfur oxide emissions by about 50 percent by the end of the century. Even with controls on sulfur emissions, long-term atmospheric deposition of nitrate- and ammonia-nitrogen remains an uncontrolled threat to many aquatic (and other) ecosystems. The effects of nitrogen deposition have received far less study than the toxic effects associated with the deposition of sulfate and its associated hydrogen ions, and increased attention to this issue is warranted. Indeed, the EPA recently announced that it would not propose a new acid deposition standard to protect sensitive areas from nitrogen deposition because "scientific uncertainty" makes it "difficult to determine the appropriate level of a standard or standards at this time" (EPA, 1995). This inaction is of particular importance to individuals in New York who are concerned about protecting vulnerable areas in the Adirondack Mountains (Renner, 1995).

Along with sulfur and nitrogen, fossil fuels contain a variety of toxic metals, among them mercury, lead, and cadmium. Upon combustion these are emitted to the atmosphere on particulate matter that can be transported over long distances. Smelters add to the atmospheric loading of these metals. In addition, the combustion of fossil fuels produces a variety of polycyclic aromatic hydrocarbons that can be transported in the atmosphere by particulate matter. Some of these, such as benzo-*a*-pyrene (also present in tobacco smoke), are extremely toxic. Other organic micropollutants distributed regionally through the atmosphere are a consequence of the agricultural use of herbicides and insecticides. Still other organic micropollutants (such as dioxins and PCBs) are emitted to the atmosphere—as well as directly to aquatic ecosystems—as a result of industrial activity. Although PCBs and many chlorinated insecticides have been banned from manufacture and use in the United States, the legacy of past use still provides a source of these materials. Some banned insecticides such as DDT are still used in other countries, such as Mexico, and enter U.S. ecosystems as a result of long-range atmospheric transport.

The deleterious effects of these metal and synthetic organic compounds upon inhabitants and users of aquatic ecosystems are less well documented than those of acid rain. However, in the case of mercury and PCBs, warnings have been issued in more than 30 states concerning the consumption by humans of fish from numerous lakes and rivers, including many in the Great Lakes states. Atmospheric transport and deposition is the major source of mercury in these regions (Swain et al., 1992). Eisenreich and coworkers (1979) have shown that the principal mechanism by which PCBs enter Lake Superior is long-range atmospheric transport and deposition by rain, snow, and dryfall. Dioxins and PCBs also have been found in sediments and biota of Siskiwit Lake, an otherwise pristine water body on Isle Royale in Lake Superior (Swackhamer and Hites, 1988). Isle Royale, a U.S. national park, has no permanent human inhabitants, and the watershed for Siskiwit Lake is wholly forested. Thus, there are no sources of organochlorine compounds in the watershed; long-range atmospheric transport of these compounds is the only logical explanation for their presence in the lake. In Canada, airborne contamination of lakes has resulted in consumption advisories for fish throughout Ontario and in regions as remote as the Yukon Territories (Kidd et al., 1995b) (see Box 3-9).

The presence of metals and synthetic organic compounds is a concern because many of them may cause birth defects, cancer, or immunological and reproductive disorders (Colborn and Clement, 1992; Colborn et al., 1993). Many bioaccumulate to such high levels that they are toxic to the end-members of aquatic food chains. In the case of mercury, rules of the U.S. Food and Drug Administration state that fish cannot be sold if the concentration exceeds 1 part per million (ppm). In Canada, the corresponding threshold for commercial sale is 0.5 ppm; that value is exceeded in many areas, most notably in reservoirs (Rosenberg et al., 1995). The Minnesota Department of Health recommends that pregnant women eat no more than one meal per month of fish containing more than 0.15 ppm of mercury. Minnesota has a lake water standard for mercury concentra-

---

BOX 3-9 LONG-RANGE TRANSPORT OF TOXIC COMPOUNDS TO LAKE LABERGE, YUKON TERRITORIES

Lake Laberge is widely regarded as a symbol of remote northern wilderness as the result of the popular poem "The Cremation of Sam McGee," written by Robert W. Service during the Yukon gold rush. The catchment of the lake is still largely uninhabited except for the city of Whitehorse, above the lake on the Yukon River. The lake is large (200 km²) and served as an important source of protein for local indigenous peoples.

In 1992, the fishery of the lake was closed because of high toxaphene concentrations in the fish. Concentrations of DDT and PCBs also were higher than in other large lakes of the Yukon. Initially, investigators hypothesized that the contamination was caused by surreptitious dumping of toxaphene in the lake. However, detailed studies of toxaphene concentrations and stable isotopes of nitrogen in Laberge and other lakes of the area, which had lower toxaphene concentrations, showed that the higher toxaphene concentrations in Laberge fish may be explained by the fact that food chains were one step longer in Laberge than in other lakes of the area (Kidd et al., 1995a,b). The toxaphene entered the lake via contaminated rainfall from the United States and Eurasia and then accumulated in the fatty tissues of aquatic organisms, with higher levels of toxaphene found at higher levels of the food chain (see Figure 3-3).

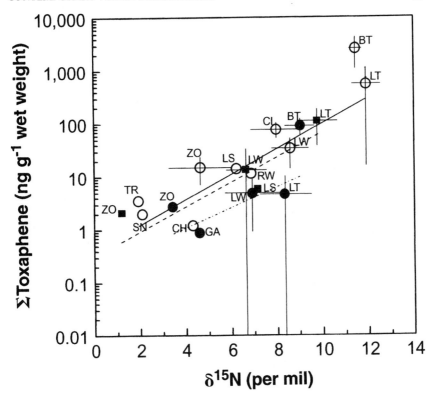

FIGURE 3-3 Toxaphene concentrations in species at various levels of the food chain in three Yukon lakes: Laberge (○, solid line), Fox (●, dash-dot line), and Kusawa (■, dashed line). The stable nitrogen isotope ($\delta^{15}N$) value shown on the horizontal axis is a surrogate measure for position on the food chain; $\delta^{15}N$ increases from prey to predator by an average of 3.4 per trophic level. The data show that species at the highest level of the food chain accumulate the largest amount of toxaphene and that lake trout and burbot in Lake Laberge are nearly a trophic level higher (and thus several times more contaminated) than in other lakes.

Organisms shown on the figure are as follows: LT, lake trout; BT, burbot; LW, lake whitefish; RW, round whitefish; CL, least cicso; LS, longnose sucker; ZO, zooplankton; GA, *Gammarus* sp.; CH, chironomid (subfamilies Tanypodinae, Prodiamesinae, and Chironomidae); SN, snail (family Lymnaeidae); and TR, tricopteran (family Limnephilidae).

SOURCE: Reprinted, with permission, from Kidd et al. (1995b). © 1995 by *Science*.

tion of only 7 nanograms (ng) per liter, intended to protect humans from the consumption of contaminated fish. In parts of the state, this standard may be inadequate because in some lake waters with mercury levels of only 2 ng per liter, the concentration in fish exceeds 0.45 ppm—three times the consumption advisory for pregnant women. Larger, older predatory fish at the top of the aquatic food chain are of particular concern because they accumulate the highest levels of mercury. Figure 3-4 provides an example of advisory recommendations for an individual lake in northern Minnesota contaminated by atmospheric deposition of mercury.

Wildlife, of course, cannot respond to fish advisories, nor do all humans heed the warnings. Birds such as belted kingfishers, common loons, ospreys, and bald eagles consume large amounts of fish, as do mink and river otters among the mammals. In Minnesota there is evidence that mercury accumulation may be impairing reproduction by common loons. There is also an indication that walleye reproduction may be affected. Elevated concentrations of mercury have been detected in some mink and river otters. Researchers attributed the death of a Florida panther—a rare and endangered species—in the Everglades in 1989 to elevated levels of mercury found in the animal's liver, leading to concern about mercury pollution throughout the south Florida ecosystem (Jordan, 1990). In the Great Lakes and below pulp mills, organochlorine compounds delivered via the air and other routes have been linked with reproductive

		Fish Size (inches)				
Location	Species	5-15	15-20	20-25	25-30	30+
Sand Point *(St. Louis Co., N of Crane Lake)*	Northern Pike		◒	◒	◒	◒
	Walleye	◯	◒	◒	●	
	Sauger	◒				
	Smallmouth Bass	◯				
	Black Crappie	◯				
	Yellow Perch	◯				
	Cisco	◯				

Mercury	◯	◯	◒	●
Vacation	unlimited	unlimited	1 meal/week	1 meal
Season	unlimited	2 meals/week	2 meals/month	1 meal/month
Annual	unlimited	1 meal/week	1 meal/month	do not eat

FIGURE 3-4   Minnesota fish consumption advisory, 1994. "Vacation" refers to those who eat the particular fish species only during short vacations; "season" refers to those who consume the fish seasonally; and "annual" refers to those who consume the fish year-round. SOURCE: Minnesota Department of Health, 1994.

problems—including sterility, birth defects, unsuccessful hatching, and deformities in offspring—in fish-eating birds (Gilbertson et al., 1991). In northern Canada, indigenous peoples who eat large quantities of aquatic organisms often have unacceptably high concentrations of mercury in their blood and hair (Lockhart, 1995).

The likelihood of synergistic interactions among diverse pollutants from atmospheric sources needs far more investigation. For example, a number of organochlorine compounds are now known to have similar effects on enzyme and hormone systems, indicating that they may have additive or synergistic effects (Colborn and Clement, 1992). Another example of such synergism is mercury and acid rain. Bioaccumulation of mercury along food chains depends in the first instance on its transformation by microbes from the inorganic form to methylmercury, which has greatly increased mobility and biological uptake. Several studies have shown that the rate of methylation increases upon lake acidification (Xun et al., 1987; Ramlal et al., 1993); this may provide an explanation for the typically higher concentrations of mercury observed in fish from lakes with low pH (Wiener et al., 1990).

Limnologists, in association with a variety of other scientists (fisheries biologists, toxicologists, aquatic chemists, hydrologists, atmospheric scientists), have been centrally involved in the study and solution of problems caused by acid rain and airborne toxic compounds. Table 3-1 chronicles some of the key scientific discoveries concerning the effects of acid rain and toxic air pollutants on aquatic ecosystems.

BIOLOGICAL CHANGES IN AQUATIC ECOSYSTEMS

All of the physical and chemical changes described above can markedly affect the populations that inhabit inland waters. Physical changes (such as dam construction and global warming) and chemical inputs (from acid rain, runoff, direct waste discharges, airborne toxic compounds, etc.) can change the structure of the aquatic food web and create conditions in which native species cannot survive. In addition to causing biological changes via physical and chemical changes, humans have affected aquatic biota in more direct ways: by introducing exotic species, altering aquatic communities to support game fish, and causing the extinction of native species.

Exotic Species Introduction

Sometimes intentionally and sometimes inadvertently, humans have substantially expanded the geographic ranges of many species. For example, about 20 percent of the 5,523 vascular plant species of eastern North America have been introduced (Fernald, 1950). In many cases, species

TABLE 3-1 Some Significant Discoveries Concerning the Effects of Atmospheric Deposition On Aquatic Ecosystems, 1957–1995

Date	Reference	Topic
1957	Mackereth	Showed the degree of acidification of English lakes to be related to substrate geology
1959	Dannevig	Recognized the relationship among acid rain, surface water acidity, and disappearance of fish in Norway
1967	Woodwell et al.	Reported biological magnification of DDT in aquatic food chains
1971	Winchester and Nifong	Indicated atmospheric precipitation as an important source of trace metals in Lake Michigan
1974	Almer et al.	Showed a reduction of biodiversity in phytoplankton, zooplankton, and fish in acidified Swedish lakes
1977	Murphy and Rzeszuko	Found atmospheric deposition to be an important source of PCBs to Lake Michigan
1978	Ferguson et al.	Implicated air pollution and associated acid deposition in the disappearance of *Sphagnum* mosses from northern English bogs during the industrial age
1979	Cronan and Schofield	Showed toxicity to fish of aluminum released from soils by acid deposition
1980	Davis and Berge	Inferred pH profiles in dated cores of lake sediment from diatom stratigraphy
1985	Schindler et al.	Demonstrated that adverse effects of acid deposition on fish can occur via harm to food organisms
1992	Wright et al.	Showed rapid recovery of catchments following reduced acid loading
1995	Muir et al.	Demonstrated high concentrations of organochlorine contaminants in arctic marine fauna due to long-range transport and bioaccumulation

have been introduced into regions to which they are foreign. Other cases involve the shifting of species distributions within narrower geographic regions. In either situation, introduced species can have substantial effects on aquatic habitats; they may shift the occurrence or abundance of native species, change important ecosystem functions, and interfere with desired human uses of the water body. Limnologists are important in developing assessments of the consequences of inadvertent introductions. They also can play a key role in the development of realistic predictions for intentional introductions. Finally, limnologists can help create management strategies to minimize the impacts of introductions that have already taken place.

Many exotic species, including the sea lamprey, alewife, and zebra mussel, have invaded the Great Lakes over the past century. In response, a variety of nonnative fish species have been introduced intentionally, in some cases to control invading species and in other cases to replace fish eliminated by the invaders. Many of the invading species entered the lakes after large canals and locks were constructed to allow oceangoing commercial ship traffic in the upper lakes. In essence, the Great Lakes have served as an unintentional laboratory for investigating the effects of invading and introduced exotic species because so many have established themselves there (Mills et al., 1994).

One exotic organism that has received considerable recent attention is the zebra mussel, *Dreissenia polymorpha* (Herbert et al., 1991). Zebra mussels were introduced to the Great Lakes in 1986, when a ship from Europe discharged its ballast water into the St. Clair River between Lakes Huron and Erie (Roberts, 1990). The zebra mussel was not detected until 1988, when two Canadian students discovered the species in a sample from Lake St. Clair. By that time, the mussel had already moved downstream into Lake Erie, where its population exploded. Since then, the mussel has spread throughout the Great Lakes and into other waterways, including the Mississippi and Hudson rivers.

Zebra mussels have a wide variety of effects on invaded water bodies. They filter vast quantities of water to obtain phytoplankton for their diet. Zebra mussel filtering is likely responsible for a recent doubling in water clarity in Lake Erie (Holland, 1993). Not all effects are positive, however. The National Biological Service estimates that this invader will cost $5 billion to industries, municipalities, and private citizens by the year 2000 because of its ability to attach to almost any hard surface. Negative effects range from clogging of water intake pipes for industries and water treatment plants to loss of important native mussels and clams that cannot compete with the invader. For example, at a water treatment plant on Lake Erie, a combination of zebra mussels and ice completely blocked the intake, so that the city of Monroe, Michigan, had to impose emergency water conservation measures and close businesses and schools. Reopening the water intake cost $250,000; completely retrofitting the system to prevent such problems in the future will cost $6 million (Roberts, 1990). Zebra mussels also threaten native clams and fish spawning beds because they can literally cover them and thereby get first access to the food supply. The mussels' phenomenal numbers (a mature, thumbnail-sized female can produce a million eggs) and prodigious appetite have been found to produce three striking effects on waters they inhabit: (1) precipitous drops in phytoplankton populations, (2) increased water transparency, and (3) greatly enhanced growths of filamentous algae. The net effect is a shift in algal community structure, which can impact higher trophic levels (including fish) in aquatic food webs.

Since 1991, the federal government has appropriated $10.3 million to the Sea Grant program for university-based research on zebra mussels and public education efforts to help control their spread; various water management agencies at the federal and state levels (such as the U.S. Army Corps of Engineers) also have research programs related to this pest. Sea Grant's zebra mussel research has been conducted in six areas: (1) biology and life history, (2) effects on ecosystems, (3) socioeconomic analyses, (4) control and mitigation, (5) prevention of new introductions, and (6) reduction of spread. Limnologists at academic institutions have been involved in a wide range of research efforts within these categories. For example, limnologists have found *Dreissenia polymorpha* to be genetically highly diverse; this may allow it to spread and adapt to new environments. Laboratory-reared larvae have been observed to postpone settlement and attachment and to prolong their free-swimming period for an additional seven weeks beyond their normal time in this life form. These delays could have profound effects on the dispersal characteristics of zebra mussels. Although it is not possible to eliminate zebra mussels from water bodies where they have become established, limnologists are studying a variety of physical, chemical, and biological control measures. These include the use of ultraviolet radiation to kill larval forms of zebra mussels, application of high-voltage fields on water intake pipes, use of various chemicals to kill adults or prevent larval attachment, and isolation of bacteria that inhibit larval attachment or cause disease in zebra mussels.

Another invader of the Great Lakes from Europe via oceangoing vessels is *Bythotrephes cederstroemi*, a large crustacean zooplankton. This species is less conspicuous to the public eye than the zebra mussel but is nevertheless capable of causing significant economic damage. A voracious predator of other zooplankton, *Bythotrephes* was first observed in Lake Huron in December 1984, detected soon after that in Lake Erie, and present in all of the Great Lakes by 1988. It spread rapidly throughout Lake Michigan after its first appearance in the north end in 1986, and it has moved into lakes connected to the Georgian Bay of Lake Huron and into northern Minnesota, which is outside the drainage basin of the Great Lakes. Several groups of limnologists are investigating the role of *Bythotrephes* in the Great Lakes ecosystem (Lehman, 1991).

Changes in the zooplankton population of Lake Michigan immediately followed the increase of *Bythotrephes*. The abundance of *Daphnia* decreased sharply, with changes occurring in body size that would be expected from selective predation by *Bythotrephes*. *Daphnia* is an important food for several species of game fish, so the change may threaten the vitality of important fish populations. *Bythotrephes* is a poor food for small game fish because it has a stiff tail spine, a centimeter long in the largest individuals. However, it is eaten freely by the alewife (*Alosa pseudoherengus*), an undesirable exotic fish species in the Great Lakes. Since the production of a full-grown *Bythotrephes* involves the consumption of many *Daphnia*, the net effect on the production of game fish could be negative. Apart from its indirect effect on fisheries, *Bythotrephes* is a direct nuisance to net fisherman. Its long tail spines stick to nylon nets, and animals accumulate on nets in slimy masses, making the nets difficult and unpleasant to handle. There is no obvious technology to control *Bythotrephes*, and there appears to be no predator in the Great Lakes capable of controlling it, except perhaps the alewife, itself an undesirable exotic species whose population is now controlled by large predatory nonnative fish (salmon) that have been stocked in the lakes.

Whereas the zebra mussel and *Bythotrephes* were introduced to the Great Lakes accidentally, introductions of nonnative species are often intentional. Stocking of fish, and in some cases fish food, in particular, is a common management practice. More than 25 percent of inland fish caught in the United States are nonnative stocks (Moyle et al., 1986). The effects of such introductions often are poorly anticipated and adverse to native populations. Spencer et al. (1991) described a case in which the negative effects of the stocking the opossum shrimp (*Mysis relicta*), a forage species for fish, extended out of Flathead Lake, Montana, one of the largest natural lakes in the western United States, to eagles and bears that inhabit the area around it. Eagle and bear populations declined markedly as the introduced shrimp reduced the number of salmon, on which the animals had depended as a major food source. The salmon decline was linked to direct competition between juvenile salmon and opossum shrimp, which ultimately depend on the same prey species.

Another major example of the purposeful introduction of nonnative species is the widespread stocking of salmonid fish into lakes in the mountains of western North America. Fishless alpine lakes were stocked with eastern brook trout, European brown trout, Atlantic salmon, arctic char, smallmouth bass, and many other species. Up to 80 percent of the fishless alpine lakes in the U.S. Rocky Mountains and 20 percent of the lakes in Canadian mountain national parks have been stocked with nonnative populations (Donald, 1987; Bahls, 1992). Limnologists have demonstrated that in fishless lakes, the original invertebrate predators, such as large calanoid copepods, benthic crustaceans, and midge larvae, were extirpated by stocked fish (Lamontagne and Schindler, 1994; Paul and Schindler, 1994); this set off changes in communities that increased algal biomass in lakes, even in pristine areas (Leavitt et al., in press). In lakes where fish were present, introduced species often displaced native stocks, in some cases eliminating them completely (see Box 3-10). In other cases, native species were eliminated prior to stocking by deliberately treating the lake with rotenone or toxaphene (Miskimmin and Schindler, 1993, 1994; Miskimmin et al., 1995).

Pacific salmon have also been introduced into the Great Lakes. The salmon were initially intended to replace native lake trout that were lost largely because of the effects of the sea lamprey, which entered the Great Lakes from the Atlantic Ocean through ship canals built in the nineteenth and early twentieth centuries (Christie, 1974). The salmon are maintained only through active management programs. A substantial portion of their

---

**BOX 3-10 EFFECTS OF FISH STOCKING IN
NORTH AMERICAN ALPINE LAKES**

Even alpine lakes at high elevations in national parks of the West have not
been spared the introduction of exotic species. In the early twentieth century,
brown trout from Europe, brook trout and Atlantic salmon from eastern North
America, rainbow trout from several regions, golden trout from the southwestern United States, arctic char, rainbow trout from several locations, and several
races of cutthroat trout, especially the Yellowstone cutthroat, were widely
introduced into waters where they were not native, including naturally fishless
alpine lakes (Donald, 1987; Bahls, 1992).

Although there were few studies of aquatic communities done in conjunction with the stocking, the few contemporary studies (Anderson, 1980) and
later paleoecological studies (Miskimmin and Schindler, 1993; Lamontagne
and Schindler, 1994) showed the impoverishment of invertebrate communities
by stocked fish. In some cases, the eliminated species did not return (Paul and
Schindler, 1994). Similar effects, dating to pre-World War I times, have been
identified by paleolimnological studies (Pechlaner, 1984). Mosquitofish introduced in 1924 to control biting insects and other tropical fish released by
hobbyists into the hot springs of Banff National Park eliminated the rare Banff
longnose dace. The native bull trout is now endangered or eliminated from
much of the West by interbreeding with introduced brook trout. The hybrids
are sterile.

---

forage base involves two other species that invaded the Great Lakes, the
alewife and the rainbow smelt, each of which has had large-scale effects
on the lakes' food webs. The mechanism for the alewife's introduction to
the Great Lakes is unclear (Scott and Crossman, 1973). Rainbow smelt
were successfully introduced into Crystal Lake, a small lake connected
to Lake Michigan by a stream, in 1912 after several failed attempts at
direct establishment in the Great Lakes (Evans and Loftus, 1987). This
population eventually expanded throughout the Great Lakes. Once established, rainbow smelt and alewife ultimately caused a substantial reduction in populations of native fish species (such as bloaters, lake whitefish,
and cisco) through the combined effects of competition and predation
(Crowder, 1980; Evans and Loftus, 1987; Crossman, 1991; McClain, 1991).
Rainbow smelt are currently spreading through inland lakes throughout
the Midwest with as yet unknown but likely important consequences for
native species and food webs (Evans and Loftus, 1987).

Important invasions of aquatic ecosystems have not been limited to
animal species. Several important cases, dating back many decades,
involve the spread of plant species. The North American aquatic plant
Elodea canadensis was introduced into Europe in the early 1800s and had

enormous economic effects (Hutchinson, 1975). Its populations extended widely throughout canals that were major arteries for shipping industries at the time and made their navigation by horse-drawn barges very difficult. The favor was returned when the Eurasian water milfoil, *Myriophyllum spicatum*, was introduced to North America in the early twentieth century. This species has proliferated throughout shallow water regions of many lakes, limiting their recreational use and leading to the development of extensive and costly aquatic weed control programs. It also has caused substantial reductions in the populations of many native plants. A range of problems has been associated with the human-caused spread of the water hyacinth, *Eichhornia crassipes*, from South America throughout much of the world (Sculthorpe, 1967). This free-floating plant proliferates quickly and has seriously clogged many waterways and lakes in the southeastern United States. Similar problems have followed the introduction of other floating plant species, such as *Pistia stratiotes* (water lettuce) and *Salvinia auriculata*, which also can cause public health problems by providing refuges for disease-bearing mosquitoes. Purple loosestrife, *Lythrum salicaria*, which was introduced to North America from Eurasia in the early 1800s, has monopolized wetlands and displaced native plant species throughout the United States and Canada (Stuckey, 1980; Thompson et al., 1987). The plant has attractive flowers that, along with their use as a pollen source for honeybees, are probably responsible for the plant's spread. Community changes caused by such nonnative plants have important deleterious effects on many waterfowl and insect species, which lose access to plant species that had provided their forage base; thus, they are the subjects of active limnological research.

An important lesson to be derived from exotic species invasions and introductions is that aquatic communities often are delicately balanced. Shifts caused by invading species can have substantial effects on ecosystem and community structure, as well as on fundamental ecosystem processes, and often these shifts have impacts that are not desirable. For example, Lamarra (1975) showed that common carp, perhaps the most widely distributed nonnative fish in North America, vastly enhances phosphorus return from lake sediments, enhancing eutrophication. Limnologists, along with fisheries biologists, are essential in evaluating such effects and predicting the occurrence and results of future invasions. Addressing the problems caused by exotic species will require a fundamental understanding of a wide range of aquatic ecosystem features, including factors dictating community structures and ecosystem processes that are the basis for much limnological research.

Species Extinction

Taken to the extreme, the net impact of the varied forms of human effects on inland aquatic ecosystems can threaten entire species with extinction. The Nature Conservancy classifies at least 30 percent of all North American fish species as rare, extinct, or at risk of extinction (Master, 1990). The proportion of species in these categories is even higher for crayfish (65 percent) and unionid mussels (73 percent) (Master, 1990). Some aquatic insect species face similar risks of extinction. About one-third of rare, threatened, and endangered species in the United States are associated with wetlands (Niering, 1988).

In general, the danger of extinction for a species is related to its distribution. Threats of extinction due to human-caused environmental effects are most severe for species that occur in only a small number of habitats. In some cases, however, threatened species can be distributed fairly widely. Causes of extinction include physical disruption of habitats, the generation of chemical stress across ecosystems, and the introduction of species that are major predators, parasites, or competitors of native species. These various mechanisms sometimes interact. For example, the impact of invaders on native fish species in California streams is more severe where the natural flow regime in a stream has been modified by humans (Baltz and Moyle, 1993). Introduction of nonnative species, often through fisheries management practices, appears to be a widespread cause of threatened extinctions (Lodge, 1993).

Limnologists are addressing the problem of species extinction in two ways. First, they are gathering information on the distribution of native species and on factors that influence their densities in their habitats. This information can be used to develop management practices that minimize both the occurrence of extirpation within particular habitats and, ultimately, overall species extinctions. Second, limnologists are working to develop an understanding of aquatic ecosystems sufficient to predict the effects of the loss of a species on the entire community and on ecosystem processes. Although some ecosystem processes are fairly immune to the loss of a single species (Frost et al., 1995), in other situations fundamental ecosystem patterns and processes are influenced strongly by the presence or absence of a single species (Huntly, 1995). Developing general understanding of linkages between ecosystems and species is currently an area of substantial research interest in aquatic and terrestrial ecology (Jones and Lawton, 1995).

ENHANCING THE ROLE OF LIMNOLOGY IN FRESHWATER ECOSYSTEM MANAGEMENT

In summary, limnologists have played important roles in the development of scientific knowledge needed to manage and protect freshwater ecosystems. Limnologists have provided scientific input on questions such as how land use affects water quality, how wetlands influence global warming and the global cycling of carbon, how wastewater discharges affect water quality and aquatic communities, how fossil fuel emissions lead to the degradation of aquatic systems, and what mechanisms lead to the deterioration of lakes and rivers via cultural eutrophication. Advances in knowledge achieved by limnologists and aquatic scientists in related fields have led to plans for long-term management of important aquatic ecosystems (see, for example, Box 3-11).

In limnology as in other fields of science, answering some questions reveals additional complexities of the systems under study and creates

---

BOX 3-11 ROLE OF LIMNOLOGISTS IN GREAT LAKES MANAGEMENT

Limnologists have played important roles in cleanup and management of the Great Lakes. In the early 1960s, pollution was rampant in the lower Great Lakes. Al Beeton, a limnologist at the University of Michigan, showed from historical records how much the chemistry and biology of the lower Great Lakes had changed since the early part of the twentieth century (Beeton, 1965). In response to this and other expressions of concern, the governments of the United States and Canada wrote a formal "letter of reference" to the International Joint Commission (IJC) asking the commission to examine whether the lower Great Lakes were polluted and, if so, to identify the causes and cures.

Richard Vollenweider, a Swiss limnologist who moved to Canada in 1969, had earlier identified phosphorus and nitrogen as the principal causes of eutrophication, which was occurring in the lower Great Lakes. Vollenweider's model of eutrophication became the scientific basis for the phosphorus control program that led to restoration of Great Lakes water quality (Vollenweider, 1968). Jack Vallentyne, a Canadian limnologist, was instrumental in focusing on the need to remove phosphates from detergents as part of the control program (Vallentyne, 1970).

This work led directly to the signing in 1972 of the Great Lakes Water Quality Agreement between Canada and the United States. However, by 1978 it was apparent that water quality could not be controlled by looking at water alone. Vallentyne, working as a head of the Ecosystem Committee of the IJC Great Lakes Science Advisory Board, proposed that an ecosystem approach to water quality be used—one that examined water in the context of all the interacting components of air, water, soil, and living organisms in the Great Lakes basin (Great Lakes Research Advisory Board, 1978). Subsequently, the IJC incorporated an ecosystem approach to environmental management into the Great Lakes Water Quality Agreement of 1978 (Vallentyne and Beeton, 1988).

The ecosystem approach had been used by limnologists since Lindeman (1942) applied it to Cedar Creek Bog, Minnesota. However, it was not until the signing of the 1978 Great Lakes agreement that the approach entered the political domain (Vallentyne and Beeton, 1986). Now, the ecosystem approach is accepted as management policy in the United States, Canada, and Europe.

new uncertainties. For freshwater ecosystems, major knowledge gaps remain in areas such as preservation of ecosystems downstream of dams, restoration of damaged ecosystems, synergistic interactions among pollutants, and preservation of native aquatic organisms, as described in this chapter and in more detail in Naiman et al. (1995). Limnologists will need to play key roles on the scientific teams working to advance knowledge about how aquatic systems function and to help the public understand how a myriad of human activities can affect the health of rivers, lakes, and wetlands.

REFERENCES

Abelson, P. H. 1985. Electric power from the north. Science 228:1487.

Almer, B., W. Dickson, C. Ekström, and E. Hörnström. 1974. Effects of acidification on Swedish lakes. Ambio 3:330–336.

American Society of Limnology and Oceanography and North American Benthological Society. 1994. Symposium Report: Regional Assessment of Freshwater Ecosystems and Climate Change in North America. Boulder, Colo.: U.S. Geological Survey, Water Resources Division.

Anderson, R. S. 1980. Relationships between trout and invertebrate species as predators and the structure of the crustacean and rotiferan communities in mountain lakes. Pp. 635–641 in Evolution and Ecology of Zooplankton Communities, W. C. Kerfoot, ed. Hanover, N.H.: University Press of New England.

Andrews, E. D. 1991. Sediment transport in the Colorado River Basin. Pp. 40–53 in Colorado River Ecology and Dam Management. Washington, D.C.: National Academy Press.

Assel, R. A., and D. M. Robertson. 1994. Changes in winter air temperatures near Lake Michigan during 1851–1993, as determined from regional lake-ice records. Presented at the American Society of Limnology and Oceanography and North American Benthological Society symposium: Regional Assessment of Freshwater Ecosystems and Climate Change in North America, Leesburg, Va., October 24–26.

Assel, R. A., and D. M. Robertson. In press. Changes in winter air temperatures near Lake Michigan during 1851–1993, as determined from regional lake-ice records. Limnol. Oceanogr.

Bahls, P. 1992. The status of fish populations and management of high mountain lakes in the western United States. Northwest Sci. 66:183–193.

Baker, L. A., and P. L. Brezonik. 1988. Dynamic model of internal alkalinity generation: Calibration and application to precipitation-dominated lakes. Water Resour. Res. 24:65–74.

Baker, L. A., P. L. Brezonik, and C. D. Pollman. 1986. Model of internal alkalinity generation-sulfate retention component. Water Air Soil Pollut. 30:89–94.

Baker, J. P., J. Böhmer, A. Hartmann, M. Havas, A. Jenkins, C. Keey, S. J. Ormerod, T. Paces, R. Putz, B. O. Rosseland, D. W. Schindler, and H. Segner. 1994. Group report: physiological and ecological effects of acidification on aquatic biota. Pp. 275–312 in Acidification of Freshwater Ecosystems: Implications for the Future, C. E. W. Steinberg and R. W. Wright, eds. Chichester, England: John Wiley and Sons.

Baltz, D. M., and P. B. Moyle. 1993. Invasion resistance to introduced species by a native assemblage of California stream fishes. Ecol. Appl. 3(2):246–255.

Barrie, L. A. 1986. Background pollution in the arctic air mass and its relevance to North American acid rain studies. Water Air Soil Pollut. 30:765–777.

Bartlett, K. B., and R. C. Harriss. 1993. Review and assessment of methane emissions from wetlands. Chemosphere 26:261–320.

Baumann, R. H., J. W. Day, Jr., and C. A. Miller. 1984. Mississippi deltaic wetland survival: Sedimentation versus coastal submergence. Science 224:1093–1095.

Bayley, S. E., D. W. Schindler, B. R. Parker, M. P. Stainton, and K. G. Beaty. 1992a. Effect of forest fire and drought on acidity of a base-poor boreal forest stream: Similarities between climatic warming and acidic precipitation. Biogeochemistry 17:191–204.

Bayley, S. E., D. W. Schindler, K. G. Beaty, B. R. Parker, and M. P. Stainton. 1992b. Effects of multiple fires on nutrient yields from streams draining boreal forest and fen watersheds: Nitrogen and phosphorus. Can. J. Fish. Aquat. Sci. 49(3):584–596.

Bedford, B. L. In press. The need to define hydrologic equivalence at the landscape scale for freshwater wetland mitigation. Ecol. Appl.

Beeton, A. M. 1965. Eutrophication of the St. Lawrence Great Lakes. Limnol. Oceanogr. 10:240–254.

Bennett, E. B. 1986. The nitrifying of Lake Superior. Ambio 15:272–275.

Billings, W. D. 1987. Carbon balance of Alaskan tundra and taiga ecosystems: Past, present, and future. Quat. Sci. Rev. 6:165–177.

Blinn, D. W., and G. A. Cole. 1991. Algal and invertebrate biota in the Colorado River: Comparison of pre- and post-dam conditions. Pp. 75–101 in Colorado River Ecology and Dam Management. Washington, D.C.: National Academy Press.

Blumberg, A. F., and D. M. DiToro. 1990. Effects of climatic warming on dissolved oxygen concentrations in Lake Erie. Trans. Am. Fish. Soc. 119(2):210–223.

Bodaly, R. A., R. E. H. Hecky, and R. J. P. Fudge. 1984. Increases in fish mercury levels in lakes flooded by the Churchill River diversion, northern Manitoba. Can. J. Fish. Aquat. Sci. 41:682–691.

Bosch, J. M., and J. D. Hewlett. 1982. A review of catchment experiments to determine the effect of vegetation changes on water yield and evapotranspiration. J. Hydrol. 55:3–23.

Bourassa, R. 1985. Power from the North. Scarborough, N.J.: Prentice-Hall.

Boutron, C., and R. Delmas. 1980. Historical record of atmospheric pollution revealed in polar ice sheets. Ambio 9:210–215.

Brezonik, P. L. 1976. Nutrients and other biologically active substances in atmospheric precipitation. Proceedings of the Symposium on Effects of Atmospheric Inputs to the Chemistry and Biology of Lakes, International Association for Great Lakes Research Special Symposium 1:166–186.

Brezonik, P. L., L. A. Baker, and T. E. Perry. 1987. Mechanisms of alkalinity generation in acid-sensitive softwater lakes. Pp. 229–260 in Chemistry of Aquatic Pollutants, R. Hites and S.J. Eisenreich, eds. Adv. Chem. Ser. 216. Washington, D.C.: American Chemical Society.

Brezonik, P. L., J. G. Eaton, T. M. Frost, P. J. Garrison, T. K. Katz, C. E. Mach, J. H. McCormick, J. A. Perry, W. A. Rose, C. J. Sampson, B. C. L. Shelley, W. A. Swenson, and K. E. Webster. 1993. Experimental acidification of Little Rock Lake, Wisconsin: chemical and biological changes over the pH range 6.1 to 4.7. Can. J. Fish. Aquat. Sci. 50:1101–1121.

Brock, T. D. 1985. A Eutrophic Lake: Lake Mendota, Wisconsin. Number 55 in the Ecological Studies Series. New York: Springer-Verlag.

Carbyn, L. N., S. M. Oostenbrug, and D. W. Anions. 1993. Wolves, Bison, and the Dynamics Related to the Peace-Athabasca Delta in Canada's Wood Buffalo National Park. Edmonton, Alberta: Canadian Circumpolar Institute, University of Alberta.

Carlson, R. E. 1977. A trophic state index for lakes. Limnol. Oceanogr. 22:361–369.

Carpenter, S. R., and J. F. Kitchell, eds. 1993. The Trophic Cascade in Lakes. Cambridge, England: Cambridge University Press.

Carpenter, S. R., J. F. Kitchell, and J. R. Hodgsen. 1985. Cascading trophic interactions and lake productivity. BioScience 35:634–639.

Chapra, S., and R. Canale. 1991. Long-term phenomenological model of phosphorus and oxygen for stratified lakes. Water Res. 25:707–715.

Charles, D. F., M. W. Binford, E. T. Furlong, R. A. Hites, M. J. Mitchell, S. A. Norton, F. Oldfield, M. J. Paterson, J. P. Smol, A. J. Uutala, J. R. White, D. R. Whitehead, and R. J. Wise. 1990. Paleoecological investigation of recent lake acidification in the Adirondack Mountains, New York. J. Paleolimnol. 3:195–241.

Cheslak, E., and J. Carpenter. 1990. Compilation Report on the Effects of Reservoir Releases on Downstream Ecosystems. Report REC-ERC-90-1. Denver: U.S. Bureau of Reclamation.

Christie, W. J. 1974. Changes in fish species composition of the Great Lakes. J. Fish. Res. Bd. Can. 31:827–854.

Colborn, T., and C. Clement. 1992. Chemically-Induced Alterations in Sexual and Functional Development: The Wildlife/Human Connection. Princeton, N.J.: Princeton Scientific Publishing Co., Inc.

Colborn, T., F. S. Vom-Saal, and A. M. Soto. 1993. Developmental effects of endocrine-disrupting chemicals in wildlife and humans. Environ. Health Perspect. 101:378–384.

Cook, R. B., and D. W. Schindler. 1983. The biogeochemistry of sulfur in an experimentally acidified lake. Ecol. Bull. 35:115–127.

Cook, R. B., C. A. Kelly, D. W. Schindler, and M. A. Turner. 1986. Mechanisms of hydrogen ion neutralization in an experimentally acidified lake. Limnol. Oceanogr. 31:134–148.

Coutant, C. C. 1990. Temperature-oxygen habitat for freshwater and coastal striped bass in a changing climate. Trans. Am. Fish. Soc. 119(2):240–253.

Croley, T. E., II. 1990. Laurentian Great Lakes double-CO_2 climate change hydrological impacts. Clim. Changes 17:27–47.

Cronan, C. S., and C. L. Schofield. 1979. Aluminum leaching response to acid precipitation: Effects on high-elevation watersheds in the Northeast. Science 204:304–306.

Crossman, E. J. 1991. Introduced freshwater fishes: A review of the North American perspective with emphasis on Canada. Can. J. of Fish. and Aquat. Sci. 48(Suppl. 1):46–57.

Crowder, L. B. 1980. Alewife, rainbow smelt and native fishes in Lake Michigan: Competition or predation? Environ. Biol. Fish. 5:225–233.

Dannevig, A. 1959. Nedb™rens innflytelse på vassdragenes surhet, og på fisk bestanden. Jeger og Fisker 3:116–118.

Davies, P. H. 1991. Synergistic effects of contaminants in urban runoff. In Effects of Urban Runoff on Receiving Streams: An Interdisciplinary Analysis of Impact, Monitoring, and Management. New York: American Society of Civil Engineers.

Davis, R. B., and F. Berge. 1980. Atmospheric deposition in Norway during the last 300 years as recorded in SNSF lake sediments, II: Diatom stratigraphy and inferred pH. Pp. 270–271 in Ecological Impact of Acid Precipitation, D. Drablös and A. Tollan, eds. Oslo-Ås, Norway: SNSF Project.

Dillon, P. J., and W. B. Kirchner. 1975. The effects of geology and land use on the export of phosphorus from watersheds. Water Res. 9:135–148.

Dillon, P. J., and F. H. Rigler. 1975. A simple method for predicting the capacity of a lake for development based on lake trophic status. J. Fish. Res. Bd. Can. 32:1519–1531.

Dillon, P. J., N. D. Yan, and H. H. Harvey. 1984. Acidic deposition: Effects on aquatic ecosystems. Pp. 167–194 in Critical Reviews in Environmental Control, vol. 13, C. P. Straub, ed. Boca Raton, Fla.: CRC Press Inc.

Donald, D. B. 1987. Assessment of the outcome of eight decades of trout stocking in the mountain national parks, Canada. N. Am. J. Fish. Manage. 7:545–553.

Edmondson, W. T. 1970. Phosphorus, nitrogen and algae in Lake Washington after diversion of sewage. Science 169:690–691.

Edmondson, W. T. 1991. The uses of ecology: Lake Washington and beyond. Seattle: University of Washington Press.

Edmonson, W. T. 1994. What is limnology? Pp. 547–553 in Limnology Now: A Paradigm of Planetary Problems, R. Margalef, ed. New York: Elsevier.

Eisenreich, S. J., G. J. Hollod, and T. C. Johnson. 1979. Accumulation of polychlorinated biphenyls (PCBs) in surficial Lake Superior sediments. Environ. Sci. Technol. 13:569–573.

Environmental Protection Agency (EPA). 1982. Results of the Nationwide Urban Runoff Program, Volume 2 Appendices. PB84-185560. Springfield, Va.: National Technical Information Service.

Environmental Protection Agency (EPA). 1994. National Water Quality Inventory: 1992 Report to Congress. EPA 841-R-94-001. Washington, DC: EPA, Office of Water.

Environmental Protection Agency (EPA). 1995. Acid Deposition Standard Feasibility Study Report to Congress (Draft for Public Comment). EPA-430-R-95-001. Washington, D.C.: EPA, Office of Air and Radiation.

Erwin, K. L. 1991. An Evaluation of Wetland Mitigation in the South Florida Management District, Vol. 1. West Palm Beach: South Florida Management District.

Evans, D. O., and D. H. Loftus. 1987. Colonization of inland lakes in the Great Lakes region by rainbow smelt, Osmerus mordax: Their freshwater niche and effects on indigenous fishes. Can. J. Fish. Aquat. Sci. 44(Suppl. 2):249–266.

Feger, K. H. 1994. Influence of soil development and management practices on freshwater acidification in central European forest ecosystems. Pp. 67–82 in Acidification of Freshwater Ecosystems: Implications for the Future, C. E. W. Steinberg and R. F. Wright, eds. Chichester, England: John Wiley and Sons Ltd.

Ferguson, P., J. A. Lee, and J. N. B. Bell. 1978. Effects of sulphur pollutants on the growth of Sphagnum species. Environ. Pollut. 16:151–162.

Fernald, M. L. 1950. Gray's Manual of Botany. New York: American Book Company.

Firth, P., and S. G. Fisher, eds. 1992. Global Climate Change and Freshwater Ecosystems. New York: Springer-Verlag.

Frayer, W. E. 1991. Status and Trends of Wetlands and Deepwater Habitats in the Contermi-nous United States, 1970s to 1980s. Houghton, Mich.: Michigan Technological University.

Frederick, K. D. 1991. Water resources: Increasing demand and scarce supplies. Pp. 23–80 in America's Renewable Resources: Historical Trends and Current Challenges, K. D. Frederick and R. A. Sedjo, eds. Washington, D.C.: Resources for the Future.

Fremling, C. R. 1989. Hexagenia mayflies: Biological monitors of water quality in the upper Mississippi River. J. Minn. Acad. Sci. 55(1):139–143.

Frey, D. G. 1955. Langsee: A history of meromixis. Mem. 1st. Ital. Idrobiol. Suppl. 8:141–161.

Frost, T. M., S. R. Carpenter, A. R. Ives, and T. K. Kratz. 1995. Species compensation and complementarity in ecosystem function. Pp. 224–239 in Linking Species and Ecosystems, C. G. Jones and J. H. Lawton, eds. New York: Chapman and Hall.

Gilbertson, M. 1988. Epidemics in birds and mammals caused by chemicals in the Great Lakes. Pp. 134–152 in Toxic Contaminants and Ecosystem Health: A Great Lakes Focus, M. S. Evans, ed. New York: John Wiley & Sons.

Gilbertson, M., T. Kubiak, J. Ludwig, and G. Fox. 1991. Great Lakes embryo mortality, edema, and deformities syndrome (GLEMEDS) in colonial fish-eating birds: Similarity to chick edema disease. J. Toxicol. Environ. Health 33:455–520.

Goldman, C. R. 1981. Lake Tahoe: Two decades of change in a nitrogen deficient oligotrophic lake. Proceedings of the International Association for Theoretical and App. Lim-nol. 21:45–70.

Goldman, C. R. 1985. Lake Tahoe: A microcosm for the study of change. Bull. Ill. Nat. Hist. Surv. 33:247–60.

Goldman, C. R. 1988. Primary productivity, nutrients, and transparency during the early onset of eutrophication in ultra-oligotrophic Lake Tahoe, California-Nevada. Limnol. Oceanogr. 33:1321–1333.

Goldman, C. R. 1989. Lake Tahoe: Preserving a fragile ecosystem. Environment 31(7):7–11.

Gonzalez, M. J., and T. M. Frost. 1994. Comparisons of laboratory bioassays and a whole-lake experiment: Rotifer responses to experimental acidification. Ecol. Appl. 4:69–80.

Gordon, N. D., T. A. McMahon, and B. L. Finlayson. 1992. Stream & Hydrology: An Introduction for Ecologists. Chichester, England: John Wiley and Sons.

Gorham, E. 1995. The biogeochemistry of northern peatlands and its possible responses to global warming. Pp. 169–187 in Biotic Feedbacks in the Global Climate System, G. M. Woodwell and F. T. Mackenzie, eds. New York: Oxford University Press.

Great Lakes Research Advisory Board. 1978. The Ecosystem Approach: Scope and Implications of an Ecosystem Approach to Transboundary Problems in the Great Lakes Basin. Windsor, Ontario: Great Lakes Regional Office, International Joint Commission.

Greeson, P. E., J. R. Clark, and J. E. Clark, eds. 1979. Wetland Values and Functions: The State of Our Understanding. Technical Publication TPA 79-2. Minneapolis, Minn.: American Water Resources Association.

Harr, R. D., and R. L. Fredriksen. 1988. Water quality after logging small watersheds within the Bull Run Watershed, Oregon. Water Resourc. Bull. 24:1103–1111.

Harriman, R., G. E. Likens, H. Hultberg, and C. Neal. 1994. Influence of management practices in catchments on freshwater acidification: Afforestation in the United Kingdom and North America. Pp. 83–101 in Acidification of Freshwater Ecosystems: Implications for the Future, C. E. W. Steinberg and R. F. Wright, eds. Chichester, England: John Wiley & Sons.

Harriss, R., K. Bartlett, S. Frolking, and P. Crill. 1993. Methane emission from high-latitude wetlands. Pp. 449–486 in Biogeochemistry of Global Change: Selected Papers from Tenth International Symposium on Environmental Biogeochemistry, San Francisco, 1991, R. S. Oremland, ed. New York: Chapman and Hall.

Heinselman, M. L. 1963. Forest sites, bog processes, and peatland types in the glacial Lake Agassiz region, northern Minnesota. Ecol. Monogr. 33:327–374.

Hendry, C. D., E. S. Edgerton, and P. L. Brezonik. 1981. Atmospheric deposition of nitrogen and phosphorus in Florida. Pp. 199–206 in Atmospheric Pollutants in Natural Waters, S. J. Eisenreich, ed. Ann Arbor, Mich.: Ann Arbor Science Press.

Henriksen, A., and D. F. Brakke. 1988. Increasing contributions of nitrogen to the acidity of surface waters in Norway. Water Air Soil Pollut. 38:183–201.

Herbert, P. D., C. C. Wilson, M. H. Murdoch, and R. Lazar. 1991. Demography and ecological impacts of the invading mollusc Dreissenia polymorpha. Can. J. Zool. 69:405–409.

Herricks, E. E., I. Milne, and I. Johnson. 1994. Balancing technology: Considering receiving system impacts caused by episodic events. Pp. 5-1–5-4 in A Global Perspective for Reducing CSOs: Balancing Technologies, Costs, and Water Quality. Alexandria, Va.: Water Environment Federation.

Holland, R. E. 1993. Changes in planktonic diatoms and water transparency in Hatchery Bay, Bass Island area, western Lake Erie since the establishment of zebra mussel. J. Great Lakes Res. 19:617–624.

Hondzo, M., and H. G. Stefan. 1991. Three cases studies of lake temperature and stratification response to warmer climate. Water Resourc. Res. 27:1837–1846.

Hondzo, M., and H. G. Stefan. 1993. Regional water temperature characteristics of lakes subjected to climate change. Clim. Change 24:187–211.

Houghton, J. T., L. G. Meira Filho, J. Bruce, H. Lee, B. A. Callendar, E. Haites, N. Harris, and K. Maskell, eds. 1995. Climate Change 1994. Intergovernmental Panel on Climate Change. Cambridge, England: Cambridge University Press.

Huntly, N. 1995. How important are consumer species to ecosystem functioning? Pp. 72–83 in Linking Species and Ecosystems, C. G. Jones and J. H. Lawton, eds. New York: Chapman and Hall.

Hutchinson, G.E. 1941. Limnological studies in Connecticut IV: Mechanism of intermediate stratification in stratified lakes. Ecol. Monogr. 11:21–60.

Hutchinson, G. E. 1969. Eutrophication, past and present. Pp. 17–26 in Eutrophication: Causes, Consequences, Correctives. Washington, D.C.: National Academy Press.

Hutchinson, G. E. 1975. A Treatise on Limnology, Vol. III: Aquatic Botany. New York: John Wiley & Sons.

Hutchinson, G. E., E. Bonatti, U. M. Cowgill, C. E. Goulden, E. A. Gerenthal, M. E. Mallott, F. Margaritoria, R. Patrick, A. Racek, S. A. Roback, E. Stella, J. B. Ward-Perkins, and T. R. Wellman. 1970. Ianula: An account of the history and development of Lago di Monterosi, Latium, Italy. Trans. Am. Philoso. Soc. 60:1–178.

Johnson, D. K., and P. W. Aasen. 1989. The metropolitan wastewater treatment plant and the Mississippi River: 50 years of improving water quality. J. Minn. Acad. Sci. 55(1):134–138.

Jones, C. G., and J. H. Lawton, eds. 1995. Linking Species and Ecosystems. New York: Chapman and Hall.

Jordan, D. 1990. Mercury contamination: Another threat to the Florida panther. Endangered Spec. Tech. Bull. 15(2):1,6.

Kellerhals, R., and M. Church. 1989. The morphology of large rivers: Characterization and management. Spec. Publ. Fish. Aquat. Sci. 106:31–38.

Kelly, C. A., J. W. M. Rudd, R. H. Hesslein, D. W. Schindler, P. J. Dillon, D. Driscoll, S. A. Gherini, and R. E. Hecky. 1987. Prediction of biological acid neutralization in acid-sensitive lakes. Biogeochemistry 3:129–140.

Kidd, K. A., D. W. Schindler, R. H. Hesslein, and D. C. G. Muir. 1995a. Correlation between stable nitrogen isotope ratios and concentrations in biota from a freshwater food web. Sci. Tot. Environ. 160/161:381–390.

Kidd, K. A., D. W. Schindler, D. C. G. Muir, W. L. Lockhart, and R. H. Hesslein. 1995b. High toxaphene concentrations in fish from a subarctic lake. Science 269: 240–242.

Kitchell, J. F. 1992. Food Web Management: A Case Study of Lake Mendota. New York: Springer-Verlag.

Kivinen, E. 1980. New statistics on the utilization of peatlands in different countries. Pp. 48–51 in Proceedings of the Sixth International Peat Congress, Duluth, Minn. Jyska. Finland: International Peat Society.

Kreuzer, K. 1994. The influence of catchment management processes in forests on the recovery in fresh waters. Pp. 325–344 in Acidification of Freshwater Ecosystems: Implications for the Future, C. E. W. Steinberg and R. F. Wright, eds. Chichester, England: John Wiley & Sons.

Kusler, J. A., and J. E. Kentula. 1989. Wetland Creation and Restoration: The Status of the Science, Vols. 1 and 2. EPA 600/3-89/638. Corvallis, Ore.: Environmental Research Laboratory.

Lamarra, V. A. 1975. Digestive activities of carp as a major contributor to the nutrient loading of lakes. Int. Ver. Theor. Angew. Limnol. Verh. 19:2461–2468.

Lamontagne, S., and D. W. Schindler. 1994. Historical status of fish populations in Canadian Rocky Mountain lakes inferred from subfossil Chaoborus (Diptera: Chaoboridae) mandibles. Can. J. Fish. Aquat. Sci. 51:1376–1383.

Larson, E. B., E. Bonebrake, B. Schmidt, and J. Johnson. 1980. Fisheries investigations of the Flaming Gorge tailwater. Salt Lake City: Utah Department of Natural Resources.

Lazaro, T. R. 1990. Urban Hydrology: A Multidisciplinary Perspective. Lancaster, Pa.: Technomic Publishing Company.

Lazerte, B. D. 1993. The impact of drought and acidification on the chemical exports from a minerotrophic conifer swamp. Biogeochemistry 18:153–175.

Leavitt, P. R., D. E. Schindler, A. J. Paul, A. K. Hardie, and D.W. Schindler. In press.

Fossil pigment records of phytoplankton in trout-stocked alpine lakes. Can. J. Fish. Aquat. Sci.

Lehman, J. T. 1991. Causes and consequences of cladoceran dynamics in Lake Michigan: Implications of species invasion by *Bythotrephes*. J. Great Lakes Res. 17: 437–445.

Lewis, H. T., and T. A. Ferguson. 1988. Yards, corridors and mosaics: How to burn a boreal forest. Hum. Ecol. 16:57–77.

Likens, G. E., ed. 1972. Nutrients and eutrophication. American Society of Limnology and Oceanography Special Symposia, Vol. 1. Lawrence, Kans.: Allen Press.

Likens, G. E., F. H. Bormann, N. M. Johnson, D. W. Fisher, and R. S. Pierce. 1970. The effect of forest cutting and herbicide treatment on nutrient budgets of the Hubbard Brook watershed-ecosystem. Ecol. Monogr. 40:23–47.

Lindeman, R. L. 1942. The trophic-dynamic aspect of ecology. Ecology 23:399–418.

Lockhart, W. L. 1995. Implications of chemical contaminants from aquatic animals in the Canadian Arctic: Some review comments. Sci. Tot. Environ. 160/161:631–641.

Lodge, D. M. 1993. Species invasions and deletions: Community effects and responses to climate and habitat change. Pp. 367–387 in Biotic Interactions and Global Change, P. M. Kareiva, J. G. Kingsdver, and R. B. Huey, eds. Sunderland, Mass.: Sinauer Associates.

MacDonald, G. M., T. W. D. Edwards, K. A. Moser, R. Pienitz, and J. P. Smol. 1993. Rapid response of treeline vegetation and lakes to past climate warming. Nature 361: 243–246.

Mackereth, F. J. H. 1957. Chemical analysis in ecology illustrated from Lake District turns and lakes. Proc. Linn. Soc. Lond. 167:161–175.

Magnuson, J. J., J. D. Meisner, and D. K. Hill. 1990. Potential changes in thermal habitat of Great Lakes fish after global climate warming. Trans. Am. Fish. Soc. 119(2): 254–264.

Master, L. 1990. The imperiled status of North American aquatic animals. Biodiversity Network News 3:1–8.

McClain, A. S. 1991. Conceptual and Empirical Analyses of Biological Invasion: Nonnative Fish Invasion into North Temperate Lakes. Ph.D. dissertation. University of Wisconsin, Madison.

McCormick, M. J. 1990. Potential changes in thermal structure and cycle of Lake Michigan due to global warming. Trans. Am. Fish. Soc. 119(2):183–194.

McHenry, M., ed. 1992. The New Encyclopedia Britannica, Vol. 26. Chicago: Encyclopedia Britannica.

Meisner, J. D., J. E. Goddier, H. A. Regnier, B. J. Shuler, and W. J. Christie. 1987. An assessment of the effects of climate warming on Great Lakes Basin fishes. J. Great Lakes Res. 13(3):340–352.

Mills, E. L., J. H. Leach, J. T. Carlton, and C. L. Secor. 1994. Exotic species and the integrity of the Great Lakes. BioScience 44:666–676.

Minckley, W. L. 1991. Native fishes of the Grand Canyon region: An obituary? Pp. 124–177 in Colorado River Ecology and Dam Management. Washington, D.C.: National Academy Press.

Minnesota Department of Health. 1994. Minnesota Fish Consumption Advisory. Minneapolis: Minnesota Department of Health.

Minns, C. K., J. E. Moore, D. W. Schindler, and M. L. Jones. 1990. Assessing the potential extent of damage to inland lakes in eastern Canada due to acidic deposition. IV. Predicting the response of potential species richness. Can. J. Fish. Aquat. Sci. 47:821–830.

Minns, C. K., J. E. Moore, D. W. Schindler, P. G. C. Campbell, P. J. Dillon, J. K. Underwood, and D. M. Whelpdale. 1992. Expected reduction in damage to Canadian lakes under legislated and proposed decreases in sulphur dioxide emission. Report. 92-1. Ottawa: Committee on Acid Deposition, Royal Society of Canada.

Miskimmin, B. M., and D. W. Schindler. 1993. Reconstructing the effect of toxaphene treatment and fish stocking on invertebrates in a western Canadian lake. Verh. Int. Verein. Limnol. 25:1929–1939.

Miskimmin, B. M., and D. W. Schindler. 1994. Long-term invertebrate community response to toxaphene treatment in two lakes: 50-year records reconstructed from lake sediments. Can. J. Fish. Aquat. Sci. 51:923–932.

Miskimmin, B. M., P. R. Leavitt, and D. W. Schindler. 1995. Fossil record of the response of cladoceran and algal assemblages to fishery management practices. Freshwater Biol. 34:177–190.

Mitchell, J. F. B., T. C. Johns, J. M. Gregory, and S. F. B. Tett. 1995. Climate responses to increasing levels of greenhouse gases and sulphate aerosols. Nature 376: 501–504.

Mitsch, W. J., and V. G. Gosselink. 1993. Wetlands, 2nd ed. New York: Van Nostrand.

Mortimer, C. H. 1941–1942. The exchange of dissolved substances between mud and water in lakes. J. Ecol. 29:280–329, 30:147–201.

Moyle, P. B. 1986. Fish introductions into North America: Patterns and ecological impact. In Ecology of Biological Invasions of North America and Hawaii. New York: Springer-Verlag.

Muir, D. C. G., R. J. Nordstrom, and M. Simon. 1988. Organochlorine contaminants in arctic marine food chains: Accumulation of specific polychlorinated biphenyls and chlordane-related compounds. Environ. Sci. Technol. 22:1071–1079.

Muir, D. C. G., N. P. Grift, W. L. Lockhart, P. Wilkinson, B. N. Billeck, and G. J. Brunskill. 1995. Spatial trends and historical profiles of organochlorine pesticides in Arctic lake sediments. Sci. Tot. Environ. 160/161:447–457.

Murphy, M. L., J. Heifetz, S. W. Johnson, D. V. Koski, and J. F. Thedinga. 1986. Effects of clear-cut logging with and without buffer strips on juvenile salmonids in Alaskan streams. Can. J. Fish. Aquat. Sci. 43:1521–1533.

Murphy, T. J., and C. P. Rzeszuko. 1977. Precipitation inputs of PCBs to Lake Michigan. J. Great Lakes Res. 3:305–312.

Naiman, R. J., J. J. Magnuson, D. M. McKnight, and J. A. Stanford. 1995. The Freshwater Imperative: A Research Agenda. Washington, D.C.: Island Press.

National Research Council. 1969. Eutrophication: Causes, Consequences, Correctives. Washington, D.C.: National Academy Press.

National Research Council. 1981. Atmosphere-Biosphere Interactions: Toward a Better Understanding of the Ecological Consequences of Fossil Fuel Combustion. Washington, D.C.: National Academy Press.

National Research Council. 1987. River and Dam Management: A Review of the Bureau of Reclamation's Glen Canyon Environmental Studies. Washington, D.C.: National Academy Press.

National Research Council. 1992. Restoration of Aquatic Ecosystems. Washington, D.C.: National Academy Press.

National Research Council. 1995. Wetlands: Characteristics and Boundaries. Washington, D.C.: National Academy Press.

Niering, W. A. 1988. Endangered, Threatened and Rare Wetland Plants and Animals of the Continental United States, Vol. 1, D. D. Hook, ed. Portland, Ore.: Timber Press.

Nilsson, J., and P. Grennfelt. 1988. Critical loads for sulphur and nitrogen. Report from a workshop held at Skokloster, Sweden, March 19–24.

Novotny, V., and G. Chesters. 1981. Handbook of Non-Point Pollution: Sources and Management. New York: Van Nostrand Reinhold.

Paul, A. J., and D. W. Schindler. 1994. Regulation of rotifers by predatory calanoid copepods (subgenus: *Hesperodiaptomus*) in lakes of the Rocky Mountains. Can. J. Fish. Aquat. Sci. 51:2520–2528.

Peakall, D. B. 1993. DDE-induced eggshell thinning: An environmental detective story. Environ. Rev. 1:13–20.

Pechlaner, R. 1984. Historical evidence for the introduction of arctic char into high-mountain lakes of the Alps by man. Pp. 549–557 in Biology of the Arctic Char, L. Johnson and B. L. Burns, eds. Winnipeg, Canada: University of Manitoba Press.

Petts, G. E. 1984. Impounded Rivers: Perspectives for Ecological Management. Chichester, England: John Wiley & Sons.

Phelps, E. B. 1944. Stream Sanitation. New York: John Wiley & Sons.

Ramlal, P. S., C. A. Kelly, J. W. M. Rudd, and A. Furutani. 1993. Sites of methyl mercury production in remote Canadian Shield lakes. Can. J. Fish. Aquat. Sci. 50: 972–979.

Renner, R. 1995. "Scientific uncertainty" scuttles new acid rain standard. Environ. Sci. Technol. 29:464A–466A.

Richardson, C. J. 1989. Freshwater wetlands: Transformers, filters, or sinks? Pp. 25–46 in Freshwater Wetlands and Wildlife, R. R. Sharitz and J. W. Gibbons, eds. CONF-8603101. Washington, D.C.: Department of Energy, Office of Health and Environmental Research.

Richardson, C. J. 1994. Ecological functions and human values in wetlands: A framework for assessing forestry impacts. Wetlands 14:1–9.

Roberts, L. 1990. Zebra mussel invasion threatens U.S. waters. Science 249:1370–1372.

Robertson, D. M. 1989. The Use of Lake Water Temperature and Ice Cover as Climatic Indicators. Ph.D. dissertation. University of Wisconsin, Madison.

Rosenberg, D. M., R. A. Bodaly, and P. J. Usher. 1995. Environmental and social impacts of large scale hydro-electric development: Who is listening? Global Environ. Change 5:127–148.

Rosenberg, R., R. Elmgren, S. Fleischer, P. Jonsson, G. Persson, and H. Dahlin. 1990. Marine eutrophication case studies in Sweden. Ambio 19:102–108.

Rosseland, B. O., and M. Staurnes. 1994. Physiological mechanisms for toxic effects and resistance to acidic water: An ecophysiological and ecotoxicological approach. Pp. 275–312 in Acidification of Freshwater Ecosystems: Implications for the Future, C. E. W. Steinberg and R. W. Wright, eds. Chichester, England: John Wiley & Sons.

Rudd, J. W. M. 1995. Sources of methyl mercury to freshwater ecosystems: A review. Water Air Soil Pollut. 80:697–713.

Rudd, J. W. M., C. A. Kelly, V. St. Louis, R. H. Hesslein, A. Furutani, and M. Holoka. 1986. Microbial consumption of nitric and sulfuric acids in acidified north temperate lakes. Limnol. Oceanogr. 31:1267–1280.

Rudd, J. W. M., C. A. Kelly, D. W. Schindler, and M. A. Turner. 1988. Disruption of the nitrogen cycle in acidified lakes. Science 240:1515–1517.

Rudd, J. W. M., C. A. Kelly, D. W. Schindler, and M. A. Turner. 1990. A comparison of the acidification efficiencies of nitric and sulfuric acids by two whole-lake addition experiments. Limnol. Oceanogr. 35:663–679.

Rudd, J. W. R., R. Harris, C. A. Kelly, and R. E. Hecky. 1993. Are hydroelectric reservoirs significant sources of greenhouse gases? Ambio 22:246–248.

Schertzer, W. M., and A. M. Sawchuk. 1990. Thermal structure of the lower Great Lakes in a warm year: Implications for the occurrence of hypolimnion anoxia. Trans. Am. Fish. Soc. 119(2):195–209.

Schindler, D. W. 1974. Eutrophication and recovery in experimental lakes: implications for lake management. Science 184:897–899.

Schindler, D. W. 1977. Evolution of phosphorus limitation in lakes: Natural mechanisms compensate for deficiencies of nitrogen and carbon in eutrophied lakes. Science 195:260–262.

Schindler, D. W. 1986. The significance of in-lake production of alkalinity. Water Air Soil Pollut. 30:931–944.

Schindler, D. W., G. J. Brunskill, S. Emerson, W. S. Broecker, and T. H. Peng. 1972. Atmospheric carbon dioxide: Its role in maintaining phytoplankton standing crops. Science 177:1192–1194.

Schindler, D. W., H. Kling, R. V. Schmidt, J. Prokopowich, V. E. Frost, R. A. Reid, and M. Capel. 1973. Eutrophication of Lake 227 by addition of phosphate and nitrate, Part 2: The second, third, and fourth years of enrichment. J. Fish. Res. Bd. Can. 30: 1415–1440.

Schindler, D. W., R. N. Newbury, K. G. Beaty, J. Prokopowich, T. Ruszczynski, and J. S. Dalton. 1980. Effects of windstorm and forest fire on chemical losses from receiving streams. Can. J. Fish. Aquat. Sci. 37:328–334.

Schindler, D. W., K. H. Mills, D. F. Malley, D. L. Findlay, J. A. Shearer, I. J. Davies, M. A. Turner, G. A. Linsey, and D. R. Cruikshank. 1985. Long-term ecosystem stress: The effects of years of experimental acidification on a small lake. Science 228: 1395–1401.

Schindler, D. W., K. G. Beaty, E. J. Fee, D. R. Cruikshank, E. D. DeBruyn, D. L. Findlay, G. A. Linsey, J. A. Shearer, M. P. Stainton, and M. A. Turner. 1990. Effects of climatic warming on lakes of the central boreal forest. Science 250:967–970.

Schindler, D. W., S. E. Bayley, P. J. Curtis, B. R. Parker, M. P. Stainton, and C. A. Kelly. 1992. Natural and man-caused factors affecting the abundance and cycling of dissolved organic substances in Precambrian Shield lakes. Hydrobiologia 229: 1–21.

Schindler, D. W., S. E. Bayley, B. R. Parker, K. G. Beaty, D. R. Cruikshank, E. J. Fee, E. U. Schindler, and M. P. Stainton. In press, a. The effects of climatic warming on the properties of boreal lakes and streams at the Experimental Lakes Area, Northwestern Ontario. Limnol. Oceanogr.

Schindler, D. W., P. J. Curtis, B. Parker, and M. P. Stainton. In press, b. Consequences of climatic warming and lake acidification for UV_b penetration in North America. Nature.

Scott, W. B., and E. B. Crossman. 1973. Freshwater Fishes of Canada. Bulletin 184. St. John's, Canada: Fisheries and Oceans Canada.

Sculthorpe, C. D. 1967. The Biology of Vascular Plants. New York: St. Martins Press.

Shapiro, J. 1995. Lake restoration by biomanipulation: A personal view. Environ. Rev. 3:83–93.

Shapiro, J., and D. I. Wright. 1984. Lake restoration by biomanipulation. Freshwater Biol. 14:371–383.

Shapiro, J., V. Lamarra, and M. Lynch. 1975. Biomanipulation: An ecosystem approach to lake restoration. Pp. 85–96 in Proceedings of a Symposium on Water Quality Management Through Biological Control, P. L. Brezonik and J. L. Fox, eds. Gainesville: University of Florida.

Simons, D. B. 1979. Effects of stream regulation on channel morphology. Pp. 95–112 in The Ecology of Regulated Streams, J. V. Ward and J. A. Stanford, eds. New York: Plenum Press.

Sjörs, H. 1961. Surface patterns in boreal peatlands. Endeavor 20:217–224.

Solbe, J. F. de L. G., ed. 1986. Effects of Land Use on Fresh Waters: Agriculture, Forestry, Mineral Exploitation, Urbanisation. Chichester, England: Ellis Horwood Limited.

Soranno, P. A., S. L. Hubler, S. R. Carpenter, and R. C. Lathrop. In press. Phosphorus loads to surface waters: A simple model to account for spatial pattern of land use. Ecol. Appl.

Spencer, C. N., B. R. McClelland, and J. A. Stanford. 1991. Shrimp stocking, salmon collapse, and eagle displacement. BioScience 41(1):14–21.

Stefan, H. G., M. Hondzo, J. G. Eaton, and J. H. McCormick. 1995. Predicted effects of global climate change on fishes in Minnesota lakes. Pp. 57–72 in Climate Change and Northern Fish Populations, R. J. Beamisk, ed. Can. Spec. Publ. Fish. Aquat. Sci. 121. Ottawa: National Research Council Canada.

Streeter, H. W., and E. B. Phelps. 1925. Public Health Bulletin 146. Washington, D.C.: U.S. Public Health Service.

Stuckey, R. D. 1980. Distributional history of Lythrum salicaria (purple loosestrife) in North America. Bartonia 47:3–20.

Swackhamer, D. L., and R. A. Hites. 1988. Occurrence and bioaccumulation of organochlorine compounds in fishes from Siskiwit Lake, Isle Royale, Lake Superior. Environ. Sci. Technol. 22:543–548.

Swain, E. B., D. R. Engstrom, M. E. Brigham, T. A. Henning, and P. L. Brezonik. 1992. Increasing rates of atmospheric mercury deposition in midcontinental North America. Science 257:784–787.

Thienemann, A. 1925. Die Binnengewasser Mitteleuropas. Die Binnengewasser 1. Stuttgart: Scheizerbartsche Verlagsbuchhandlung. 255 pp.

Thompson, D. Q., R. L. Stuckey, and E. B. Thompson. 1987. Spread, Impact, and Control of Purple Loosestrife (Lythrum salicaria) in North American Wetlands. Fish and Wildlife Research No. 2. Washington, D. C.: U.S. Fish and Wildlife Service.

Vallentyne, J. R. 1970. Phosphorus and the control of eutrophication. Can. Res. Dev. 3 (March/April):36–43, 49.

Vallentyne, J. R., and A. M. Beeton. 1988. The ecosystem approach to managing human uses and abuses of natural resources in the Great Lakes Basin. Environ. Conservation 15:58–62.

van der Leeden, F., F. L. Troise, and D. K. Todd. 1990. The Water Encyclopedia. Chelsea, Mich.: Lewis Publishers.

Vollenweider, R. A. 1968. Water Management Research. Scientific Fundamentals of the Eutrophication of Lakes and Flowing Waters with Particular Reference to Nitrogen and Phosphorus as Factors in Eutrophication. Paris: Organization for Economic Cooperation and Development, Committee for Research Cooperation.

Vollenweider, R.A. 1969. Möglichkeiten und Grenzen der elementarer Modelle der Stoffbilanz von Seen. Arch. Hydrobiol. 66:1–36.

Vollenweider, R.A. 1975. Input-output models with special reference to the phosphorus loading concept in limnology. Schweiz. Z. Hydrol. 37:53–84.

Vollenweider, R. A. 1976. Advances in defining critical loading levels for phosphorus in lake eutrophication. Mem. 1st. Ital. Idrobiol. 33:53–83.

Ward, J. V., and J. A. Stanford. 1979. Ecological factors controlling stream zoobenthos with emphasis on thermal modification of regulated streams. Pp. 35–56 in The Ecology of Regulated Streams, J. V. Ward and J. A. Stanford, eds. New York: Plenum Press.

Webster, K. E., T. M. Frost. C. J. Watras, W. A. Swenson, M. Gonzalez, and P. J. Garrison. 1992. Complex biological responses to the experimental acidification of Little Rock Lake, Wisconsin, USA. Environ. Pollut. 78:73–78.

Wein, E. E., J. H. Sabry, and F. T. Evers. 1991. Food consumption patterns and use of country foods by native Canadians near Wood Buffalo National Park, Canada. Arctic 44:196–205.

Welch, H. E., D. C. G. Muir, B. N. Billeck, W. L. Lockhart, G. J. Brunskill, H. J. Kling, M. P. Olson, and R. M. Lemoine. 1991. Brown snow: A long-range transport event in the Canadian Arctic. Environ. Sci. Technol. 25:280–286.

Wiener, J. G., W. F. Fitzgerald, C. J. Watras, and R. G. Rada. 1990. Partitioning and bioavailability of mercury in an experimentally acidified Wisconsin lake. Environ. Toxicol. Chem. 9:909–918.

Winchester, J. W., and C. D. Nifong. 1971. Water pollution in Lake Michigan by trace elements from pollution aerosol fallout. Water Air Soil Pollut. 1:50–64.

Wolf, E. W., and D. A. Peel. 1985. The record of global pollution in polar snow and ice. Nature 313:535–540.

Woodwell, G. M., C. F. Wurster, Jr., and P. A. Isaacson. 1967. DDT residues in an east coast estuary: A case of biological concentration of a persistent insecticide. Science 56:821–824.

Woodwell, G. M., F. T. MacKenzie, R. A. Houghton, M. J. Apps, E. Gorham, and E. A. Davidson. 1995. Will the warming speed the warming? Pp. 393–411 in Biotic Feedbacks in the Global Climate System, G. M. Woodwell and F. T. Mackenzie, eds. New York: Oxford University Press.

Wright, H. E., Jr., B. Coffin, and N. Aaseng, eds. 1992. Patterned Peatlands of Northern Minnesota. Minneapolis: University of Minnesota Press.

Xun, L., N. E. R. Campbell, and J. W. M. Rudd. 1987. Measurements of specific rates of net methyl mercury production in the water column and surface sediments of acidified and circumneutral lakes. Can. J. Fish. Aquat. Sci. 44:750–757.

4

Education in Limnology: Current Status and Recommendations for Improvement

Will society be capable of better managing its water resources to ensure a high quality of life for future generations? The answer to this question will depend in large part on the strength of the educational system for freshwater science. Knowledge is transferred to the practitioners of the future primarily at colleges and universities, and it is there that much of the new knowledge needed to solve problems of inland aquatic ecosystems is developed.

Colleges and universities need to provide more opportunities for future water managers to learn about limnology during their undergraduate years. Society also has to ensure that the science of limnology continues to advance via a consistently funded research program because limnology is a critical component of the scientific foundations necessary for understanding the risks and benefits of water management decisions. The challenge for colleges and universities is to meet these needs by training well-qualified professional limnologists and by providing training in limnology within other academic programs that produce aquatic resource managers, planners, and field technicians. Also needed are leaders and citizens familiar with the basic principles of limnology. Educational programs in limnology therefore must meet three goals: (1) education of future research scientists who will work to advance the understanding of aquatic ecosystems, (2) education of future leaders who will determine policies and actions that affect freshwater ecosystems, and (3) education of responsible citizens who will value freshwater ecosystems and understand water resource issues. This chapter describes the current educational system in limnology and recommends strategies for improving it to meet these three goals.

LIMNOLOGY UNDER THE CURRENT EDUCATIONAL SYSTEM

Limnology programs at American colleges and universities are highly fragmented across departments. Ironically, the main cause of this fragmentation is derived from one of the strengths of limnology as a science: it is highly interdisciplinary. As a result, limnology lacks a single "home" in the university system. Instead, limnologists and the limnology-related courses they teach are housed in a variety of departments, such as biology, civil engineering, geology, zoology, and many others. For example, 69 universities surveyed for this report indicated that faculty working in limnology and related disciplines are housed in 23 different types of departments (see Chapter 2 and Appendix A). The diversity in academic homes for limnologists applies within individual universities as well as among universities. For example, at the University of Minnesota, limnologists are found in at least eight departments housed in five different colleges on the Twin Cities campus. Only two universities surveyed for this report have departments that include the name "limnology": one is a department of marine science and limnology; the other is a department of zoology and limnology. While some interdisciplinary fields, such as oceanography, soil science, and forestry, have departments or schools of their own, academic limnology remains scattered and lacks well-defined degree programs.

One result of the scattering of limnology courses across departments and the lack of named degree programs is that limnology graduates produced by the various "nonlimnology" departments often lack knowledge of some critical subfields of limnology (such as physical limnology) or of the characteristics of the full range of aquatic ecosystems (streams and wetlands as well as lakes). Another result is that limnology faculty may not be replaced upon retirement if the department in which they are housed chooses to shift its emphasis elsewhere. This redirection may diminish the breadth of a university's offerings in limnology or even eliminate the university's limnology program altogether.

Undergraduate Education in Limnology
Under the Current System

The fragmentation of limnological studies is especially a problem at the undergraduate level. Colleges and universities in the United States generally do not offer formal undergraduate majors specifically identified as "limnology." Students who enter this field often stumble across it accidentally by signing up for a limnology course as an elective within another major. A few schools, such as the University of Wisconsin–Stevens Point (see Box 4-1), offer a limnology emphasis within a water resources major. In general, however, undergraduate students' options for studying limnology are limited to enrolling in aquatic science courses while pursu-

---

BOX 4-1 LIMNOLOGY AND FISHERIES AT THE UNIVERSITY OF WISCONSIN–STEVENS POINT

The University of Wisconsin–Stevens Point, offers B.S. degrees in water resources with an option to specialize in limnology and fisheries. The limnology and fisheries program is part of the nondepartmentalized College of Natural Resources, which employs faculty to support major curricula in five areas: water resources, forestry, wildlife management, soils, and natural resources. The integrated nature of the program makes access to faculty across various disciplines more convenient than if the college were divided into departments. The curriculum in limnology and fisheries is as follows:

• *Courses to meet university general degree requirements (22 percent of course work):* English; history; communications; humanities; social science.

• *Courses in common with all students in the College of Natural Resources (17 percent of course work):* introduction to natural resources; introduction to water resources; introduction to soil resources; introduction to forestry; introduction to wildlife ecology; land surveying; field experiences in forest measurement; soil conservation and watershed inventory techniques; field experience in soil inventory techniques; field experience in aquatic ecosystem evaluation; field experience in wildlife management techniques; integrated resource management seminar.

• *Courses in common with other water resource majors (29 percent of course work):* limnology; water chemistry and analysis; general chemistry; applied physics; botany; zoology; general ecology; computer applications.

• *Professional courses in limnology and fisheries (32 percent of course work):* hydrology; pollution ecology; limnology and fisheries research; fisheries management; resource economics or principles of macroeconomics; organic chemistry survey; genetics; animal or human physiology; aquatic invertebrate zoology or aquatic insects; ichthyology; calculus; statistics.

• *Electives (one course):* algology; aquatic insects; ecology of freshwater benthic organisms; ground water geochemistry; aquatic vascular plants; aquatic invertebrate zoology; life history and population dynamics of fish.

The university offers one senior-level limnology course per semester. It is a capstone course that explores the chemistry, physics, and biology of inland aquatic systems.

N. Earl Spangenberg, Professor
University of Wisconsin–Stevens Point

ing another major such as biology, botany, ecology, geology, zoology, geography, or (more recently) environmental studies. Because of the lack of comprehensive limnology programs, students who develop an interest in limnology during their undergraduate careers may lack guidance on the appropriate mix of courses to prepare them for further work in this field.

Where limnology classes are offered and visible to students, they are usually popular. For example, at the University of Colorado, Boulder (see Box 4-2), demand for the limnology class always exceeds the enrollment limit of 75 students. Of the 69 universities surveyed for this report, 38 responded that student interest in limnology is increasing, 17 responded that interest is holding steady, and only 2 reported declining interest (see Appendix A).

Opportunities for undergraduate students to participate in field and laboratory research in limnology typically are very limited. At some universities, field and laboratory programs are essentially nonexistent, eliminating one of the key ways that students have been attracted to limnology. Of the 69 U.S. universities surveyed for this report, 40 reported having access to research or field stations (see Appendix A), but the degree to which these research sites are available to undergraduate students is often limited. Universities rarely allocate sufficient funding and laboratory space to offer laboratory and/or field work for all interested undergraduate students. For example, because of lack of laboratory space and funding at the University of Colorado, only a few students can enroll in the laboratory portion of the limnology course (see Box 4-2). In contrast, Canada's Trent University (see Box 4-3) has invested heavily in field experiences for limnology students, offering a two-semester limnology course in which the second semester is devoted to field experiments. Recently, Trent University has been forced to cut back on its field and laboratory program: although the course had an enrollment of 120 students in 1993, the university limited enrollment to 48 in the following year because of staff reductions.

In addition to having limited opportunities for field and laboratory research, undergraduates lack opportunities to learn about streams and wetlands, at least in most introductory limnology courses. For example, in five major limnology textbooks published over six decades, about 6 percent of the pages are devoted to wetlands, and about 7 percent cover streams (see Table 4-1).

Creating a major in limnology may not be necessary for every university interested in increasing its strength in this science. For example, the University of Wisconsin–Madison, recognized as one of the leading U.S. graduate schools of limnology, does not offer a formal undergraduate degree in limnology. Instead, students gain exposure to limnology through well-supported interdisciplinary programs and highly visible introductory limnology courses, in which enrollment typically exceeds

---

BOX 4-2 LIMNOLOGY AT THE UNIVERSITY OF COLORADO, BOULDER

The University of Colorado's teaching program in limnology extends back to 1940, when Robert W. Pennak, a student of Chancey Juday (see Chapter 2), joined the faculty. Pennak developed an introductory course in limnology that served approximately 25 students at the time of his retirement in 1974. In addition, he developed a course in stream biology and initiated a graduate program in limnology. Complementary courses were offered by an invertebrate zoologist, John Bushnell, and a fish biologist, John T. Windell. These three positions constituted the biology department's investment in aquatic studies, out of a total faculty of 20.

Pennak's course originally had a laboratory. The enrollment pressure grew after 1974, however, and because the university was unable to provide teaching assistants or equipment for enlargement of the laboratory, the laboratory and lecture were split. Enrollment in the lecture is limited to 75; the laboratory is taught separately and reaches fewer students.

Today, the biology department still has one limnologist and an invertebrate zoologist on the tenure-track faculty, as well as an open position that could go to a fish biologist but could also be transmuted into something else. In terms of aquatic faculty strength, the university thus is more or less where it stood in 1950, although the biology department has now grown to 46 faculty. The university's formal course work offerings in limnology, especially the laboratory component, are still inadequate. The limnology course serves both graduate and undergraduate students; graduate students receive additional personal contact and are asked to complete additional readings and written work. Although this is not an ideal pedagogical device, the approach works reasonably well for many graduate students.

In 1985, the university created the Center for Limnology to recognize its strength in limnological research. The center, which has a modest staff and several postdoctoral associates, provides a home for limnologists, including undergraduates. In about 1980, the university initiated an internship program that allows recent graduates with B.S. degrees who have a professional interest in limnology to be employed for one or two years in a technical capacity by the Center for Limnology.

William M. Lewis, Jr., Professor
University of Colorado, Boulder

---

120 (see Box 4-4). Although a limnology major may not be a necessary option at every school or even at most schools, the visibility and cohesiveness of teaching in limnology nevertheless need to be increased so that undergraduate students are aware of the opportunity to specialize in freshwater studies. Otherwise, students too often will learn about limnology by stumbling across it accidentally, instead of recognizing it early as a possible profession.

BOX 4-3 LIMNOLOGY AT TRENT UNIVERSITY

Limnology, offered in the third year of undergraduate studies, is a core course for the Departments of Biology and Environmental Resource Studies at Trent University in Peterborough, Ontario. The course is offered annually and runs for the full year. There are two weekly presentations, one of which is given by a guest lecturer. In addition to several faculty with aquatic interests, Trent has access to the Ontario Ministry of Environment and the National Water Research Institute, with adjunct faculty from both groups.

Every second week, the course includes laboratories that alternate between field trips and analytical sessions. Students sample lakes, streams, ponds, and rivers in fall, winter, and spring. The university's location, on the Otonabee River in the heart of the Kawartha Lakes area, makes this feasible despite large enrollments. (At the course's peak, each field trip and laboratory was repeated six times.) Students perform analytical work in teams, and all members have access to the team's collective data set. Manuscript-style reports are required following each analytical session.

In spring, students carry out their own projects, pursuing their own hypotheses with their own experimental design and writing up a final project report. Students are required to use the primary literature in designing, interpreting, and discussing their results. In addition, in their fourth year, students are allowed to operate research equipment and have the opportunity to pursue research-oriented honors projects, which typically result in an honors thesis.

Prerequisites for the limnology course include general biology, inorganic and organic chemistry, physics, and math (either calculus or statistics) in the first year and ecology in the second year. A number of electives in chemistry, math, geology, and biology are also available, and students are encouraged to select broadly based programs rather than narrowly specialized ones. A high proportion of Trent graduates receives fellowships to pursue graduate work.

David Schindler, Professor
University of Alberta, Edmonton

Graduate Education in Limnology Under the Current System

Limnologists currently receive advanced degrees through a variety of departments—most often biology but also a range of other departments, as discussed above and documented in Appendix A. For example, at the University of Wisconsin–Madison (Box 4-4), students may obtain M.S. and Ph.D. degrees in aquatic science through three interdepartmental, interdisciplinary graduate programs; and at least nine traditional academic departments offer advanced degrees with specialization in some aspect of aquatic science. At the University of Alabama, students obtain graduate degrees in aquatic ecology primarily through the biology depart-

TABLE 4-1 Coverage of Streams and Wetlands in Five Limnology Texts

Author	Date	Title	Pages Total Text	Streams	Wetlands
Welch	1935	*Limnology*	394	30	4
Ruttner	1963	*Fundamentals of Limnology*	249	22	6
Cole	1983	*Textbook of Limnology*	412	26	3
Wetzel	1983	*Limnology*	753	25	91
Horne and Goldman	1994	*Limnology*	520	51	24
Total			2,260	154	128
Percentage of total			100	6.8	5.7

NOTE: Material judged to cover streams and wetlands excludes occasional brief mention in sections emphasizing lakes.

ment, which has strong connections with the geology department. At Dartmouth, limnology graduate study is through the environmental studies, biology, or geology departments. At the University of Minnesota, limnology graduate students reside in departments of ecology, geology, and civil engineering and the College of Natural Resources, as well as occasionally in other departments or colleges (such as plant biology and public health). Limnology graduate students at Arizona State are housed in the zoology department. As a consequence of this range of departmental settings, limnology graduate programs are quite diverse, both among universities and within them (depending on which department within the university grants the degree).

Although the varied nature of the departments in which limnologists pursue their graduate degrees reflects the multidisciplinary nature of the science, it also poses problems with regard to integration of knowledge across disciplines and types of aquatic ecosystems. Individual programs in discipline-based departments are not likely to cover all of the subdisciplines of limnology. For example, students who obtain limnology degrees through biology departments might have strong training in fields such as organismal physiology but much weaker knowledge of water chemistry and physical limnology. On the other hand, limnology students with degrees from civil and environmental engineering departments may have strong training in water chemistry or hydrology but relatively little knowledge of the organisms that inhabit aquatic systems. Limnology students typically do not receive adequate training across the major types of aquatic ecosystems (wetlands, streams, and lakes).

Another problem resulting from the fragmentation of graduate programs in limnology among departments is that their success often hinges on the presence of one or a few strong professors, housed in departments whose focus is not limnology or even aquatic or environmental science. This reliance on one or a few individuals leaves graduate programs at

---

BOX 4-4 LIMNOLOGY AT THE UNIVERSITY OF WISCONSIN–MADISON

The University of Wisconsin–Madison, provides extensive opportunities for the study of limnology, building on a long history in this field. For graduate studies, there are three programs—Oceanography and Limnology, Water Resources Management, and Water Chemistry—that provide degrees in aquatic sciences, along with at least nine traditional departments (including the departments of zoology, botany, meteorology, and entomology) that offer specialization in some aspect of aquatic science in their advanced degree programs. On average, more than 15 graduate degrees are awarded in limnology across campus annually through the three aquatic programs and the traditional departments. For undergraduate students, there are no formal degree programs in limnology aside from a certification in environmental studies. Undergraduate students can, however, specialize to a large extent in aquatic sciences in several degree programs.

More than 40 faculty members in at least nine departments at the university are specialists in aquatic science. More than 30 limnology-related courses, mostly at advanced levels, are taught on campus each year. Topics include water chemistry, wetland vegetation, aquatic organisms (fish, insects, phytoplankton, zooplankton, bacteria), geology, and physical limnology.

Introductory-level limnology courses are taught each year during the fall semester and also during most summer sessions. Both courses are popular; the fall lecture course routinely attracts more than 120 students. Lectures are taught along with an optional, smaller-enrollment laboratory course. The introductory courses are intended to provide a basic introduction to limnology and cover geological, physical, chemical, and biological aspects of aquatic investigations with an emphasis on lake studies and human impacts on aquatic ecosystems. Students with strong career interests in limnology are encouraged to enroll in more advanced courses and to participate in research programs at the university and its field facilities.

Limnology at the University of Wisconsin has long had a tradition of emphasizing the interdisciplinary nature of the science. Collaboration among departments and between engineering and science programs is facilitated by the organization of the university, which provides strong support for the work of limnologists, as demonstrated most recently by the creation in 1992 of the Center for Limnology.

Thomas M. Frost, Associate Director
University of Wisconsin, Center for Limnology

risk of being closed when a professor retires or moves (see Chapter 2 and Box 4-5).

Improving Education in Limnology: The Basis for Reform

In summary, the current status of limnological education in North America is unsatisfactory. Individual programs generally are very small

---

BOX 4-5 LIMNOLOGY AT THE UNIVERSITY OF WASHINGTON

Prior to 1986, graduate students at the University of Washington who wanted to do research in limnology for a Ph.D. or M.S. degree took a course called Limnology in the Department of Zoology and assembled supporting courses from a variety of departments. The course was also taken as background by many graduate students in the College (now School) of Fisheries who were working on projects in fresh water. Undergraduates also could use Limnology as one of the optional courses for their major or for distribution credits; the course was required for fisheries undergraduates in the freshwater option.

A course on limnology was taught at Washington for the first time in 1923 by Trevor Kincaid; mine started in 1949. The course included three lectures and two three-hour laboratory sessions per week for a quarter for five credits. Later, it was split into a lecture course with no limit on enrollment and a laboratory course with a nominal limit of 20.

Students at all levels had plenty of supporting courses to choose from, including a graduate seminar, Topics in Limnology. Appropriate courses were also offered by the Departments of Oceanography, Fisheries, Botany, Civil Engineering, and Statistics. Later, an interdisciplinary Center for Streamside Studies was created.

After retirement in 1986, the Department of Zoology took the occasion to review its educational responsibilities to the College of Arts and Sciences. It was recognized that the department was outstanding in the field of ecology, including limnology, but some other fields needed to be updated to maintain the quality of the department as a whole. To keep limnology as a part of the departmental program until a faculty position could be reinstated, the faculty decided to offer the course at intervals with visiting professors. Negotiations for a fourth session were postponed in 1993 after a general decrease in state funding. Through 1995, there was no course at the University of Washington that gave the kind of one-package overview that the original limnology course did, but courses in various aspects of freshwater aquatic ecology were available in the departments mentioned above. In 1996, the university hired a new, full-time assistant professor to resume the limnology course.

W. T. Edmondson, Emeritus Professor
University of Washington

and spread in an uncoordinated fashion among several departments in a given university. Each program deals with the field in only a fragmentary manner. Taken as a whole, these programs can be criticized as representing less than the sum of the component parts. Overall, the present status is reflective neither of the needs of society nor of the vitality and comprehensiveness of this science.

To provide the training essential for limnologists to deal with future water problems, limnological education must be expanded and improved. Undergraduate programs should be developed, and graduate programs should be strengthened. Curricular enhancements are needed to ensure that students are exposed to the broad range of subdisciplines comprising this field and that they are encouraged to integrate this knowledge into a holistic understanding of how inland aquatic ecosystems function. In addition, administrative changes are necessary to improve coordination among the limnologists housed in a variety of departments and colleges of major universities or to bring them together under a single program or department.

STRENGTHENING LIMNOLOGY THROUGH ADMINISTRATIVE REFORMS

Enhancement of limnological education first and foremost is a matter of curriculum reform, and most of the remainder of this chapter deals with such issues. As indicated in previous chapters of this report, educational programs in limnology need to be broad, encompassing the essentials of the many scientific disciplines that contribute to an understanding of how aquatic ecosystems function; they also need to be integrative and rigorous. A diverse faculty committed to these ideals is prerequisite to the establishment and maintenance of programs that combine these features, but the practical realities of modern universities also must be taken into account if such programs are to succeed. In particular, the administrative structures in which academic programs are housed can be critical to their success and even their survival. Large research universities do not operate as completely amorphous entities in which students (and their advisers) are free to design individual degree programs cafeteria-style from course catalogs. Rather, university curricula are structured around degree programs that most commonly are associated with departments, which are further organized by broad disciplines into schools and colleges. University resources—faculty, funds for teaching assistants and support staff, classrooms, laboratories, and space for research and offices—generally are allocated along collegiate and departmental lines.

In designing better educational programs for limnology, faculty limnologists (and university administrators) thus need to consider what sorts of administrative structures will promote and maintain the necessary curricular reforms and provide the financial resources to maintain high-quality programs. Two broad administrative paths are (1) to establish departments of aquatic science in which limnology is a major focus at a few American universities on a regional basis and (2) to create strong interdisciplinary, interdepartmental programs in aquatic science at other universities. Each option has certain advantages and limitations, and the approach that is chosen will depend on the faculty and leadership of a university, as well as on existing strengths and circumstances within the university.

Limnologists and university leaders will face challenges in pursuing either of the paths. The political strength in most universities lies within established departments, and existing departments may object to the creation of new departments or programs that are perceived as siphoning off their resources. Nevertheless, the interdisciplinary approach of limnology is consistent with current trends in modern science, which is becoming more and more interdisciplinary as the fundamental principles of individual sciences become increasingly understood and as scientists discover critical links among disciplines. Educational institutions need to find ways to support interdisciplinary programs and reward those who engage in interdisciplinary work, even if it challenges the existing system. Scientists in interdisciplinary fields need to find ways to convince their colleagues that their programs and work are worth supporting.

Departments of Aquatic Science

Creating strong academic departments of aquatic science in one or a few universities in each region of the country would serve to create centers of strength for this science. Such departments could provide comprehensive majors in limnology, as well as other fields of aquatic science, for both undergraduate and graduate students (M.S. and Ph.D.). Although a few universities already have strong interdepartmental graduate aquatic science programs, very few have undergraduate programs. Although interdepartmental undergraduate majors in other fields can be found at some major universities, by far the majority of bachelor's degree programs are housed within traditional departments. The interdepartmental approach, which is inherently less structured, may be less satisfactory for undergraduate degree programs than for graduate programs. Moreover, the geographic coverage of existing interdepartmental graduate programs is highly inadequate, and even the best of the existing programs tends to focus on producing research-oriented specialists in specific subareas of a field rather than the broadly trained practitioners and managers that represent the bulk of the human resource needs in water resource management.

Several European universities have "institutes of limnology" that are essentially the equivalents of academic departments, granting degrees at the undergraduate, master's, and doctoral levels (see Boxes 4-6 and 4-7). These institutes maintain faculty with specialties in the subdisciplines of limnology, conduct large research programs, have their own budgets, and

---

BOX 4-6 LIMNOLOGY AT EUROPEAN INSTITUTES OF LIMNOLOGY

Several European universities, such as Uppsala University in Sweden, have institutes of limnology; and others, such as the Swiss Federal Institute of Technology in Zurich (see Box 4-7), have comprehensive institutes for aquatic or environmental science that include strong limnological components. Uppsala's institute, like others in Europe, is effectively a department in the North American tradition; it grants degrees at the undergraduate, master's, and doctorate levels; maintains faculty with various specialties in instruction and research; and has its own budget and building.

Several conspicuous differences emerge in both the philosophy and the methods of training in limnology within the European system in comparison to that prevailing in North America. Students are rigorously selected for entrance into the universities. Commitments to a discipline are made early, usually during the second year of undergraduate studies. Course requirements and sequences are set forth rigorously to provide the foundations for professional training. Generally, the undergraduate program extends for a five-year period. During the fourth and particularly the fifth years, nearly all training is in the institute of limnology. Detailed, specific training is given with intensive field-oriented problem solving on natural river, lake, and reservoir ecosystems. Most students are involved directly in ongoing research projects and receive training in the scientific method and in scientific publication. Graduation with an undergraduate degree from these programs approximately equals training obtained after the master's degree in North America.

A select minority of graduates of the institute continues for advanced degrees. Nearly all master's and Ph.D. students are actively involved in teaching undergraduates. Ph.D. research is often brought into instructional programs, affording students exposure to the realities of complex ecological research.

Nearly all graduates at all levels of training are incorporated into the professional work force of environmental evaluation and regulation in regional and national government agencies, environmental protection groups, and industries. Only a small fraction, likely less than 10 percent, enters the higher education field in universities.

Robert G. Wetzel, Professor
Department of Biological Sciences, University of Alabama and
Erlander Professor, Institute of Limnology, Uppsala, Sweden

BOX 4-7 ENVIRONMENTAL SCIENCE PROGRAM AT THE SWISS FEDERAL INSTITUTE OF TECHNOLOGY IN ZÜRICH

In 1987, the Swiss Federal Institute of Technology initiated a new program in response to the growing awareness that many questions dealing with the environment call for an interdisciplinary approach and for a new kind of scientist who is not only broad in his or her scientific approach but is also able to build bridges to the social sciences. The curriculum leads to both M.S. and Ph.D. degrees in environmental sciences. Rather than choosing among the classical disciplines (physics, chemistry, biology), the student focuses his or her curriculum on an environmental system, such as soil-terrestrial, atmosphere, or hydrosphere. For those choosing the hydrosphere as their principal domain, the curriculum corresponds to a multidisciplinary education in limnology in its broadest sense.

The program lasts five years, including the time needed to write a master's thesis. Course work is the same for all students in the program during the first two years, and it sets a solid base in mathematics and all natural sciences. Some first courses in the social sciences are included as well. During the last three years, the courses are focused on the system that is selected by the student. During this time, students also receive practical training in applied, multidisciplinary problem solving, and they spend at least four months with a company, federal agency, or other institution that works in the chosen field.

Experiences with the first graduates of the program, who completed their work at the end of 1993, are encouraging. Based on questionnaires filled out by former students and representatives of industry, both the former students and their employers consider interdisciplinary training important to meeting their future needs in a time when knowledge changes quickly and new fields no longer fit within the boundaries of the classical disciplines.

Dieter Imboden
Swiss Federal Institute of Technology,
Environmental Physics Department

---

often have their own buildings. Norway, Sweden, Germany, Switzerland, Poland, Italy, and France all have such institutes within one or more major universities or field stations closely connected with universities.

Some may argue that the highly stressed financial situation within most American universities provides an unlikely climate for the development of new departments. However, current unrest in academia involves much more than retrenchment caused by financial considerations. As is true of many societal institutions at the close of the twentieth century, American universities are seeking to redefine their missions and their structures. Current departmental and collegiate structures, which in many cases represent a legacy from early in this century or even from the nineteenth century, are not necessarily optimal for science education as the twentieth century comes to a close. To the extent that limnologists can make effective arguments that existing programs and administrative arrangements for education in limnology, or more generally in aquatic science, are inefficient and inadequate to satisfy national needs, proposals to develop new departments may find a more receptive audience among university leaders than ever before.

Interdisciplinary Programs in Aquatic Science

Within the recent past, some universities have strengthened their aquatic science offerings by establishing interdepartmental programs. Limnologists and university leaders may encounter less resistance to forming new interdepartmental programs than to establishing new departments. In addition, these programs have the advantage of encouraging collaborative work among faculty from diverse departments rather than creating new divisions. Such broad programs may offer students access to large numbers of formal courses that are related to aquatic science but might not necessarily belong in an aquatic science department.

One example of an interdepartmental aquatic science program is Kent State University's Water Resources Research Institute (see Box 4-8). Through this institute, students may obtain M.S. and Ph.D. degrees in aquatic science from four departments that cross-list their courses, encouraging students to bridge departmental boundaries; faculty members may also receive joint appointments in the four departments. Another example of a cross-departmental program is Utah State University's Watershed Science program (see Box 4-9). This program, established in 1990, offers B.S., M.S., and Ph.D. degrees in interdisciplinary water science with options to emphasize hydrology, management, or ecology. Similarly, a recently approved (1995) interdisciplinary graduate program in water resources science at the University of Minnesota provides broad training through core courses in surface and ground water hydrology, limnology, aquatic chemistry, water quality management, and water resource policy and law. Several areas of specialization are possible beyond the common core, including biological, chemical, and geological limnology and watershed management.

Potential drawbacks of interdepartmental programs revolve around the current reward structure for university faculty, which is tied closely to existing departments. For example, faculty tenure status and salary increases are decided at the departmental level, and funds for laboratory equipment and new faculty positions also reside primarily within departments. Faculty priorities thus tend to be directed first toward fulfilling intradepartmental needs. In some departments and some universities,

---

BOX 4-8 LIMNOLOGY AT KENT STATE UNIVERSITY

Limnology at Kent State University has grown from two faculty members in 1967 to a present-day multidepartment program (Departments of Biological Sciences, Geology, Chemistry, and Geography) with 14 full-time faculty members. The interdepartmental limnology program is housed within the Water Resources Research Institute, a unit with nonacademic departmental status that has a line item in the budget of the College of Arts and Sciences. The university's support of the institute has practically eliminated problems normally associated with interdepartmental research and teaching. Departments share overhead costs, cross-list courses, and may make joint faculty appointments. Through the institute, students may obtain M.S. and Ph.D. degrees in each of the four departments. The number of graduate student applicants to the limnology program doubled between 1984 and 1994: from 15 to 17 applicants per year in 1984–1990 time period to 32 applicants in 1994.

Following are the graduate-level courses in limnology and related disciplines offered in the four departments of the Water Resources Research Institute:

- *Biological sciences:* limnology, microbial ecology, limnological techniques, dynamics of aquatic communities, ecosystem ecology, lake management, experimental limnology, population and community ecology, vertebrate zoology, phycology, palynology, stream ecology.
- *Chemistry:* advanced analytical chemistry, environmental chemistry.
- *Geology:* introductory hydrogeology, hydrogeochemistry, engineering geology, geochemistry, hydrology, hydrogeological systems, advanced hydrogeology, advanced geochemistry, paleoecology, soil mechanics.
- *Geography:* remote sensing, geographical information systems, earth imagery, soils.

G. Dennis Cooke, Professor
Kent State University

---

faculty receive strong support for participation in extradepartmental activities such as interdisciplinary graduate programs, but in other cases, faculty may encounter disincentives to such participation. Leaders of interdepartmental aquatic science programs may find the problem of divided faculty loyalties to be a challenge in developing a strong and cohesive faculty and program. Furthermore, support from all of the involved departments is important; reduced staffing or funding by one department may be highly detrimental to the success of interdisciplinary programs. The challenge of balancing loyalties and developing incentives and rewards for interdisciplinary activities of faculty is not unique to aquatic science programs but is true of interdisciplinary, interdepartmental academic programs in general.

BOX 4-9 WATERSHED SCIENCE AT UTAH STATE UNIVERSITY

Until 1990, Utah State University (USU) offered limnology through the Department of Fisheries and Wildlife, whose faculty members had expertise in biological aspects of limnology. Undergraduate students could emphasize aquatic ecology within the fisheries curriculum, and the department offered M.S. and Ph.D. degrees in aquatic ecology in addition to degrees in fisheries. This model is similar to that in most universities that provide training in limnology, which usually is offered as an emphasis within a biology or fisheries degree. Such training usually focuses on one aspect of limnology, such as biology, while neglecting other areas.

In 1990, the College of Natural Resources at USU expanded a small, interdepartmental program in watershed science as a means of offering students the opportunity to pursue a degree in water science that bridged the physical, chemical, biological, and social sciences. More than 20 faculty members in the Departments of Geography and Earth Resources, Fisheries and Wildlife, Forest Resources, and Range Science participate in the Watershed Science Program. Faculty members have expertise in hydrology, geomorphology, restoration ecology, biogeochemistry, remote sensing, geographic information systems, statistical analysis, modeling, and water policy in addition to aquatic ecology.

The Watershed Science Program offers students interested in the interdisciplinary study of water the opportunity to pursue B.S., M.S., and Ph.D. degrees with emphasis on hydrology, management, or ecology. The undergraduate degree requires students to complete courses in calculus, physics, chemistry, biology, geology, economics, and policy, in addition to courses in their area of emphasis. Undergraduates interested in interdisciplinary aspects of limnology typically would take courses in aquatic ecology, fisheries, aquatic chemistry, water quality analysis, hydrology, and fluvial geomorphology.

Course work for graduate degrees is tailored to the specific career objectives of students, but all students are expected to demonstrate competence in six core areas of watershed science: hydrology, geomorphology, biogeochemistry, aquatic ecosystems, terrestrial ecosystems, and water resource policy or watershed management. Graduate students are also expected to explore thesis topics that bridge departmental disciplines.

Charles Hawkins, Director
Watershed Science Program, Utah State University

The approach taken for the environmental science program at the Swiss Federal Institute of Technology (see Box 4-7) may be useful in overcoming these problems. In this case, a core of faculty has appointments (including tenure and salary lines) in the environmental science program. These faculty members serve as program leaders and links to a much larger group of faculty members who are associated with the environmental science program but whose tenure and salary lines are located within traditional, discipline-oriented faculties.

Scientists in cross-departmental programs report that strong support from the university administration is essential for success. For example, strong university commitment to Kent State's water resources program has been essential to eliminate potential problems in dividing overhead costs among departments, cross-listing courses, and approving joint appointments. Faculty at the University of Wisconsin–Madison, which has three interdepartmental programs providing graduate study in aquatic science, also report strong support from the university administration. At the University of Alabama (see Box 4-10), which also has interdisciplinary programs in limnology, the administration promotes interdepartmental work by housing a Center for Freshwater Studies and aquatic scientists from various departments in a single building.

DESIGNING EDUCATIONAL PROGRAMS IN LIMNOLOGY: CURRICULAR ISSUES

A wide range of educational institutions exists in North America, and each institution has its own philosophy, mission, strengths, peculiarities, and imperfections. Therefore, the administrative route to strengthening programs in limnology, whether through developing interdepartmental programs, forming new departments, or modifying existing ones, cannot be prescribed specifically. Nonetheless, the fundamentals of limnology and the career possibilities for students remain the same regardless of the institution. In establishing programs in limnology, universities should follow some common curricular themes, described below.

Curricula for Undergraduate Students

Undergraduate students—science and nonscience majors alike— should have the opportunity to be introduced to limnology early in their education. Those with strong science skills and strong interest in limnology then should have the opportunity to pursue a curriculum that provides the foundations for further work in limnology. Undergraduate programs at all universities therefore must meet two needs. First, they must provide students having differing backgrounds and career plans with opportunities to learn about the field of limnology and its role in the protection of water resources. Second, they must provide an adequate foundation in limnology and related sciences for those who wish to pursue this science.

BOX 4-10 LIMNOLOGY AT THE UNIVERSITY OF ALABAMA

The interdisciplinary program in freshwater sciences at the University of Alabama receives unusual university-wide administrative support. Although the instructional and research programs in limnology are affiliated predominantly with the Department of Biological Sciences, where 15 faculty members have their primary research and teaching activities in aquatic ecology, research collaborations and interdisciplinary graduate research programs extend between several departments. Collaboration is particularly strong between the Departments of Biological Sciences and Geology, in which six faculty members specialize in hydrogeology, aqueous geochemistry, and hydrology. Nearly all of the ecosystem-oriented biological limnologists and the hydrogeologists and geochemists are co-located in a building completed in 1994. In addition to major centralized analytical facilities, the building houses a glasshouse experimental mesocosm facility and experimental wetland growth facilities.

Alabama offers 54 courses in limnology and related subjects in aquatic ecology in the Department of Biological Sciences and more than 50 additional courses in hydrogeology, aqueous geochemistry, hydrology, water resources, and applied limnology in allied departments. Further aquatic offerings are found in the environmental sciences undergraduate program, focused largely on water resources, in the Department of Geography. The Schools of Business and Law offer courses in environmental economics and law, particularly associated with water and wetlands. The Departments of Chemical Engineering and Environmental Engineering offer undergraduate and graduate courses in hydrology, applied aquatic chemistry, and water treatment.

A University of Alabama Water Resources Network fosters information transfer among research and instructional programs. In addition, the Center for Freshwater Sciences facilitates interdisciplinary research and education in freshwater ecosystems.

Robert G. Wetzel, Professor
University of Alabama

Introductory Course in Limnology

The goal of introductory undergraduate courses in limnology should be to provide a broad view of the science for students with differing backgrounds and interests. Such courses should be widely available at U.S. colleges and universities and should be designed not just for those who plan to become aquatic scientists but for all undergraduates with an interest in lakes, rivers, and wetlands. The course should educate students about the nature, origin, and development of inland aquatic ecosystems, their interactions with the surrounding environment, and their importance for human society. The instructional objective of such courses should be to convey an understanding of how these ecosystems are structured and how they function. Topics to be covered should include productivity of aquatic biota; interactions among organisms at various levels of the food web; cycling of water, nutrients, and toxic materials; decomposition of organic matter; biodiversity; and ecosystem development over time. The course should emphasize problems of management for human purposes, including case studies of degradation, restoration, creation, and protection of diverse inland aquatic ecosystems. Throughout, the course should describe the history of scientific advances in limnology and major questions to be answered by future research.

Although introductory limnology courses typically are taught as upper division (junior or senior level) courses at present, the fact that limnology is a highly multidisciplinary subject also makes it an attractive option for a general science course at the freshman or sophomore level. Such a course would have to be broadly integrative and include all types of inland aquatic ecosystems within a watershed perspective. Box 4-11 describes one possible scheme for organizing an introductory limnology course.

Undergraduate Curriculum in Limnology

Students who wish to pursue education in limnology beyond an introductory class will need direction about courses that are necessary for further study in this field. Details of how this direction is dispensed may vary with the university, but in general students should receive direction through brochures and catalogs that describe a set of prescribed courses and electives for a specialization in limnology, combined with one-on-one advice from a faculty member.

Whether the ultimate goal of these students is a B.S. degree in a field that emphasizes freshwater ecosystem science, an undergraduate minor in this subject, or a graduate degree in limnology, the student's curriculum should have depth and breadth in mathematics and science and should integrate the study of biological, physical, and chemical processes in aquatic systems with an overarching theme of the ecosystem. Further, the curriculum should provide exposure to detailed characteristics of and critical concepts for lakes, streams, and wetlands. Finally, it should include social science courses such as natural resource economics and environmental policy to provide students with perspectives on the many complex social issues that influence stewardship of freshwater ecosystems.

The curriculum may be similar for students planning to complete their education at the B.S. level and go on to employment and for those planning to continue for either an M.S. or a Ph.D. degree. An M.S.-level education or higher is usually preferable in limnology, however, because it is difficult to obtain sufficient depth of knowledge in the many subjects necessary to understand aquatic systems during a four-year undergraduate program.

Students wishing to seek employment immediately after a B.S. degree may need a somewhat greater complement of practically oriented (technician-level) courses to be attractive to prospective employers than would students expecting to continue their education in advanced degree programs. Conversely, students expecting to continue into graduate programs may need a greater complement of rigorous basic science and mathematics courses at the undergraduate level. The danger in designing a terminal B.S. degree in limnology that satisfies the interests of prospective employers is that practice-oriented courses may be included at the expense of depth, rigor, and breadth in the basic sciences and mathematics, so that recipients of such degrees may not have adequate preparation for graduate studies. This problem is not unique to limnology; it is common in integrative disciplines such as the natural resource sciences. To the extent that disciplines such as forestry and soil science have resolved these problems in developing a full range of degree programs (B.S. to Ph.D.), it should be possible to do the same in limnology.

Many U.S. college students do not realize their academic potential or learn about the possibilities for advanced graduate studies until late in their undergraduate careers. Therefore, it is important to provide adequate counseling to undergraduates, especially early in their student careers, and to design sufficiently rigorous B.S. curricula to provide maximum flexibility for advanced study, whether immediately after the B.S. degree or at some later time in the person's career.

Box 4-12 presents an example of an undergraduate curriculum in limnology. This curriculum could fit within an interdisciplinary aquatic science program or within a limnology major in a department of aquatic science. In order to leave students with as many options as possible after graduation, it attempts to achieve a balance between generalization and specialization.

Synthesis Course in Limnology

The undergraduate curriculum shown in Box 4-12 includes an advanced limnology course covering all types of freshwater ecosystems and taught through case studies. In this capstone course, students would learn about current water resource problems at the global, national, and local levels and about how limnology can be brought to bear to solve them. By studying these problems, students would learn directly about the interconnectedness of water resources with human activities and land uses.

Ideally, this course would be team-taught by a group of limnologists and other aquatic resource specialists with knowledge of hydrology, hydrodynamics, aquatic biology, aquatic chemistry, and geology. At the beginning of the course, the teaching team would describe a problem and review the key factors contributing to it, tools needed to assess it,

BOX 4-11 COMPONENTS OF AN
INTRODUCTORY LIMNOLOGY COURSE

An introductory limnology course might include the following major sections:

• *Introduction*: The course could begin by presenting definitions and examples of the various types of inland aquatic ecosystems and their importance to human society. It could introduce lakes, streams, and wetlands as open ecosystems that process solar energy and exchange materials with the environment—that is, as models for other ecosystems, including the oceans and the biosphere as a whole. For example, a wetland might be examined in terms of its role in the cycles of greenhouse gases, fixing carbon dioxide in peat but emitting methane to the atmosphere.

• *Geography, origin, development, and classification*: This section could introduce the environmental factors—geology, geomorphology, climate, and hydrology—that determine how aquatic ecosystems are formed. Students could learn why regions as different from one another as bedrock-dominated landscapes on the Canadian Shield and prairie pothole landscapes on glacial drift of the Great Plains both are rich in lakes and wetlands (but of different kinds), in contrast to desert and semidesert regions with very limited (and predominantly riverine) aquatic ecosystems. These topics could be followed by a discussion of how aquatic ecosystems develop over time. Such a discussion could lead to consideration of various schemes for classifying inland aquatic ecosystems, for instance, the classification of lakes in terms of their geologic origins (glacial lakes, volcanic crater lakes, solution lakes in "karst" limestone, oxbow lakes, and so on) versus classifications based on nutrient status (oligotrophic, mesotrophic, eutrophic, hypereutrophic).

• *Ecological and biogeochemical functions and dynamics*: Under these topics students could examine productivity and decomposition in different categories of lakes, streams, and wetlands and the excess of production over decomposition, which leads to storage of nutrients (carbon, nitrogen, and phosphorus) and toxic materials such as (lead, DDT, polychlorinated biphenys (PCBs), and others. This section of the course also could describe dynamics of food webs. Students could learn about species interactions and their roles in biodiversity and ecosystem dynamics.

• *Landscape interaction*: This section could describe the interactions (1) among inland aquatic ecosystems, (2) between aquatic and terrestrial ecosystems, and (3) between aquatic ecosystems and the atmosphere. For example, the role of geomorphology in developing and differentiating flowing water systems would be described in the context of the River Continuum Concept.

• *Ecosystem development*: Students could learn about patterns of ecosystem succession caused by both external environmental factors and processes inherent in ecosystems themselves. For example, they might learn how topography and climate influence peatland development and how, in turn, peat accumulation leads to changes in biotic communities.

• *Human uses*: Under this heading might come a series of case studies of constructive and destructive uses of inland aquatic ecosystems. Examples of the former might include maintenance or introduction of species to enhance sport fishing in streams and lakes, management of prairie wetlands for duck production, or use of riparian wetlands for purifying wastewater discharges. Examples of the latter might include the cultural eutrophication of lakes by urban and agricultural runoff or the damage to sport fisheries and ecosystem integrity in streams receiving sewage or acid mine drainage.

• *Management, restoration, and creation:* In this section of the course, students could learn about restoration, for instance, of streams dammed but no longer used for hydropower, lakes subjected to cultural eutrophication, or streams affected by acid mine drainage. Management of reservoirs created for water supply (and recreational use) and wetlands constructed for wastewater control or to replace wetlands lost to development, forestry, and agriculture also could be described.

• *Major research questions:* Finally, students could be introduced to key research areas such as (1) the effects of climate change, acid rain, and ozone depletion on aquatic systems; (2) the use of remote sensing and geographic information systems to assess ecosystem structure and function on landscape, regional, and global scales; (3) the roles of lakes, rivers, and wetlands in the global circulation of carbon, nitrogen, sulfur, trace metals, and xenobiotic molecules of human origin; and (4) the degree to which gradual changes in the environment owing to human activities may lead to sudden, drastic, and often unwanted changes in aquatic ecosystems.

approaches for correcting or reducing it, and legal and socioeconomic issues pertaining to it. After the class assessed two or three such problems in detail, students could divide into subdisciplinary groups to study a local problem. Each group would be led by a course instructor. For two or three sessions, the subgroups would meet separately to identify key issues, relevant information, and possible solutions to the problem. The full class would then reconvene, and each subgroup would present its analysis. The class could then be subdivided again into different groups and repeat the approach so that each student would have an opportunity to learn about different subdisciplines.

Representatives of state regulatory agencies, environmental groups, and others active in water resource problem solving could be invited to speak to the class about their work. A mock legal or permit hearing might be included as part of the class, with students presenting the cases of various water users (such as developers or discharge applicants) and citizen groups. This type of activity would illustrate the social, legal, and ethical issues surrounding water resource management.

Laboratory and Field Experience for Undergraduates

Undergraduate limnology programs must include laboratory and field components. A good example of a limnology course providing extensive

---

BOX 4-12 CURRICULUM FOR AN UNDERGRADUATE LIMNOLOGY MAJOR

Recommended Prerequisites

- math: calculus, differential equations, statistics
- physics: one year, with calculus
- chemistry: three semesters of introductory and organic chemistry
- biology: one-year introductory course and ecology course with field lab
- humanities, social science, and general requirements for B.S. degree

Recommended Core Courses

- introductory limnology
- advanced limnology (synthesis course covering lakes, streams, and wetlands, with attention to case studies)
- geology
- hydrology
- aquatic botany or zoology (including systematics and taxonomy)
- aquatic chemistry and biogeochemistry
- intensive summer field study

Recommended Additional Courses

- natural resource economics
- ecosystem ecology
- geomorphology
- aquatic microbiology
- soil science
- ground water hydrology
- environmental fluid mechanics
- geographic information systems, remote sensing methods
- environmental policy or law

---

laboratory and field experience is the Trent University class described in Box 4-3. Another example, requiring fewer resources but providing less comprehensive field experience, is the limnology course taught by W. T. Edmondson at the University of Washington before his retirement (see Box 4-13). Two examples of texts for teaching laboratory and field methods in limnology are those by Wetzel and Likens (1991) and Hauer and Lamberti (1996). Field and laboratory work requires a large investment in time for students and resources on the part of the university. Hands-on experience is essential, however, to provide adequate exposure to real aquatic ecosystems, so that students' knowledge will extend beyond the textbook.

BOX 4-13 NOTES ON THE LABORATORY COURSE IN LIMNOLOGY AT THE UNIVERSITY OF WASHINGTON

The laboratory course offered in conjunction with the introductory limnology course at the University of Washington illustrated in 18 three-hour sessions a selection of the kinds of information on which the lectures (Box 4-5) were based, and it gave students personal experience with ways of getting such information. It started with two field trips. The first was on a fisheries boat in Lake Washington, observing demonstrations of field equipment and procedures in a smaller research boat. The second trip was to a bog lake with a *Sphagnum* mat, where the students took sediment cores.

In the laboratory, physical limnology was illustrated by experiments on thermal stratification in tanks of water. Most of the laboratory time was spent working in various ways with organisms in a survey of plankton and benthos. Students were given the opportunity to observe live material whenever possible. The emphasis with zooplankton was on feeding, life history, and adaptations to planktonic existence. A morphological study of *Daphnia* and *Diaptomus* showed how to get information needed for species identification and to understand the results of feeding experiments. Experiments with visual and tactile predators feeding on planktonic and benthic organisms made the concept of selective predation meaningful.

The course was not intended for detailed instruction in methods, but a selection was made to illustrate the problems of getting different kinds of information. Methods of plankton counting were explained, and students counted both phytoplankton and zooplankton samples. Sediment cores were studied for paleolimnological data. The final session was a demonstration of different kinds of chemical and other methods to show the principles involved (colorimetry, spectrophotometry, titration, and gravimetry, for example). Students who needed to develop skills in chemical analytical techniques could take courses in the Department of Chemistry or the Department of Oceanography. Those who wanted more field experience were invited on research trips and allowed to help in the lab afterward.

W. T. Edmondson, Emeritus Professor
University of Washington

In addition to field exercises within regular courses, undergraduate field sessions or summer field camps should be a key component of education in limnology. One innovative approach to providing field experience for students is Dartmouth College's course on tropical ecosystems. Students who enroll in this course spend a semester in the field in Costa Rica, Jamaica, and Panama learning in detail about the surrounding ecosystems. Another example is Wayne State University's summer field pro-

University of Wisconsin limnology class on a field trip to Little Rock Lake. SOURCE: Thomas M. Frost, University of Wisconsin, Trout Lake Station.

gram at Fish Lake, during which students investigate in detail special problems pertaining to the lake. Such sustained field experiences provide unique opportunities to observe and learn field techniques and to develop an enthusiasm for limnology. The acquisition of detailed knowledge of an ecosystem may be more effective than a lecture course in conveying the ecosystem and landscape perspectives. Field sessions should strive to involve scientists from agencies, consulting firms, and industry to broaden the "real-world" experience for students.

Programs for the Master of Science Degree

A master of science degree can serve one of three purposes. First, it can provide rigorous training in limnology for those who did not have adequate opportunity to study aquatic science as undergraduates. Second, it can provide advanced training for those with undergraduate degrees in limnology or aquatic science to prepare them for jobs requiring more sophisticated expertise. Third, it can provide training for entry into a Ph.D. program. The course requirements for all three purposes are similar, but the details may differ. In particular, students entering limnology graduate programs who did not study aquatic science in obtaining their bachelor's degree probably will need some introductory limnology and aquatic science courses that are not necessary for students with aquatic science backgrounds. All students will require a balanced exposure to stream, lake, and wetland ecosystems and a broad knowledge of physical, chemical, and biological processes in aquatic systems.

The formal course component for a master's degree should cover the following broad topics directly related to inland aquatic ecosystems:

1. limnology, integrating physical, chemical, and biological processes across lakes, streams, and wetlands;
2. aquatic biology;
3. aquatic chemistry and biogeochemistry; and
4. hydrology, geomorphology, and physical limnology.

The specific mix of courses appropriate for a given student will depend on the student's educational background. For example, students with undergraduate degrees in biology may need introductory courses in aquatic chemistry and physical limnology, and those with degrees in hydrology or environmental engineering may need introductory courses in aquatic biology. Even as more universities create strong interdisciplinary undergraduate programs in aquatic science, and even if some universities create limnology majors for undergraduates, master's degree programs will have to maintain flexibility to accommodate the varied backgrounds of students. Some of the most prominent limnologists have stumbled on this field from unusual pathways, and master's degree programs should be open to accepting students who do not "discover" limnology until the end of their undergraduate years.

Although aquatic science courses should be the key focus of a master's program in limnology, students should be encouraged to select supporting courses that provide additional technical expertise and insights about the societal context for limnology. Such courses include, for example, environmental law, decisionmaking, management, policy, and economics, as well as advanced courses in chemistry, biology, statistics, and geology. The master's program should also provide training in special skills such as experimental design, modeling, and trend analysis. Finally, the development of good communication skills (writing and public speaking) is essential. Limnologists must be able to communicate effectively not only with other scientists but also with the concerned public. M.S. programs must provide a variety of opportunities for students to develop these skills.

At the master's level, there are several possible lines of further specialization. One specialization would be in aquatic biology, with advanced courses in subjects such as biodiversity, taxonomy of a major group of organisms, or trophic dynamics. Another common specialization would be in chemical limnology, with advanced study in biogeochemical cycling, surface chemistry, photochemistry, or contaminant transformations. Master's degree programs should include a field component. This requirement will ensure that students have mastered a range of basic field skills for collecting samples and making measurements such as stream flow, water transparency, and dissolved oxygen levels. Field experience also provides students with an appreciation for the difficulty of collecting data and a healthy skepticism for data that they may encounter in future jobs. Experience at a field station is an excellent way to obtain this training, but it can also be accomplished as part of a laboratory for a graduate-level limnology course.

A thesis or practical project is a highly important component of a master's degree because it provides students the experience of initiating, carrying out, and completing a project of their own. A thesis is valuable even for students who do not plan to become researchers because it provides an opportunity to learn about the nature of research and a perspective for evaluating studies conducted by others. On the other hand, these students could benefit equally from being given a practical problem to solve, applying the skills they have developed in formal classes to develop a solution, and writing and presenting their suggested course of action. Such experience also might be gained through a formal internship, part of which would be dedicated to developing a written solution to a practical problem at the organization in which the student is interning. Exceptions to the requirement for a thesis or practical project may be appropriate for M.S. students who already have substantial project or research experience through postbaccalaureate work experience, but in general, "course-work-only" graduate programs that are mere extensions of an undergraduate experience should not be promoted in limnology.

Programs for the Ph.D. Degree

Doctoral training is necessary for those who will lead or manage scientific research investigations, either independently or as part of a team. Jobs that generally require a Ph.D. include faculty positions in academic institutions, many research scientist positions in government agencies or private research centers, and program officers in federal agencies such as the National Science Foundation, Environmental Protection Agency, Department of Agriculture, and National Biological Service. Many consulting firms also hire Ph.D.-level scientists for the expertise and added credibility they bring to the job.

Ph.D. programs should require broad knowledge of all the subdisciplines of limnology—that is, at least the equivalent of the knowledge base expected of M.S. students, including field and laboratory experience. To distinguish themselves, Ph.D. students should develop an area of specialization within their major field. Also valuable is expertise in an analytical specialty such as sediment dating, genetic analysis (using, for example, the polymerase chain reaction), or model development (using, for example, geographic information systems combined with mathematical models).

As in any other scientific discipline, the major component of a Ph.D. program in limnology is the thesis research project. The Ph.D. student must propose an appropriate area of research (within funding constraints), develop a plan for conducting the research, carry out the study, analyze the data, and write a detailed account of the project in a dissertation. The final step of the dissertation is publication of key results in scientific journals in order to communicate findings to the broad community of aquatic scientists.

Ph.D. students also must develop skills in written and oral communication and in collaborative research. Experience in giving technical presentations to describe research findings is essential. Teaching experience is valuable because the ability to teach and convey information is an increasingly important skill in all job markets. Professors should provide opportunities for their Ph.D. students to conduct a portion of their thesis research as part of a team, perhaps including students working in related disciplines. Finally, Ph.D. students should have training in writing and reviewing research proposals.

More so than in educating B.S. and master's degree candidates, mentoring plays a critical role in the training of Ph.D. scientists. Many of the skills required of high-level scientists, such as identifying promising research areas and developing proposals to pursue this research, cannot be taught in the classroom but are learned by apprenticeship. The stream ecology program at Arizona State University, described in Box 4-14, provides one example of a strong mentoring system. In addition to mentoring, however, Ph.D. programs need structure within the academic institution; a program that relies solely on the presence of one or two professors who serve as mentors is at risk of being eliminated if these professors retire or move to another institution.

Postdoctoral Programs

For those who have completed a Ph.D. and plan to pursue a research career, a common next step is a postdoctoral position before an academic faculty or government appointment. Ideally, postdoctoral work should be completed at an institution where the graduate can broaden his or her skills and gain new perspectives.

Mentoring relationships may continue to be important for postdoctoral researchers, but the obligation of senior scientists to provide a supportive learning environment recedes to some extent. The postdoctoral researcher has responsibility for developing research opportunities and broadening scientific collaborations and interactions. Postdoctoral scientists may accept positions to carry out research programs, or some part of research programs, that have been developed and funded by senior scientists. These positions could provide an opportunity to develop management

---

BOX 4-14 EDUCATION THROUGH MENTORING: STREAM ECOLOGY AT ARIZONA STATE UNIVERSITY

Training of Ph.D. students in stream ecology at Arizona State University relies on a mentoring system. My colleague, Nancy Grimm, and I spend one evening a week with students at our home. We cook and eat dinner with the students, discuss papers in an organized way, plan research, and contemplate issues such as where to get funding, appropriate ethics in science, barriers to minorities, and career options.

We encourage students of stream ecology to take courses in limnology, ecosystems, fluvial geomorphology, and statistics and to select a broad range of courses in ecology (for example, physiological ecology and evolutionary ecology). In addition, we encourage students to participate regularly in topical and departmental seminars. However, as with many small universities (or in this case, small programs in large universities), there is no written, formal aquatic curriculum. Faculty advisers approve student course selections based on (1) the expectation that they will become broadly trained ecologists and (2) their research needs. To be successful, students need to learn how to identify and solve problems, how to stay informed, and where to go to gain new, relevant skills. Highly structured curricula will not automatically provide this without sound mentoring.

The program is predicated on the assumption that there are two issues in the education of Ph.D. scientists: technical training and idea development. The former should be enhanced and will respond to curricular revision. The latter is as important and will not necessarily respond to a longer list of required formal courses. The future leaders of the field will continue to come from those places where students are engaged by active, demanding, caring mentors.

Stuart Fisher, Professor
Arizona State University

---

skills that might not have been obtained in an individually designed Ph.D. project.

At universities, postdoctoral positions, which are often vital to research programs, are typically supported by grants to faculty members. The National Research Council (NRC) postdoctoral program has been effective in bringing young scientists to federal research facilities. These scientists benefit the agencies by bringing new or advanced techniques to federal laboratories, from which they may be disseminated further for application in operating offices of the agencies. Postdoctoral scientists also enhance interactions between federal agencies and academic institutions. Funding for NRC postdoctoral positions is often vulnerable to tight budgets within the agencies, however. Federal agencies with responsibilities for water resources need to invigorate their NRC postdoctoral programs with an emphasis on inland aquatic ecosystem research.

ROLE OF FIELD RESEARCH SITES IN AQUATIC SCIENCE EDUCATION

As emphasized in this chapter, field research is a critical part of education in limnology. Although the principles of how aquatic ecosystems operate can be learned in classrooms, hands-on experience is the only way to gain a true appreciation for the complexities of aquatic systems. Field experience is essential not only for graduate students, who may themselves become scientific researchers and aquatic resource managers, but also for undergraduate students, who benefit greatly from a firsthand understanding of how aquatic systems operate, from learning scientific research methods, and from the atmosphere at field research sites, which is distinctly different from the on-campus atmosphere. Students can become completely immersed in their field study and can interact with scientists as members of a team. Many employers place a high value on field experience.

It is highly unfortunate that field experience in limnology is extremely limited for undergraduates at U.S. universities. Opportunities are relatively limited even for graduate students to obtain experience with a diversity of aquatic ecosystems at field stations or in organized field courses.

Field research sites and stations have contributed significantly to the collection of long-term limnological data, interdisciplinary research, and experimental manipulation of large-scale aquatic ecosystems. Efforts to expand the educational mission of such sites and stations should be encouraged. Encouraging a mix of research and educational activities at a field station may provide a more stable funding base to maintain the station than would exist if it focused solely on one of these activities.

Some field stations do offer opportunities for student research experience (see the background paper "The Role of Major Research Centers in the Study of Inland Aquatic Ecosystems" at the end of this report). A few examples include the Trout Lake Station, associated with the University of Wisconsin–Madison; the Lake Itasca Forestry and Biological Station, affiliated with the University of Minnesota; the University of Notre Dame Environmental Research Center; the H. J. Andrews Experimental Forest, which has a major stream study component and is associated with Oregon State University and the U.S. Forest Service; the Center for Great Lakes

Studies (see Box 4-15), part of the University of Wisconsin–Milwaukee; and Canada's Experimental Lakes Area (described in Chapter 5), which is affiliated with U.S. as well as Canadian universities.

According to the Organization of Biological Field Stations, 103 biologi-

---

BOX 4-15 RESEARCH AND TEACHING AT THE CENTER FOR GREAT LAKES STUDIES

The Center for Great Lakes Studies, part of the University of Wisconsin-Milwaukee, is housed in a large warehouse that has been converted to a high-tech research facility on Milwaukee Harbor, one of the major Great Lakes ports. The center operates the 71-foot research vessel *Neeskay*, as well as several smaller boats for estuarine and near-shore sampling. It is also home port to the *R/V Roger Simons* and the winter port of the *Lake Guardian*, both operated by the Environmental Protection Agency's Great Lakes National Program Office. Specific research capabilities include instrumentation for water analysis; a radioisotope laboratory for dating sediment cores; modern molecular biological facilities; and a state-of-the-art, computer-interfaced video microscopic system for behavioral studies of plankton. Two full-time chemical and instrument technicians are responsible for day-to-day maintenance and the training of new users.

Research at the center involves undergraduate and graduate students in all activities, from field collections, to laboratory analyses, to interpretation. The facilities used for these research projects are available for undergraduate independent research projects as well as graduate dissertation research. Many of the research labs are also used for demonstrating techniques in laboratory courses, giving students exposure to modern equipment and techniques.

Examples of research projects at the center include the following:

• *Coastal dynamics*: Center researchers continuously measure and plot the spatial distribution of chemicals and organisms in the water. These data can be compared with satellite images of the area in a geographic information system to provide a broad picture of the regional distribution of these variables. Access to such real-time data provides a powerful teaching tool for field courses.

• *Underwater robotics*: Center researchers have been involved in developing instrumentation and robots for obtaining underwater samples and in conducting underwater experiments.

• *Microorganisms in sediments*: Center researchers are studying the activity and genetic characteristics of microorganisms in Lake Michigan sediments that play important roles as sinks for atmospheric methane and as potential consumers of environmental contamination.

Arthur Brooks, Professor
Center for Great Lakes Studies

Fish sampling done as part of routine analyses of a group of LTER lakes. SOURCE: Thomas M. Frost, University of Wisconsin, Trout Lake Station.

cal field stations either currently have, or have the potential for, aquatic science research and education. In addition, the National Science Foundation supports a network of field sites, listed in Table 4-2, through its Long Term Ecological Research (LTER) program; most of these sites include an aquatic research component, and four (North Temperate Lakes for lakes and Hubbard Brook, Coweeta, and Andrews for streams) have long histories of advancing knowledge in limnology.

Educational opportunities at field stations should be improved in several ways:

• More undergraduates should be encouraged to take advantage of field stations or other outdoor laboratories as part of their general education in aquatic science. For example, more stations could develop short courses or workshops specifically for undergraduates. Universities with on-campus aquatic resources (lakes, streams, or wetlands) could establish their own outdoor laboratory programs for undergraduates.

• The range of ecosystem types and the geographic distribution of field stations should be expanded to encompass a broad range of aquatic ecosystems that occur across the continent. For example, the LTER sites listed in Table 4-2 reflect neither the range of habitat types nor the latitudes at which those habitats occur, especially for wetland and lake ecosystems. Funding resources should be focused on developing additional sites representing lake and wetland ecosystems over the range of climatic conditions

TABLE 4-2 Sites in the Long-Term Ecological Research Network

Site	Aquatic Emphasis
H. J. Andrews Experimental Forest	Forest-stream interactions
Arctic Tundra	Lake and stream studies
Bonanza Creek Experimental Forest	Wetland processes
Cedar Creek Natural History Area	Wetland processes, plant communities, paleoecology
Central Plains Experimental Range	No aquatic emphasis
Coweeta Hydrologic Laboratory	Stream and riparian zone studies
Harvard Forest	Paleolimnology
Hubbard Brook Experimental Forest	Hillslope hydrology and biogeochemistry
Jornada Experimental Range	Water transport and playa lake processes
W.K. Kellogg Biological Station	Agricultural ecosystem hydrology and biogeochemistry
Konza Prairie	Prairie stream processes
Luquillo Experimental Forest	Stream processes
McMurdo Dry Valleys	Lake and stream processes
Niwot Ridge/Green Lakes Valley	Montane hydrology and biogeochemistry
North Temperate Lakes	Lake processes and watershed hydrology
Palmer Station	Marine ecosystem processes
Sevilleta National Wildlife Refuge	Watershed studies
Virginia Coast Reserve	Terrestrial-aquatic interactions in coastal habitats

in the United States and ensuring that each has a strong education and research program.

• Programs should be established to allow graduate students to conduct thesis research across a set of different sites or stations. This would provide students not only with an opportunity to increase their knowledge of diverse ecosystem types but also with an opportunity to observe the different research methods used at different sites.

• An educational program should be developed in which students from various universities would visit a set of a field stations representing a wide range of ecosystem types and investigation techniques over the course of a semester. Students would be expected to engage in research projects at each site and to develop an independent research project at the final site.

Funding for some of the above activities (particularly the last two) can come from research grants or from specific programs at universities or federal or state agencies. In addition, the National Science Foundation has a special Research Experience for Undergraduates (REU) program to provide scientists with funding specifically to support undergraduates. The REU program provides individual professors with support for one or a few students to work on an established research project and also provides groups of professors with funding to establish REU sites for large numbers of undergraduates.

Although federal funds for specific research and training initiatives are a key component in financing field station operations, the maintenance of field station facilities over the long term is primarily a responsibility of individual universities and states. In times of financial difficulties, such as universities are currently experiencing, administrators may view such facilities as luxuries that the university no longer can afford. Limnologists need to argue convincingly against this attitude. Field stations should be regarded as essential for teaching and research in limnology just as teaching and research hospitals are essential for medical schools, experimental farms for agricultural colleges, observatories for academic astronomers, and structural testing laboratories for civil engineering departments. For this attitude to take hold, however, limnologists must make field activities a much more integral part of undergraduate and graduate education in limnology than typically has been the case in the past.

RECOMMENDATIONS FOR RESTRUCTURING EDUCATIONAL PROGRAMS IN LIMNOLOGY

In summary, education in limnology must serve three purposes: (1) educating responsible citizens about stewardship of aquatic resources; (2) educating future policymakers and managers of these resources; and (3) training the next generation of scientists who will help develop the knowledge necessary to reverse the damage done to the world's lakes, streams, and wetlands in order to preserve their usefulness for the future. As awareness of the value of freshwater resources increases, so will the importance of establishing strong educational programs that meet all three of these needs. Indeed, as shown in Appendix A, student interest in limnology is increasing at many universities, reflecting the increasing importance that society is placing on protecting its freshwater resources.

Following are steps that limnologists, universities, government agencies, and private-sector companies can take to strengthen educational opportunities in limnology.

Strengthening Limnology Within the University System

• Establish departments of aquatic science with a strong focus in limnology at one or a few universities in each region of the United States to provide the full range of training (B.S. to Ph.D.) in limnology.

• At universities that do not create departments of aquatic science but have significant faculty expertise in aquatic science in various departments, establish interdepartmental programs in aquatic science with a strong focus on limnology; such programs are particularly suitable for educating graduate students but may not provide sufficient structure for educating undergraduate students.

• Ensure that these programs provide adequate coverage of all types of freshwater ecosystems: lakes, streams, and wetlands.

Educating Responsible Citizens

• Develop general introductory limnology courses (lower division or freshman-sophomore level) that are accessible to all types of students, with the goal of conveying how freshwater ecosystems respond to various human activities in a watershed context.
• Provide enough faculty support to allow all interested students to enroll in introductory limnology courses.
• Ensure that introductory limnology courses include coverage of wetlands and streams as well as lakes.

Educating Future Water Managers

• Provide students with increased opportunities to gain exposure to practical problems such as the management of freshwater systems in urban areas.
• Increase student internship opportunities in federal and state agencies and the private sector.
• Strengthen NRC postdoctoral programs in federal agencies with water resource management responsibilities.
• Provide all interested students with an opportunity to gain laboratory and field experience, at either on-campus aquatic ecosystems, nearby aquatic ecosystems, or formal field stations.

Educating Future Limnologists

• Provide undergraduates showing special interest in limnology with better guidance on selecting a curriculum that will provide the breadth of skills and knowledge needed to solve problems of freshwater ecosystems.
• Increase the opportunities for undergraduate and graduate students to attend summer- or semester-long limnology "field camps."
• Develop rigorous and comprehensive degree programs at the M.S. and Ph.D. levels to educate limnologists who (1) are knowledgeable across the broad spectrum of freshwater ecosystem types; (2) have an integrated understanding of the physical, chemical, and biological processes operating in these ecosystems, as well as the political, economic, and cultural factors that affect aquatic ecosystems and their management; and (3) have the problem-solving and communication skills necessary to apply their knowledge in the protection, management, and restoration of freshwater resources.
• Encourage the inclusion of a research component (thesis) or practical

problem-solving component in all M.S. programs and discourage course-work-only programs.

• Establish recommendations for the knowledge base that students should master at each degree level in limnology.

The primary responsibility for carrying out the above recommendations rests with limnology-oriented faculty in universities because only they can develop the courses and degree programs described in this chapter. If reform were easy, however, many of the above recommendations already would have been enacted. Numerous barriers must be overcome to achieve the goals described, and it will be difficult for academic limnologists to overcome these barriers by themselves. University administrators, water resource managers in government and the private sector, and professional societies involving limnologists all have key roles to play. University administrators can help advance reform by supporting needed changes in university administrative structures for limnology programs and by allocating appropriate financial resources to these new structures. University alumni working as aquatic ecosystem scientists and managers can promote change by advising faculty and administrators of the essential and desirable training that students should receive for employment in their field and by developing internship opportunities for students. Professional societies involving limnologists (both researchers and managers) can organize symposia at their annual meetings or cosponsor a national conference to discuss limnological education. As individual societies, or preferably as a joint effort, they should develop initiatives to influence the direction of educational reform. Such initiatives would serve not only to guide the efforts of faculty at individual universities but also to provide encouragement and support for those efforts. In summary, to accomplish changes in the educational system limnologists in academic institutions, government, and the private sector, supported by their professional societies and by university administrators, will need to join forces.

REFERENCES

Cole, G. A. 1983. Textbook of Limnology, 3rd ed. St. Louis, Mo.: Mosby.

Hauer, F. R., and G. A. Lamberti, eds. 1996. Methods in Stream Ecology. San Diego, Calif.: Academic Press.

Horne, A. J., and C. R. Goldman. 1994. Limnology, 2nd ed. New York: McGraw Hill.

Ruttner, T. 1963. Fundamentals of Limnology, 3rd ed. (translated by D. G. Frey and F. E. J. Fry). Toronto: University of Toronto Press.

Welch, P. S. 1935. Limnology. New York: McGraw-Hill.

Wetzel, R. G. 1983. Limnology, 2nd ed. Orlando, Fla.: Harcourt Brace Jovanovich.

Wetzel, R. G., and G. E. Likens. 1991. Limnological Analyses, 2nd ed. New York: Springer-Verlag.

5

Future of Limnology:
Linking Education and
Water Resource Management

Future employment prospects for graduates are a critical consideration when evaluating the merits of educational programs. In the case of limnology, however, those designing educational programs should look beyond the specific job market for limnologists because many positions in government and the private sector, whether or not they are held by limnologists, involve actions that affect freshwater ecosystems. Regulators who oversee the implementation of environmental laws, scientists and engineers in environmental consulting firms, managers of industrial facilities located on lakes and rivers, operators of water and wastewater treatment plants, urban planners, and policymakers in legislative bodies all make decisions that affect freshwater ecosystems. Limnology programs can serve not just to educate people who will pursue careers specifically in limnology but also to educate people whose work might someday influence the long-term, sustained use of freshwater ecosystems.

This chapter describes a wide range of jobs in public- and private-sector water resource management, as well as academia and research, in which limnological skills would be valuable and limnologists of the future may find careers. It also recommends strategies for improving the education of limnologists by linking academic programs more closely with applied problems and with agencies and companies involved in the management of lakes, rivers, and wetlands. Finally, it recommends ways to enhance the involvement of limnologists in management decisions that affect freshwater ecosystems.

CAREERS IN LIMNOLOGY

Typically, limnologists with advanced degrees have pursued careers in academia or research. For example, 49 percent of the members of the North American Benthological Society, one of the primary professional societies for limnologists, are employed in academia (see Appendix B). However, in addition to typical academic and research careers, promising employment opportunities exist for limnologists in the private and public sectors.

Private-Sector Employment Opportunities

Employment options in the private sector often are overlooked by limnologists, but a variety of potential careers exists with utilities, industries, nonprofit institutions, and consulting companies. The multidisciplinary training and experience that limnologists offer are assets for many positions in the private sector. Small and large organizations alike benefit from having staff with the ability to integrate several disciplines when confronting complicated environmental issues such as watershed management, wetland restoration, or mitigation of point-source pollutant discharges. At the same time, limnologists benefit from the intellectual challenges in the private sector as they seek solutions to environmental problems that involve diverse scientific and policy issues.

Two examples illustrate private-sector opportunities for limnologists:

1. *Environmental manager of a hydroelectric facility:* The responsibilities of environmental managers at hydroelectric facilities (see Box 5-1) often include monitoring water quality for compliance with relevant regulations, determining impacts of, a dam on fish and aquatic invertebrate organisms, protecting wetland habitats influenced by the dam, and maintaining terrestrial plant and animal communities of the riparian areas surrounding the impoundment and river. Such managers must understand the physics, chemistry, and biology of the river and reservoir system—for which training in limnology is directly relevant.

2. *Watershed manager for a nongovernmental watershed conservancy:* Land, lake, and river conservancies in various parts of the United States purchase critical land, negotiate conservation easements, and establish cross-ownership management partnerships in areas they wish to protect. Watershed managers for such conservancies must develop, implement, and monitor natural resource management plans that integrate activities on conservancy lands, consider industrial activities within the watershed, and seek to improve water and wetland quality of the mainstream and tributaries. As with the hydroelectric facility operator, the conservancy watershed manager must be able to consult with science experts from disparate fields and evaluate their advice when confronting complex restoration

BOX 5-1 HYDROELECTRIC FACILITY MANAGER ON THE MENOMINEE RIVER

The environmental manager at a hydroelectric facility on the Menominee River in Michigan's Upper Peninsula has a variety of responsibilities for which training in limnology would be ideal. Many of these responsibilities are illustrated in the tasks the manager carried out to obtain a new license for the facility from the Federal Energy Regulatory Commission (FERC). These tasks included

- arranging for surveys of rare organisms;
- developing management plans to protect a population of rare wood turtles (*Clemmys insculpta*) and globally endangered oak savanna prairie remnants on company property;
- monitoring bald eagle use of the impoundment and river, and identifying and protecting potential nest trees;
- conducting fish entrainment studies to determine the impact of the facility on the fishery;
- surveying aquatic insects and mollusks in the impoundment and river to determine the health of these communities and provide a baseline against which to monitor ecosystem changes;
- mapping all Menominee River wetlands influenced by the project and estimating their water-level fluctuations;
- evaluating optimal water flows for whitewater kayakers and rafters on rapids downstream of the facility and reconciling these uses with the needs of downstream fisheries and the demands for electric power.

In most cases, the environmental manager contracted services to conduct these studies but required broad-based expertise in aquatic science such as is provided by limnology to prepare requests for proposals, select consultants, and evaluate and interpret reports.

or environmental impact challenges—responsibilities for which broad, interdisciplinary training in limnology is valuable.

In addition to direct opportunities in water management, the private sector offers limnologists opportunities to conduct in-depth studies of challenging practical questions. Examples of private-sector questions that limnologists can address include the following:

- How wide should a vegetated buffer zone be to protect stream quality from adjacent landscape activities?
- What habitat changes result from construction of roads over small streams?
- How can downstream fisheries be maintained despite dramatic water flow fluctuations caused by a hydroelectric facility?

- Can a borrow pit pond be transformed into a productive wetland ecosystem?
- How can herbicides be applied to pine plantations without damaging an adjacent wetland or stream community?
- How can stormwater runoff be managed or treated to protect the recreational values of urban lakes?
- How can a tailing pond be designed to protect aquatic habitats while allowing mining operations to proceed?

Some of these private-sector research questions can be addressed by the application of existing research. Addressing others will require studies tailored to a geographic location. Still others will require original research in limnology and other aquatic sciences.

Public-Sector Employment Opportunities

At all levels of government, managers make critical decisions that affect aquatic ecosystems. Examples of such decisions include how close to a lakeshore construction should be permitted, whether to allow alterations to a wetland, what water level to maintain in a reservoir, and what concentration of chemical constituents to allow in a wastewater discharge. Inclusion of limnologists on the teams addressing water management problems will enhance the potential for successful outcomes.

A variety of employment opportunities exists for limnologists in government agencies. Examples include the following:

- In Michigan, the Department of Natural Resources (DNR) employs more than 150 environmental quality analysts and land and water specialists, many of whom have training in limnology, to assess water quality problems. Some of these analysts work with city and county planners to determine the relative impacts of point and nonpoint sources of pollution on aquatic life. Approximately 50 of them are wetland specialists who analyze the impacts of human activities, such as road building, on wetlands. In addition, the Michigan DNR has positions with the titles environmental enforcement analyst, aquatic biologist, fisheries biologist, research biologist, and park interpreter—jobs for which training in limnology is highly valued. Currently, most of the employees in such positions have B.S. degrees from state universities, combined with water management work experience.
- The Wisconsin Department of Natural Resources has many opportunities for graduates of limnology and related aquatic science programs. Employment opportunities exist within its fisheries management, water resource management, wastewater management, water supply, water regulation and zoning, and water resource research bureaus. In 1994, the DNR filled approximately 35 such positions, mostly with individuals who

have master's degrees. In addition to limnologists, employees in these programs include environmental engineers, hydrogeologists, aquatic toxicologists, aquatic biologists, watershed planners and modelers, and hydrologists.

• Federal water management agencies such as the Bureau of Reclamation, the Army Corps of Engineers, and the Tennessee Valley Authority perform a large number of tasks, ranging from reservoir management to wetland delineation, that could benefit from the involvement of limnologists. For example, the Tennessee Valley Authority retains teams of scientists whose services can be contracted to monitor the condition of fish and benthic organisms in reservoirs and streams. The U.S. Geological Survey has hired limnologists in its National Water Quality Assessment Program, designed to evaluate the status of the nation's waters, trends in water quality, and the reasons for these trends. However, the hiring of limnologists for federal positions is complicated by the lack of a specific federal job description for the profession of limnology, as explained later in this chapter.

• Some Native American nations have opportunities for limnologists. For example, the Sault Ste. Marie Tribe of Chippewa Indians employs water quality specialists to develop nonpoint-source pollution management plans, delineate wetlands, and evaluate Great Lakes water quality. The tribe also employs water resource technicians for field and laboratory analysis of water quality. These specialists and technicians typically have B.S. degrees in a field of aquatic science.

Thus, numerous employment opportunities exist for limnologists in public-sector water management. The degree to which these jobs are filled by scientists with training in limnology varies. As the complexity of sound management of aquatic ecosystems becomes more widely recognized, the value of training in limnology should increase in the job market.

Career Opportunities at Research Centers

Ecological research centers offer a variety of employment opportunities for limnologists at all education and skill levels. The United States has a large number of research centers operated by organizations ranging from private foundations, to state and federal government agencies, to universities. Compared with traditional university departments, such centers can offer more opportunities for interdisciplinary research. For example, the objectives of the Maryland International Institute for Ecological Economics specifically include integration, study, and management of ecological and economic systems. Some centers, such as Notre Dame's Environmental Research Center in the upper Midwest, focus on one or a few ecosystems in one geographic region. Others cover a broad range of ecosystems and geography. For example, the University of Georgia's Institute of Ecology conducts studies on wetlands, forests, coastal waters, rivers, lakes, and agricultural systems in the southeastern United States and several tropical countries.

Research centers employ Ph.D.-level scientists full- or part-time in research, management, education, and extension positions. Types of appointments include fixed-term (temporary) postdoctoral assignments for new Ph.D.s and permanent positions. Postdoctoral positions may be funded by center-derived fellowships or research grants; alternatively, the center may simply provide on-site resources to individuals who have obtained their own fellowship support. Employment of permanent staff scientists ranges from full twelve-month appointments at large, well-funded government laboratories (such as the Savannah River Ecology Laboratory in Aiken, South Carolina) to support for nine months with the expectation that three months of funding will be generated by the scientist from research grants or contracts (as at the Jones Ecological Research Center in Newton, Georgia), to salary support dependent completely on research grants (as at the Marine Biological Laboratory in Woods Hole, Massachusetts).

Centers that also have responsibility for managing land or water resources (such as the Flathead Lake Biological Station in Polson, Montana) often have resource management positions. Centers with an extension mission (such as the U.S. Department of Agriculture's Southeast Watershed Research Station in Tifton, Georgia) have employment opportunities in education and information outreach, as well as research. For individuals holding regular faculty positions, research centers may provide sabbatical fellowships or summer support (for example, the Institute of Ecosystem Studies in Milbrook, New York) or defined contract work on a consulting basis (for example, the Environmental Sciences Department at Battelle Pacific Northwest Laboratory in Washington).

Centers also employ technical personnel with B.S. or M.S. degrees for field and laboratory research, laboratory management, resource management, computer support, and extension activities. As with Ph.D.-level positions, a variety of contractual arrangements is possible. Some centers assist in training graduate students by providing graduate stipends, fellowships, and internships, and some provide grant support that targets specific groups. Many centers also provide research experience for undergraduate students. For example, the Savannah River Ecology Laboratory regularly provides research opportunities for undergraduates with funds from the National Science Foundation's Research Experience for Undergraduates (REU) program, which is intended to introduce undergraduate students to basic research.

Career Opportunities in Academia

Despite the concern among some academic limnologists about a decline in limnology programs in North America, opportunities for academic careers in colleges and universities still exist and are likely to continue for imaginative and well-educated limnologists. An understanding of how aquatic ecosystems function and how they are influenced by a variety of human activities will remain a critical societal need for a long time. Colleges and universities will have to produce individuals with such understanding; therefore, a continuing demand for limnologists as instructors and professors can be projected.

As is true of some other fields of science, the nature of academic opportunities for limnologists has changed in recent decades (see Chapter 2). In particular, there is increasing emphasis on wetland and flowing water limnology and on expertise related to such disciplines as chemistry and geology, so that available positions frequently are in nontraditional (that is, nonbiological) departments. Graduate education in limnology must change in response to these trends if the field is to remain vital.

In contrast to graduate limnology programs of the past, which focused on producing Ph.D.-level scientists for academic and research careers, new limnology programs should recognize that the academic job market is likely to remain fairly small. The greater needs in limnology are in the production of practicing scientists at both the M.S. and the Ph.D. levels. Ph.D. students, in particular, need to be counseled about the rewarding limnological careers that exist outside academia, the job skills they will need to compete successfully for such positions, and the limited opportunities for academic positions.

COLLABORATION AMONG UNIVERSITIES, GOVERNMENT AGENCIES, AND THE PRIVATE SECTOR

In preparing this report, the Committee on Inland Aquatic Ecosystems met with water resource managers from around the United States to discuss their perspectives on education in limnology. Perhaps the most emphatic message delivered by the resource managers was that *"universities must provide the broad training necessary for solving practical management problems."* This clear statement is fundamental to delivering a successful education to limnologists who hope to practice their discipline in real-world settings. To answer questions such as the following, planners of limnology curricula need to communicate directly and frequently with the employers who form the market for graduates:

- What kind of work will limnology graduates be performing?
- What scientific knowledge is necessary for this work?
- What communication and other social skills are most valued?

In short, universities can benefit from studying the market carefully in order to prepare their limnology students for the workplace.

One way for universities to determine whether they are preparing their students adequately for the workplace is to survey former students who are working in various resource management careers to find out about any gaps in their educational backgrounds. Beyond communication with former students, however, there is a wide variety of opportunities for creating links between universities and water resource managers, as outlined below. Strengthening these links will enhance the relevance and overall quality of the educational experience for limnology students.

Collaborative Teaching and Research

Universities and environmental agencies have decidedly different missions. The role of a university is to educate students, train professionals through graduate programs, conduct research to increase society's knowledge base, and disseminate information to scientists and the public. The mission of environmental agencies is to manage and protect the environment and public health, as codified in state and federal laws. Because of these very different roles, the interactions between agencies and universities are not always as extensive as they could or should be. Nonetheless, several successful models of extensive collaboration between government agencies and university aquatic science programs provide examples that could be duplicated elsewhere.

Many of the existing successful university-government partnerships involve agency scientists working part-time as adjunct professors. Agency managers and researchers may teach semester-long university courses or short courses themselves or in conjunction with university professors. Adjunct appointments provide agency scientists with access to graduate and postdoctoral students, research laboratories, and specialized field and laboratory equipment. Such appointments also enable agency scientists to submit joint applications with regular university faculty to funding agencies such as the National Science Foundation. Adjunct appointments can benefit universities because of the access they provide to scientists working on applied resource problems. In addition, universities can gain access to state and/or federal funding sources, equipment, and field technicians. In many situations, duplication of laboratory and field equipment can be prevented, providing a savings to both institutions. For example, scientists from the University of Wisconsin and the Wisconsin Department of Natural Resources share laboratories for evaluating trace-metal concentrations in water samples.

In addition to collaboration through appointment of government scientists and managers as adjunct professors, universities and agencies may conduct joint research projects. Some of the most effective collaboration can occur at field research stations, where agency and university scientists can work together on real-world problems (see the background paper "The Role of Major Research Centers in the Study of Inland Aquatic

BOX 5-2 RESEARCH PARTNERSHIPS AT HUBBARD BROOK EXPERIMENTAL FOREST

The Hubbard Brook Experimental Forest, a 3,160-hectare reserve in Woodstock, New Hampshire, in the White Mountain National Forest, is dedicated to the long-term study of forests and associated aquatic ecosystems. Established in 1955 as a center for hydrologic research in New England, it was expanded in 1963 with the initiation of the Hubbard Brook Ecosystem Study. F. Herbert Bormann, Gene E. Likens, and Noye M. Johnson, then on the faculty of Dartmouth College, and Robert S. Pierce of the Forest Service proposed to use small watershed ecosystems at Hubbard Brook to study linkages between the hydrologic cycle and the transport of elements in forested and associated aquatic ecosystems. In addition, they proposed to study the response of ecosystems to natural and human disturbances such as air pollution, forest cutting, land-use changes, increases in insect populations, and climatic variations.

The Hubbard Brook site is operated by the Forest Service and since 1963 also has received continuous support from the National Science Foundation. Cooperative efforts among many educational institutions, government agencies, foundations, and private industries continue to produce extensive data and information on the biology, geology, and chemistry of forest and freshwater ecosystems.

Today the primary goals of the Hubbard Brook sites are to (1) advance scientific understanding of forest and associated aquatic ecosystems, (2) provide a scientific basis for the improved management of natural resources, (3) provide educational opportunities for students, and (4) promote public awareness of the importance of forest and associated aquatic ecosystems and concern for ecological issues.

Work conducted at Hubbard Brook is oriented toward applied research and management issues that are regional, national, and global in scope. Awareness and understanding gained from this research aid in the long-term management of natural resources for water supply, water quality, wildlife, timber yield, and sustained forest growth. Examples of Hubbard Brook research include the following:

• Acid rain has been studied since the early 1960s. Research at Hubbard Brook produced data that have played an influential role in U.S. and international assessments of this major environmental problem.

• Considerable research quantifying the effects of clear-cutting on northern hardwood forests has been conducted. This research has been instrumental in developing techniques to minimize the impacts of clear-cutting on forests, streams, and lakes.

• Data from Mirror Lake, located at the base of Hubbard Brook Valley, date back to the 1960s, making the lake one of the most intensively studied in the world and providing critical information for the management of lake ecosystems.

• By monitoring lead in precipitation, soil, vegetation, and streams since the mid-1970s, Yale University scientists have documented the recovery of forests from lead pollution since the U.S. phaseout of leaded gasoline.

The sustained research at Hubbard Brook has been invaluable in providing continuity in pursuit of patterns and trends, in identifying extreme observations, and in catalyzing significant new research questions. These long-term data also have influenced policy decisions regarding management of natural resources (see G. E. Likens et al., 1978; G. E. Likens, 1992). More than 100 senior scientists have worked at the site. Hubbard Brook research has generated more than 800 published papers, 5 books, 63 Ph.D. theses, 30 M.S. theses, and 22 undergraduate honor theses. A listing of all publications can be found in P. C. Likens (1994).

Gene E. Likens, Professor
Institute of Ecosystems Studies
Millbrook, New York

Ecosystems" at the end of this report). Examples of research stations, of which government and academic limnologists work in tandem include

- Hubbard Brook Experimental Forest (see Box 5-2), which was founded by scientists from Dartmouth College and is now supported by the U.S. Forest Service and the National Science Foundation;
- the Ontario Ministry of the Environment and Energy research station at Dorset, Ontario, which is supported by the Canadian government but draws scientists from the Universities of Toronto, Trent, Waterloo, and York;
- the Coweeta Hydrologic Laboratory in North Carolina, where University of Georgia and Virginia Tech scientists work closely with the U.S. Forest Service to study stream processes;
- the H. J. Andrews Experimental Forest in Oregon, which supports collaborative stream research by the Universities of Oregon and Washington and the Forest Service;
- the Freshwater Institute at Winnipeg, Manitoba, where University of Manitoba researchers work with funding from the Canadian government; and
- the University of Wisconsin Trout Lake Station in Boulder Junction, Wisconsin, which has dedicated space for research programs of the Wisconsin Department of Natural Resources and the U.S. Geological Survey.

One of the most productive partnership models to emulate has been that of the Experimental Lakes Area (ELA), near Kenora, Ontario (see Box 5-3). This field research station is operated by the Canadian government and staffed with agency and university scientists alike. From the late 1960s through the 1980s, research conducted at ELA on the causes and effects of major water pollution problems, such as lake eutrophication and acid deposition, was followed closely by agency resource managers.

BOX 5-3 COOPERATIVE RESEARCH AND EDUCATION AT THE EXPERIMENTAL LAKES AREA

Canada established the Experimental Lakes Area (ELA) in 1969, after a survey of several lake regions in Manitoba and northwestern Ontario, to perform ecosystem-scale experiments that would guide policies for controlling eutrophication in the Great Lakes and other areas of Canada (Johnson and Vallentyne, 1971). In 1973, the project mandate expanded to include whole-ecosystem experiments to address a variety of water management issues. Institutions participating in these experiments have included the Canadian Departments of Fisheries and Oceans, Environment, and Forestry; the Universities of Manitoba, Alberta, and Toronto; Lamont-Doherty Earth Observatory; and several U.S. universities.

Over the history of this program, ELA researchers have carried out a range of significant ecosystem experiments, some of which have played critical roles in policy formulation:

• ELA experiments with nutrients demonstrated the key role of the atmosphere in remedying carbon and nitrogen deficiencies, supporting the decision to control phosphorus as a first step in controlling eutrophication (Schindler et al., 1972; Schindler, 1974, 1977).

• ELA experiments with lake acidification demonstrated that much of the buffering capacity in some lakes is driven by microbiological action that can assist lakes in recovering once acid deposition is reduced, supplementing earlier theories that emphasized terrestrial and geological sources of buffering (Schindler, 1980, 1986; Kelly et al., 1982; Cook and Schindler, 1983; Cook et al., 1986; Schindler et al., 1986).

• Other ELA experiments in acidification showed that damage from acidification could occur at much higher pH levels than previously believed; the experiments demonstrated that biodiversity declined by more than 30 percent in acidification to pH 5, which some had viewed as the point at which damage to lakes began (Schindler et al., 1985; Schindler, 1990).

Measurements of precipitation, streamflow, lake and stream chemistry, and a broad suite of other biological and physical parameters made in several lakes in the ELA date back as long as 22 to 27 years. These measurements serve as references for experimental systems and show how the lakes and streams have changed over two decades of warming and drying trends.

Because an "ecosystem approach" requires a broad suite of investigations, ELA has always had strong inputs from graduate students, even though it has had no official mandate to do so. Harvey (1976), describing Canadian programs in aquatic environmental quality, called ELA ". . . our most outstanding field station for graduate work in limnology." As of 1989, 35 percent of the more than 500 scientific publications from ELA had been authored or coauthored by graduate students from several Canadian and U.S. universities. Approximately 60 graduate theses have been completed, and 18 doctoral and master's projects are under way at the site.

ELA typically employs 8 to 12 undergraduates as summer research assistants. Although there is no formal instruction beyond a weekly seminar series, a high proportion of undergraduates employed at ELA go to graduate school in limnology. Undergraduates receive continuous informal instruction because they live, work, and eat with graduate students and research scientists. ELA encourages undergraduate research projects by awarding a prize for the best undergraduate research project each summer.

As of late 1994, the ELA project was in the process of being transferred from Fisheries and Oceans Canada to Environment Canada. Despite being rated as a project of outstanding value to taxpayers by the Auditor-General of Canada in 1991, the funding base for ELA has undergone years of erosion. The facilities, which were designed as "temporary" in 1968, have fallen badly into disrepair, and the fate of the project is uncertain.

David Schindler, Professor
University of Alberta, Edmonton

University scientists working at ELA during that period were from institutions throughout North America, including the University of Alberta, Columbia University, the University of Minnesota, and the University of Manitoba. The partnership that developed between agency and university scientists advanced the understanding of mechanisms involved in pollution problems through whole-lake experimentation, and the results were valuable in the development of process-driven environmental models employed by agency managers. They also were of great value in international negotiations between the United States and Canada to reduce pollutant emissions to the atmosphere. Such research partnerships could be adopted by other universities and government agencies.

Student Internships

Internships with government agencies and private companies constitute one of the best mechanisms for providing students with exposure to practical problems in water resource management. In some cases, agencies and companies have formal internship programs. Examples of formal internship programs include the following:

• In Maryland, the governor's office awards summer fellowships to students for internships in a variety of agencies, including the Department of Environment and the Department of Natural Resources. Interns in these agencies have worked on projects such as developing management strategies for pollution control in specific water bodies, creating regional action plans for reducing the release of toxic materials to the environment, and ranking the environmental risk factors facing the state. Each February, the governor's office sends a request for ideas for projects that can be undertaken by summer fellows to the various departments and selects the best from among them for funding. Student fellows may come from any university.

• The South Florida Water Management District hires co-op students and summer interns for salaried positions. Students work under the guidance of a senior scientist on conducting ecological censuses, designing wetland restoration projects, evaluating nutrient removal strategies, and other tasks. At the end of their term, students are expected to produce a finished project. Interns and co-op students must be at the junior level of undergraduate work or higher.

• The Sault Ste. Marie Tribe of Chippewa Indians hires paid summer interns for a variety of projects. The tribe's Department of Fisheries and the Tribal Environmental Multimedia Department both employ interns with backgrounds in aquatic science.

• The Wisconsin DNR contracts with the University of Wisconsin for approximately 30 interns annually from aquatic science fields. Most of the interns are undergraduates, but a few are master's students. Interns work within a variety of department programs, including fisheries management, water resources, water supply, water regulation and zoning, wastewater management, and water resource research. Most work only during the summer, but some work for a full year or more. The university pays the interns through a contract with the DNR.

• The Michigan DNR has formal intern programs in its Divisions of Surface Water Quality, Fisheries, Land and Water Management, Environmental Response, and Wildlife. The department has a predetermined pay scale that applies specifically to interns and is based on years of education. Recently, however, the department has been forced to scale back internship opportunities because of state budget cuts.

In other cases, internship arrangements are less formal. For example, the Electric Power Research Institute (EPRI), which runs substantial research programs in water quality management for the electric utility industry, has no formal intern program but employs aquatic science students on an ad hoc basis, depending on the needs of the individual project manager. Students in M.S. and Ph.D. programs (rather than undergraduates) are most likely to find jobs at EPRI.

In recent years, hiring student interns within federal government agencies has become more difficult because of efforts to curb the size of the federal work force. Previously, student interns did not count as part of an agency's allocation of "full-time equivalent" employees. In this way, funding could be spent supporting a short-term project carried out by a student, benefiting the agency and providing valuable experience for the student. In 1992, the government reversed this policy; the full-time equivalents associated with student appointments now count against an agency's full-time equivalent allotments, which have decreased for numerous agencies. The result has been fewer student appointments and the development of indirect mechanisms (involving contractors) for supporting students; the opportunities for students to gain exposure to practical water management problems and to learn field sampling and measurement techniques, which often are not taught in universities, have decreased.

The U.S. Geological Survey (USGS) recently developed an indirect mechanism to support interns via a program operated by the university-based Water Resources Research Institutes (WRRI) program, which the USGS administers for the federal government. The WRRI program helps to support a water resource center (or institute) at the land-grant university in all 50 states. Among other activities, these centers sponsor small grant programs for faculty and student research on water resource problems in their states. In the new USGS internship program, students will be employed by a university (through the local WRRI) via a contract between the USGS and the university. Other federal agencies may find this mechanism useful in filling their needs for student interns.

Intergovernmental Personnel Agreements Between Agencies and Universities

Adjunct faculty appointments provide a means of bringing qualified individuals from agencies into the university setting, thus promoting agency-university collaboration. Mechanisms also are needed to bring university faculty into agency settings. The benefits of such arrangements are similar to those mentioned above for adjunct appointments: (1) the agency can gain expertise in an area where it is lacking; (2) the faculty member can gain exposure to real-world problems and the ways in which theory is brought into practice; (3) the faculty member can learn new skills and techniques; (4) both sides benefit from enhanced interactions and from the sharing of expertise, facilities, and equipment; and (5) opportunities for additional collaborative educational, research, and outreach activities can be developed. At present, such arrangements are pursued mostly on an individual basis by faculty members seeking partial support for sabbaticals and by agencies seeking to fill a specific temporary need. The Intergovernmental Personnel Act (IPA) provides a mechanism for making such arrangements. Through this act, individuals may work on temporary assignment in a given agency, which arranges to pay their salary and benefits to their permanent agency (or university). Universities would do well to promote the use of IPA arrangements for their faculty, both as a means of enhancing agency-university ties and as a means of maintaining faculty skills and involvement in cutting-edge issues.

Professional and Continuing Education

Continuing education programs provide one of the most direct ways to enhance the two-way communication between those at the forefront of confronting and solving aquatic ecosystem problems and those at the forefront of synthesizing knowledge and imparting it to others. Through continuing education courses and programs, water professionals can develop valuable new skills, improve their comprehensive understanding of ecosystems, or learn newly developed scientific techniques. Teachers of courses oriented toward practicing professionals can learn about the questions of most importance to professionals in the field.

The "packaging" of course work for practicing professionals is an important issue that needs the attention of limnological educators. As a minimum, for widespread participation by practicing professionals, courses that are taught as a part of a university's regular semester or quarter schedule need to be offered at convenient times for such students (in the early morning or late afternoon, in the evening, or on Saturday). In general, courses taught in larger blocks of time, fewer times per week, are more accessible to off-campus students than are courses scheduled in the traditional 50- to 60-minute class period three or four times per week. Alternative formats to traditional semester- or quarter-long courses also have to be developed to meet the needs of practicing professionals. Examples include intensive short courses and workshops lasting for a few days to a week and half-day or all-day classes held over a series of Saturdays.

The University of Wisconsin–Madison and the Bureau of Research at the Wisconsin Department of Natural Resources have established a formal program for graduate-level continuing education of DNR scientists that could serve as a model for other agencies or companies and universities. Under this program, several DNR employees have pursued advanced degrees in the University of Wisconsin's Oceanography and Limnology Program, working primarily with faculty at the university's Center for Limnology. DNR scientists who undertake graduate study develop thesis research programs that are tied closely to the DNR's scientific needs. Because of the close connection between the graduate program of these employees and their bureau's mission, the DNR is able to continue full-time support for them while they pursue graduate work. This program has been very successful but to date has been limited to fewer than half a dozen participants.

Although formal graduate study by professional water managers offers one avenue for continuing education, other possibilities also exist. In many cases, the kind of two-way exchange that most benefits both the students and the teachers of continuing education programs can best be accomplished in field settings, where the boundaries between teacher and student are deliberately less rigid and discussions result in learning on both sides. An example of the field approach is the Total Ecosystem Management Strategies (TEMS) program, established in 1990 by Mead Corporation and White Water Associates, Inc. (an ecological consulting firm in Michigan). Although the program is administered by a private firm, it nonetheless illustrates a valuable model that universities could adopt. The goal of TEMS is to develop improved techniques for ecological forest management and to continue the education of forest managers in ecology. The students of TEMS are 60 Mead employees including field foresters, equipment operators, geographical information system staff, and managers. There are four instructors, one of whom is a limnologist who contributes expertise in riparian areas, wetlands, water quality, stream ecology, and vernal ponds. Among the most valuable components of TEMS are day-long field trips in which a specific block of land owned by Mead Corporation serves as a landscape for study. Three to five such trips are taken annually. Prior to the field trip, TEMS organizers query Mead's geographical information system and produce a map of its tree stands and other landscape features. The forester in charge of the landscape works with scientists from the consulting firm to develop an itinerary with stops that illustrate specific management challenges. Discussion topics include dispersal corridors, buffer zones, sensitive wetlands, water quality, erosion, and biodiversity. Following each field trip, the consulting firm produces a written report, and the area forester develops a five- to ten-year plan that incorporates concepts discussed in the field. A long-term result of the TEMS program is that practical questions posed by forest managers frequently become the focus of joint research projects between foresters and scientists.

ENHANCING THE INVOLVEMENT OF LIMNOLOGISTS IN WATER RESOURCE MANAGEMENT

As discussed in Chapter 2 and elsewhere in this report, one key reason why limnologists are not always involved in resolving water resource management problems is limited public awareness of the field of limnology. Whereas many people are aware, for example, that oceanographers can answer questions about oceans, few are aware that a limnologist is someone with expertise in freshwater ecosystems. A variety of steps can be taken to enhance the visibility of limnology and increase the involvement of limnologists in important water resource management decisions. First, professional societies involving limnologists can coordinate programs for certifying limnologists and can consider whether it is desirable to establish minimum competence levels across the broad spectrum of ecosystem types (lakes, streams, wetlands) included in limnology. Second, the federal government can develop a job classification for limnologists. Finally, limnologists themselves can work to increase public awareness of their field.

Certification of Professional Limnologists

Universities and some research institutions have long evaluated the productivity of scientists by their track records—typically in terms of number and quality of peer-reviewed publications and/or number of graduated Ph.D. students. A pragmatic evaluation or certification system for aquatic ecosystem scientists and water resource professionals might be valuable for a variety of positions in the public and private sectors, especially in consulting firms, state and federal resource and regulatory agencies, and private industry. Examples of specific positions for which certification might be useful include state water quality managers, federal agency water quality regulators, aquatic science and engineering consultants, hydroelectric facility environmental supervisors, and paper mill aquatic environmental supervisors.

There are at least three compelling but simple reasons for certification of limnologists in the private and public sectors. First, managers of aquatic ecosystems need a standard and basic set of limnological skills to make good decisions involving aquatic resources. Second, employers of aquatic resource scientists and evaluators of programs or environmental impact studies would benefit from knowing that the employee, consultant, or study author is certified as possessing certain key knowledge. In addition, having certified limnologists in some of these jobs would facilitate communication and negotiations regarding aquatic resources because a common language would be present.

Several models of certification of aquatic scientists exist. These could be adapted or combined into a broad, umbrella aquatic science certification program having several subspecialties (one of which would be limnology). Existing certification programs include the following:

• The American Institute of Hydrology certifies applicants in hydrology. The evaluation process involves review of the applicant's education and professional experience, along with a written examination. Educational requirements include a B.S. in hydrology or hydrogeology or a major in physical or natural sciences or engineering with at least 25 semester credits in hydrology or hydrogeology. Applicants must have at least eight years of experience after the B.S. degree, six years after an M.S., or four years after a Ph.D. Five letters of reference are required from individuals who know the applicant's qualifications, integrity, and professional conduct. Some of the requirements may be waived for applicants whose reputation or standing in the professional community clearly shows that the applicant qualifies for certification. Applicants are certified as "professional hydrologist," "professional hydrogeologist," or "professional hydrologist (ground water)."

- The Ecological Society of America certifies "senior ecologists," "ecologists," and "associate ecologists" based on educational background and work experience. Certification as an associate ecologist requires a bachelor's or higher degree in ecology or a related science and at least one year of postgraduate professional experience in applying ecological principles. Certification as an ecologist requires a master's or higher degree in ecology or a related science and at least two years of professional experience, or a bachelor's degree and five years of professional experience. Certification as a senior ecologist requires a Ph.D. in ecology or a related science and at least five years of professional experience, or a master's degree and at least ten years of professional experience. In addition to completing a form showing educational background and work experience, applicants must provide the names of three references. The certification program is administered by the society's Board of Professional Certification, which consists of seven certified ecologists elected by the society's general membership.

- The Society of Wetland Scientists certifies "professional wetland scientists" and "wetland professionals in training" based on educational background and work experience. Certification as a wetland professional in training requires at least 15 semester hours of biological sciences, 15 semester hours of physical sciences, and 6 semester hours of quantitative sciences. Certification as a professional wetland scientist requires an additional 12 semester hours of courses specifically related to wetlands, plus five years of work experience demonstrating application of current technical knowledge in dealing with wetland resources; some of the work experience requirement can be met by advanced degrees in related fields. In addition to documentation of meeting these requirements, applicants must submit the names of five professionals who can serve as references, plus written references from three of the professionals.

- The North American Lake Management Society (NALMS) has a certified lake manager program that was established in 1990 as a means of identifying lake management professionals within NALMS who meet certain standards. By NALMS definition, a lake manager is any person directly involved in the comprehensive management of a pond, lake, reservoir, or other body of water or its watershed. Certification candidates may apply for either "provisional" certification, which is based on college course work and requires no experience, or "professional" certification, which requires five years of professional lake management experience if the academic degree is not related to aquatic sciences or a minimum of two years of lake management experience if the B.S. degree is in one of the environmental fields (such as zoology, botany, or environmental engineering). Certified lake managers must renew their certification every two years by completing a minimum of four continuing education units or equivalent academic course work. Certification is granted by the Certified Lake Manager Board, which is appointed by the NALMS Board of Directors.

Although some limnologists argue strongly for a certification program to validate the credentials of limnologists who are working as managers of aquatic systems, others argue equally strongly against certification. Arguments against certification include the risk of not certifying highly qualified professionals because they fail to meet one or two "paper requirements" of the certification program and, at the same time, the risk of certifying a "charlatan" who meets the paper requirements but does not keep up with current developments in the field or lacks scientific ethics. Another argument against certification is that it may not account for the great variation in quality of the educational backgrounds of applicants. Some also might argue that outlining specific course requirements for certification might stifle innovation in education and limit the practice of limnology to those with narrow backgrounds. Nevertheless, certification could provide a useful resource for those wishing to hire professional limnologists.

Certification of professional limnologists and consolidation of the existing, related certification programs are issues that leaders of the various professional societies involving limnologists (see Appendix B) should evaluate. In reviewing certification programs, the societies should consider whether there is a need for limnologists who are certified as competent across the spectrum of aquatic ecosystem types (lakes, flowing waters, various types of wetlands) that are within the purviews of modern limnology, and if so, what level of education (B.S., M.S., Ph.D.) would be required.

Federal Job Classification for Limnologists

To gain employment in the federal government, professionals are hired through the Office of Personnel Management (OPM) from the federal job register, which is a listing of qualified applicants for a wide range of job classifications (such as "hydrologist" or "environmental engineer") at different levels. Currently, there is no federal job classification for "limnologist." Scientists who are trained as limnologists may be hired as hydrologists (a classification that does not necessarily require biological training), microbiologists (which does not necessarily require training in hydrology and physical and chemical sciences), fisheries biologists, or some other related position title.

For a limnologist to be hired by a federal agency, the applicant must be ranked as among the best-qualified (top five or ten) of those listed in the particular job register (hydrologist or microbiologist, for example). At higher levels of government service (GS-12 and above), the specific qualifications and expertise required for a particular job are considered in ranking candidates. However, at the lower range (especially GS-4 through GS-9), specific qualifications are not given as much weight. As a result, agencies wishing to hire a limnologist can have trouble finding candidates with suitable training.

In 1969, OPM established the hydrologist register through the efforts of Luna Leopold, who was then chief hydrologist for the USGS. Establishing this register facilitated the hiring of qualified scientists by the USGS's Water Resources Division—positions that had previously been difficult to fill because of the limitations of using the job registers for "geologist" and "hydraulic engineer" to identify qualified hydrologists. Federal agencies would benefit similarly from establishment of a limnologist category in the federal job register. Such a listing would be especially valuable for hiring staff for the USGS's National Water Quality Assessment Program and for a variety of positions in the Environmental Protection Agency, National Biological Service, Fish and Wildlife Service, National Oceanic and Atmospheric Administration, National Park Service, Forest Service, Soil Conservation Service, Bureau of Land Management, and Army Corps of Engineers.

Coordination Among Professional Societies

As described in Chapter 2 and shown in Appendix B, there are several professional societies in the field of limnology, each focusing on one aspect of the broad field. This diversity of organizations has many positive attributes. For example, the relatively small size of the societies' annual meetings (attendance well under a thousand) makes them more manageable for both organizers and participants than the large meetings (with thousands of attendees) of some societies. Small societies engender a sense of belonging and tend to have less bureaucratic structures and lower annual dues than large societies.

On the other hand, the current situation has several important drawbacks. None of the present societies represents the field of limnology as a whole, and none of their journals provides coverage of all ecosystem types and the broad spectrum of research activities encompassed within modern limnology. The small size of the individual societies means that none is able to support a professional staff to represent the interests of the society or the field to Congress, federal agencies, or the national press. Consequently, limnologists as a group do not have a strong voice in legislative and budgetary decisions that affect their field.

Limnologists collectively have invested considerable time and energy to develop the current limnological societies and journals, and they rightfully have a sense of pride in their accomplishments and a sense of ownership in these entities. Consequently, radical changes in the current organizational structure are not likely in the near term. Nonetheless, there are many steps that the societies could take to improve coordination among themselves without imposing a loss of individual identities. For example, presidents of the various societies could form a coordinating council that would meet once or twice a year to exchange information and plan joint activities, such as jointly sponsored symposia or working groups to prepare reports on pressing issues. The societies should consider organizing joint annual meetings, which could provide considerable savings in time and travel costs to many scientists who now attend several different limnological meetings each year. Over time, as the societies work more closely together and build increasing levels of trust and understanding, they may decide that some of their business functions could be consolidated in a central office. The efficiencies gained by consolidation could save money that might be used to enhance efforts at outreach and public education or other member benefits. Eventually, it may be possible and desirable to form a confederation of limnological societies in which each society would maintain its identity but business and operating functions would be handled centrally.

Public Education

As emphasized in this report, academic limnologists need to train not only their successors in science but also broad-minded citizens who are aware of the many ways in which human activities affect aquatic ecosystems. Limnologists must reach beyond university students whose focus is aquatic science to help educate the broader public. There are several ways in which academic limnologists, as well as those in the private and government sectors, can increase public awareness of limnology and the vulnerability of freshwater systems to human-caused stresses. Such outreach activities include testifying at public hearings involving proposed water management projects, speaking at public gatherings, working with K-12 teachers and participating in K-12 education initiatives, providing scientific information on local water quality issues to local media (newspapers, radio stations, and television stations), and helping community and environmental organizations address aquatic resource issues. Aquatic scientists can attract large audiences at the meetings of a variety of organizations, ranging from environmental and sporting groups to Girl Scouts, Lions Clubs, Rotary Clubs, Leagues of Women Voters, and many others. Limnology programs can also attract interest at their own open houses or at tours that they provide of wetlands, streams, or lakes. In some cases, such programs can be run through a university's extension service.

There have been numerous successful outreach programs. The Institute of Ecosystem Studies in Millbrook, New York, has extensive citizen education programs on ecosystem processes. Researchers at the Center for Limnology at the University of Wisconsin–Madison have worked with the staff of a museum to develop interactive programs on lake biology. Researchers at Wisconsin's Trout Lake Station have participated regularly in a regional lake fair and provided people with introductions to the range of aquatic organisms that occur in their region and to the features of natural systems that are monitored routinely. Trout Lake personnel also provide summer field walks of wetlands in the region through an effort coordinated by the Wisconsin Department of Natural Resources. Many members of the Organization of Biological Field Stations and the National Association of Marine Laboratories routinely offer programs that attract a wide general audience.

Citizen-based water quality monitoring programs are a rapidly growing phenomenon in many parts of the country. Academic limnologists can play useful roles as advise to such groups, helping them to select useful and practical measurement variables to ensure proper collection of high-quality data and to interpret and understand the data they are collecting. In some cases, citizen monitoring is done by school-age children under the supervision of high school science instructors. In other cases, the monitoring is done by homeowners on lakes. For example, a citizen-based program to monitor the Secchi disk transparency of lakes was started in Minnesota more than 20 years ago by Joseph Shapiro, a limnologist at the University of Minnesota, and this program has provided the state's pollution control agency with long-term data on trends in water clarity for a large number of lakes. A similar national program called the "Great Secchi Disk Dip-In" is being developed by Robert Carlson, a professor of limnology at Kent State University in Ohio.

Limnologists should increase their efforts to participate in such public outreach programs to increase general citizen awareness of the importance of inland aquatic ecosystems. Outreach activities traditionally have been an important function of land-grant universities. All too often, however, outreach activities are regarded by faculty as something to be done only by extension agents or specialists, who typically are associated with colleges of agriculture.

SUMMARY: RECOMMENDATIONS FOR IMPROVING LINKS BETWEEN LIMNOLOGY AND WATER MANAGEMENT

In limnology, traditional boundaries between basic and applied research are inappropriate. Basic understanding of how aquatic systems function is essential for devising realistic management plans. Consequently, most fundamental research on aquatic ecosystems ultimately has practical implications. At the same time, applied problems are so widespread that there are few ecosystems in which problem-solving research approaches are irrelevant. Despite the lack of clear boundaries between the applied and theoretical perspectives, limnologists, like scientists in other disciplines, sometimes have tried to draw a distinction between them. This separation has led to weak ties between academic limnologists and the government agencies responsible for protecting and managing aquatic ecosystems. The challenge for modern educators and water managers is to avoid such distinctions and to improve the links between formal educational programs in limnology, which often are considered to focus on fundamental research, and agencies and companies with practical problems to solve. As described in this chapter, these links can be improved in several ways:

• Universities should (1) survey past students to determine how to tailor limnology programs to the needs of the job market, (2) provide opportunities for senior-level water managers to serve as adjunct faculty, (3) develop better continuing education opportunities for practicing limnologists, and (4) reward faculty members for educating the public about aquatic ecosystem management.

• Scientists and water resource managers in universities, government agencies, and the private sector should seek opportunities for collaborative research.

• Government agencies and universities should expand opportunities for faculty to take leave from their normal academic responsibilities to work on solving practical aquatic resource problems in agencies.

• Government agencies and private companies should increase opportunities for student interns to work on projects related to management of aquatic ecosystems.

• The federal government should establish a limnologist classification in the federal job register to facilitate the hiring of persons with training in this field.

• Professional societies representing various types of limnologists should (1) evaluate the need for formal certification of limnologists across the spectrum of aquatic ecosystem types, (2) form a coordinating council to plan joint activities, (3) consider organizing joint annual meetings, and (4) coordinate strategies to increase their influence on policies regarding the protection and restoration of inland waters.

REFERENCES

Cook, R. B., and D. W. Schindler. 1983. The biogeochemistry of sulfur in an experimentally acidified lake. Ecol. Bull. (Stockholm) 35:115–127.

Cook, R. B., C. A. Kelly, D. W. Schindler, and M. A. Turner. 1986. Mechanisms of hydrogen ion neutralization in an experimentally acidified lake. Limnol. and Oceanogr. 31:134–148.

Johnson, W. E., and J. R. Vallentyne. 1971. Rationale, background, and development of experimental lake studies in northwestern Ontario. J. Fish. Res. Board Can. 28:123–128.

Kelly, C. A., J. W. M. Rudd, R. B. Cook, and D. W. Schindler. 1982. The potential importance of bacterial processes in regulating rate of lake acidification. Limnol. and Oceanogr. 27:868–882.

Likens, G. E. 1992. The Ecosystem Approach: Its Use and Abuse. Excellence in Ecology, Book 3. Oldendorf-Luhe, Germany: Ecology Institute.

Likens, G. E., F. H. Bormann, R. S. Pierce, and W. A. Reiners. 1978. Recovery of a deforested ecosystem. Science 199:492–496.

Likens, P. C. 1994. Publications of the Hubbard Brook Ecosystem Study. Millbrook, N.Y.: Institute of Ecosystem Studies.

Schindler, D. W. 1974. Eutrophication and recovery in experimental lakes: Implications for lake management. Science 184:897–899.

Schindler, D. W. 1977. Evolution of phosphorus limitation in lakes: Natural mechanisms compensate for deficiencies of nitrogen and carbon in eutrophied lakes. Science 195:260–262.

Schindler, D. W. 1980. Experimental acidification of a whole lake: A test of the oligotrophic hypothesis. Pp. 370–374 in Ecological Impact of Acid Precipitation: Proceedings of an International Conference, Sandefjord, Norway, March 11–14. Oslo: SNSF Project.

Schindler, D. W. 1986. Recovery of Canadian lakes from acidification. Paper presented at Effects of Air Pollution on Terrestrial and Aquatic Ecosystems: Workshop on Reversibility of Acidification, Grimstad, Norway, June 9–11.

Schindler, D. W. 1990. Experimental perturbation of whole lakes and tests of hypotheses concerning ecosystem structure and function. Oikos 57:25–41.

Schindler, D. W., G. J. Brunskill, S. Emerson, W. S. Broecker, and T.-H. Peng. 1972. Atmospheric carbon dioxide: Its role in maintaining phytoplankton standing crops. Science 177:1192–1194.

Schindler, D. W., K. H. Mills, D. F. Malley, D. L. Findlay, J. A. Shearer, I. J. Davies, M. A. Turner, G. A. Linsey, and D. R. Cruikshank. 1985. Long-term ecosystem stress: The effects of years of acidification on a small lake. Science 228:1395–1401.

Schindler, D. W., M. A. Turner, M. P. Stainton, and G. A. Linsey. 1986. Natural sources of acid neutralizing capacity in low alkalinity lakes of the Precambrian Shield. Science 232:844–847.

BACKGROUND PAPERS

Organizing Paradigms for the Study of Inland Aquatic Ecosystems

Patrick L. Brezonik
Department of Civil Engineering and Water Resources Center
University of Minnesota
Minneapolis and St. Paul, Minnesota

SUMMARY

This paper describes the major organizing principles that have driven the development of the science of limnology for lakes, flowing waters, and wetlands. Several broad themes that contribute to the understanding of how these water bodies behave as ecosystems have played important roles in integrating the component disciplines. These include the concept of lakes as microcosms, wetlands as ecotones or "gradient ecosystems," and the River Continuum Concept. In recent years, the importance of terrestrial-aquatic interactions has been widely recognized, and a paradigm common to all classes of inland aquatic ecosystems regards them as reflections or integrators of conditions in the watershed in which they are located.

INTRODUCTION

Limnology is usually defined as the science of all inland aquatic systems: lakes, reservoirs, rivers, streams, and wetlands.[1] For example, in a chapter entitled "What Is Limnology," Edmondson (1994) describes limnology as

> the study of inland waters . . . as systems. It is a multidisciplinary field that involves all the sciences that can be brought to bear on the understanding of such waters: the physical, chemical, earth, and biological sciences, and mathematics.

---

[1]Some aquatic scientists include ground water in this list, but most limnologists probably regard ground water as they do atmospheric precipitation: a potential source of water (and chemicals) to their systems of interest, which may be studied in this context, but not systems that they study in their own right.

Similarly, a recent "state of the science" assessment (Lewis et al., 1995) defined limnology as

> the integrative study of inland waters. It encompasses the biological, chemical and physical phenomena, as well as all levels of organization extending from individual chemical reactions or adaptations of individual organisms to the analysis of entire ecosystems.

Nonetheless, many aquatic scientists, as well as the general public, associate the word *limnology* exclusively with the study of lakes (and reservoirs). There are historical reasons; organized studies of lakes using techniques of modern science began much earlier (late nineteenth century) than studies of flowing waters and wetlands.[2] In addition, the discrete, semiclosed nature of lake basins, compared with the open, flowing nature of streams and the often diffuse boundaries of wetlands, makes it easier to view lakes as appropriate objects of scientific study. Lake studies have always been approached holistically to include physical, chemical, biological, and geological aspects, even though biologists conducted most of the early work.

In contrast, the physical, geological, chemical, and biological characteristics of streams have been studied primarily within separate disciplines (i.e., as separate fields of study)—hydrology, geomorphology, geochemistry, sanitary (environmental) engineering, public health biology, fisheries science. These studies generally were pursued in separate academic departments such as civil engineering, geography, geology, public health, biology, and fisheries. To a considerable extent, this fragmentation still exists today—to the extent that stream limnology as a distinct field within the broad field of limnology might be questioned. To be sure, academic lake limnology suffers from the same fragmentation, but it is still considered a discrete field.

Except for studies on wetland flora and fauna by botanists and zoologists, studies on wetlands are even more recent, especially in North America, and for the most part, comprehensive studies of wetlands as ecosystems are the product of the past 20 to 30 years.

This paper examines the various paradigms that have driven the fields and subdisciplines of limnology since its founding in the late nineteenth century. As described above, the subareas of limnology differ substantially in the length of time over which they have developed, and this has important implications concerning the driving paradigms.

---

[2]Some ecological studies on wetlands predate the beginnings of modern science in the late eighteenth and early nineteenth century (see section of this paper on Wetlands for further details), but the prevailing attitude that wetlands are wastelands limited both scientific and public interest in these ecosystems until recent years.

LAKES

Lakes come in all sizes and shapes and have a wide range of chemical and biological characteristics. Nonetheless, they all share fundamental attributes that make it useful to define them as a class of objects worthy of study. They have well-defined physical boundaries and are semiclosed systems (in the sense that water stays within a lake's boundaries long enough to develop distinctive biological and chemical characteristics compared with the water that flows into it). Moreover, lakes are important to humans for practical, recreational, and aesthetic reasons, and this, as much as any factor, led to an early development of the science of lake limnology. Major changes have occurred in the organizing paradigms for the study of lakes over the past century, reflecting advances not only in our understanding of how these complicated natural systems operate, but also in the basic sciences that contribute to the study of lakes.

Lakes as Microcosms or Integrated Ecosystems

The founders of lacustrine limnology in the latter half of the nineteenth century viewed the subject expansively as the application of all (relevant) basic sciences to the investigation of lakes as fundamental objects of study. In his seminal paper of 1887, Stephen Forbes, one of the earliest limnologists, described lakes as "microcosms," or little worlds. Although the term "ecosystem" was not introduced for another half century (by Tansley in 1935), Forbes' approach was essentially that of an ecosystem scientist. He proposed that lake studies should focus on many of the ecosystem-level processes that define the late twentieth century field of ecosystem ecology: the circulation of elements and substances (now called biogeochemistry), the production and decomposition of organic matter, food web interactions (especially predator-prey relationships) and their resulting effects on the structure of biological communities, and the effects of physical conditions and gradients on biological communities. Together, these topics define lakes as functioning, integrated systems.

The paradigm of the lake as a microcosm or integrated ecosystem has pervaded the study of lakes to this day. Of course, the concept has been refined as the science of ecology has developed, and it has been supplemented by other organizing paradigms at the same scale of organization (e.g., lakes as experimental units) and some at even larger scales of organization (the more recent paradigm of the lake as an integrator or reflection of its watershed). Current limnological studies on lakes focus on lakes as "open" systems that receive inputs of water, solar energy, and chemical substances from outside the system. As implied by the second definition of limnology given in the introduction, studies within lakes at smaller scales of organization (e.g., studies of chemical reactions or populations of organisms) also are part of the field of limnology, and as described below, numerous paradigms have been developed for these scales and the scientific subdisciplines of limnology.

François Forel was the first scientist to use the term limnology in a publication, and his three-volume *Le Leman: Monographie Limnologique*,[3] published over the period 1892 to 1904, is considered to be the first book on limnology. Encyclopedic in scope, the book is divided into 14 chapters (Table 1), 11 of which define the main supporting fields of modern limnology (Edmondson, 1994). Taken together, Forbes' and Forel's contributions provide the multidisciplinary and interdisciplinary framework that still defines the science of limnology.

The founders of academic limnology in North America, Edward Birge and his younger colleague Chancey Juday, continued this multidisciplinary, interdisciplinary tradition at the University of Wisconsin (Brooks et al., 1951; Mortimer, 1956; Frey, 1963). Birge was trained as a zoologist and was attracted to lake studies (in the context of zooplankton life cycles) even as an undergraduate at Williams College in Massachusetts during the early 1870s. Juday also was trained as a biologist and was first hired by Birge in 1897 to help conduct lake surveys for the newly established Wisconsin Geological and Natural History Survey, of which Birge served as the first director. However, Birge and Juday soon branched into the physics and chemistry of lakes as they realized that one could not understand the dynamics of plankton without knowledge of these subjects. Their studies on annual cycles of thermal stratification (Birge, 1898) and dissolved gases (Birge and Juday, 1911) are seminal works that provided limnologists with information essential for understanding virtually all biological cycles in lakes. Birge and Juday knew the limitations of their training and actively sought collaboration with physicists and chemists to study lake phenomena beyond their own field of expertise. Together, these scientists developed many new techniques to measure physical properties and processes (e.g., light transmission, heat transfer, and heat

TABLE 1 Contents of Forel's Treatise on Lac Leman (Lake Geneva)

Vol. 1 (1892)	Vol. 2 (1895)	Vol. 3 (1904)
Geography	Hydraulics	Biology
Hydrography	Temperature	History[a]
Geology	Optics	Navigation[a]
Climatology	Acoustics[a]	Fish
Hydrology	Chemistry	

[a]Not considered a subfield of modern limnology.

---

[3]Lac Leman is Lake Geneva (Switzerland/France).

budgets) and many chemical characteristics of lakes, including the first methods used to analyze the nature of dissolved organic matter.

The early Wisconsin limnologists collected a wealth of information on individual lakes and synthesized this at the regional scale. As practiced by these scientists and their contemporaries, limnology was essentially a passive or observational science, but this term is not meant (nor should it be taken) in a pejorative sense as merely qualitative and lacking in rigor. The knowledge gained was largely from sample collection, analysis, and interpretation rather than from controlled experiments. Quantitative analyses and interpretation of data were undertaken on various physical processes (e.g., heat budgets and light transmission) as far back as the 1920s.

The regional, descriptive approach to limnology, at least in part, can be regarded as an outgrowth of the lake-as-a-microcosm paradigm, in that the observations on individual lakes usually were multidisciplinary, involving physical, chemical, and biological elements and implicitly or explicitly recognizing lakes as complex organized systems. Regional efforts during the first half of the century focused on grouping lakes into major types or classes based on a multidimensional set of descriptors. For example, in the classification scheme that categorizes lakes according to trophic state (i.e., general nutritional status),[4] a wide array of indicators was developed for classification purposes, including physical measures (transparency), chemical concentrations (nutrients), and biological characteristics (species types and abundance and primary production). In this sense, classification, which is looked upon by many current scientists as a fairly unimportant or useless exercise, was a force for integration and synthesis leading to generalizations about lakes as systems.

Lakes as Experimental Systems

The idea that lakes can serve as subjects for scientific manipulation is also a product primarily of the Wisconsin school of limnology. Juday apparently was the first to treat a whole lake in this way. In the 1930s, he added fertilizer to Weber Lake, a small pond in northern Wisconsin, to study its effects on plankton production and fish populations (Juday and Schloemer, 1938). Einsele (1941) did a similar experiment on a small lake in northern Germany a few years later. However, neither of these manipulations seems to have had much impact on the development of experimental limnology, perhaps because of the disruptive influence of World War II on natural science.

---

[4]The trophic classification system groups lakes into three main categories: oligotrophic (low in nutrients and plant production), mesotrophic (intermediate in these characteristics), and eutrophic (rich in nutrients and high in plant production).

The later but better-known "before-and-after" experimental liming of Cather Lake in 1950 (Hasler, 1964) and "paired manipulation" of Lakes Peter and Paul by University of Wisconsin limnologist Arthur Hasler (Johnson and Hasler, 1954; Stross and Hasler, 1960) are sometimes cited as the first ventures in whole-lake experimentation. In the latter case, a small, colored, two-basin lake was separated into two lakes by an earthen dike. One basin was treated with lime in an effort to improve water clarity and increase primary production; the other basin was retained as a control. Hasler's group was involved in several other whole-lake manipulations during the 1950s and 1960s. Examples include the first whole-lake experimental acidification, which involved a small bog lake (Zicker, 1955); aeration-induced destratification experiments on several eutrophic lakes (Schmitz and Hasler, 1958); and the addition of short-lived radioisotopes to the water of stratified lakes to measure rates of water movement (e.g., Likens and Hasler, 1960). Hasler's group also pioneered the use of small artificial ponds (wading pool size) treated in various ways to simulate lakes.

An earlier whole-lake radiotracer experiment by G. Evelyn Hutchinson at Yale University actually predates all of the lake manipulation experiments of Hasler's group in Wisconsin. In June 1946, he added approximately 10 millicuries of ^{32}P-phosphate to Linsley Pond, a small lake in Connecticut. The distribution of radiophosphorus was determined in several strata of the water column and in littoral macrophytes on two dates over the following month (Hutchinson and Bowen, 1947).[5] Compared with the sophistication and complexity of limnological papers published today, the ^{32}P tracer experiment on Linsley Pond was extremely simple. It involved only a few crude measurements and only the simplest mathematical analysis of the data. Techniques available to measure the radiotracer in lake samples were very crude (a simple Geiger counter) compared with the sophisticated and highly sensitive tools available today. Also, the regulatory climate for use of radioisotopes in the ambient environment was much more relaxed in the 1940s than it is today.

Today, experimental limnologists conduct at least three types of whole-lake manipulations:

1. stress-response experiments, in which a whole lake (or one basin of a multibasin lake) is treated with some chemical stressor, such as excess nutrients or acid, and the responses of the lake system are studied;

2. remediation and rehabilitation manipulations, involving hydrologic manipulations such as water level fluctuations to improve littoral (near-

---

[5]According to these authors, an even earlier but unsuccessful effort was made to conduct such an experiment in 1941. A similar ^{32}P experiment was conducted by F. R. Hayes and coworkers on a small lake in Nova Scotia a few years later (Coffin et al., 1949).

shore) habitat, physical manipulations such as aeration and removal of contaminated sediments by dredging, chemical manipulations such as addition of alum to remove phosphorus or lime to neutralize acidity, or biological manipulations such as fish stocking or removal and other food web controls; and

3. tracer additions to measure rates of physical processes, such as use of radiotracers to determine water movement and sulfur hexafluoride (SF_6) and noble gases to monitor air-water gas exchange.

Not all of these manipulations are done for scientific reasons; many of the remediation and rehabilitation procedures outlined above are performed for management purposes rather than to gather new knowledge. Of course, prototype studies on these methods also are done to study their effectiveness. Most stress-response manipulations of lakes are done under the lake-as-a-microcosm paradigm in that a wide range of responses are studied: changes in chemical concentrations and chemical processes; effects on individual organisms, populations, and communities; and ecosystem-level parameters.

Whole-lake manipulations generally cannot be replicated, and some manipulation projects have not even included a control or reference lake. Premanipulation data can be used as a baseline, but because lakes are dynamic systems, there are always some uncertainties about this approach (i.e., if changes are observed following manipulation, would they have happened even if the manipulation had not occurred?). Because of this, a few scientists have argued that such manipulations are not true experiments, and in the sense of traditional statistical design, they are not. However, sophisticated statistical methods such as randomized intervention analysis (Carpenter et al., 1989) have been developed to analyze the data from such manipulations. In the literature of the past decade, there has been considerable discussion about the interpretation of large-scale manipulations with their common problem of limited replication (Hurlbert, 1984; Stewart-Oaten et al., 1986, 1992; Carpenter, 1990). Overall, however, there appears to be a strong consensus on the importance and utility of observing responses to manipulations made at the whole-system level.

Experimental limnology on a large scale (whole lakes or large enclosures) did not play a prominent role in the science until the late 1960s and 1970s, probably because of the general lack of funding to support such initiatives. Widespread concern about lake eutrophication prompted large government research programs in the United States, Canada, and several countries in western Europe in the 1960s, and these facilitated the wider use of experimental approaches in limnology. Even since then, whole-lake experiments have been relatively few in number for two practical reasons: (1) the high cost, general difficulty, and long time required for completion; and (2) the limited number of lakes available for such purposes. Whole-lake experiments of the stress-response type are practical only on small lakes with no shoreline habitation and in watersheds where public access can be limited.

For these reasons, experimental approaches on a smaller scale, such as in situ plastic enclosures (one to a few meters in diameter), often called mesocosms, limnocorrals, or limnotubes, have certain practical advantages and have been popular for field experiments both in Europe, where their use was pioneered by John Lund, and in North America. This intermediate scale of experimentation (smaller than whole lakes or lake basins but larger than laboratory microcosms) has enabled limnologists to perform experimental studies under conditions that can be replicated and subjected to reasonable control using systems more similar in complexity to whole lakes than is possible in the laboratory. Nonetheless, mesocosms cannot duplicate the complicated communities and ecosystems of whole lakes, and they are especially inadequate for studying responses of large fish over long periods (Gorham, 1992).

Despite the small number of whole-lake experiments that have been conducted over the past 30 years, they have been very important in two ways: (1) advancing understanding of fundamental limnological and ecological processes and (2) providing critical evidence for management of major pollution issues such as eutrophication[6] (e.g., Schindler, 1974) and acidification (Schindler et al., 1985, 1992; Brezonik et al., 1993). Their strengths for both purposes lie in their ability to test hypotheses (Schindler, 1990) and to provide a "platform" for related laboratory or field experiments at a range of scales. Two brief examples from eutrophication studies at the Experimental Lakes Area (ELA) in western Ontario during the early 1970s illustrate both types of advances.

One of the major issues in eutrophication during this period concerned the role of carbon in limiting primary production in lakes. Rates of carbon dioxide (CO_2) transfer across the air-water interface and the question of whether CO_2 transfer is a simple physical process or is enhanced by chemical reactions at the air-water interface were of great interest because of the potential importance of this transfer in renewing the limited supply of CO_2, especially in low-alkalinity lake waters. Field studies that included carbon mass balances on one of the ELA lakes were used by Emerson (1975) to quantify the amount of chemical enhancement in the CO_2 air-water transfer process. The transfer of CO_2 across the air-sea interface has been a major topic of research for many years in relation to global climate

---

[6]Eutrophication is the nutrient enrichment of lakes that results in an array of symptomatic changes, including an increase in primary production and in the abundance and composition of phytoplankton and other aquatic organisms, and a decrease in water clarity. Some of these changes are considered objectionable and limit the usefulness of the lake for recreational purposes (e.g., swimming, fishing) or as a drinking water supply.

change, and the ELA work also serves as an important example of the use of lakes as models for global-scale processes.

Related studies at ELA during this period also produced dramatic evidence of the key role of phosphorus in lake eutrophication. Addition of phosphorus to one side of an oligotrophic lake (which had been divided into two basins by a plastic barrier) led to the rapid appearance of a dense blue-green algal bloom (Schindler, 1974). In contrast, addition of inorganic nitrogen to the other side of the lake did not produce any change in algal biomass. Evidence for the key role of phosphorus in eutrophication had been accumulating from a wide variety of other limnological studies, including laboratory experiments, in situ limiting-nutrient bioassays, nutrient budgets, lake surveys, and application of mathematical models. Nonetheless, the whole-lake results were thought by many to be incontrovertible evidence; they received much public and scientific attention, and they played an important role in government decisions to control phosphorus inputs to lakes.

Lakes as Chronicles of Natural History and Evolving Ecosystems

Recognition that lake sediments are a repository of information about the past conditions in lakes and their watersheds—and how they have evolved over time—dates back to the early days of limnology. Most of this early "paleolimnology" was done by geologists on lithified sediments of ancient lakes. Bradley's chapter on paleolimnology (Bradley, 1963) in the compendium *Limnology in North America* deals almost exclusively with such studies. Probably the first paleolimnological studies in the United States were on Pleistocene Lake Bonneville, the much larger, freshwater ancestor of the present Great Salt Lake (Gilbert, 1891), and on Pleistocene Lake Lahontan in western Nevada (Russell, 1885).

Stratigraphic studies on the sediments of modern lakes to reconstruct historical conditions date back at least to the 1920s. For example, Nipkow (1920) was the first to observe and explain the existence of thinly laminated or "varved" sediments. He showed that the laminae resulted from an annual depositional cycle in which spring and summer photosynthesis causes precipitation of calcite, which settles and forms a thin carbonate layer on the sediments. Organic detritus deposited during the remainder of the year forms a dark (organic-rich) layer on top of the calcite layer. These thin annual laminae allow limnologists to count back in time to date individual strata of a sediment core. By studying plant and animal microfossil remains (e.g., pollen, diatom shells, remains of zoo plankton bodies) in various laminae, paleolimnologists thus could reconstruct historical conditions in the lake and/or its watershed.

Unfortunately, relatively few lakes deposit varved sediments, and it was not until the 1950s and 1960s, when radioisotope dating methods (especially ^{14}C and ^{210}Pb methods) were developed, that the field of paleo-limnology grew to become one of the important subdisciplines of limnology. The tools of paleolimnology include radiochronometers (^{14}C, ^{210}Pb) that date the time of deposit for sediment strata, pollen as an indicator of terrestrial vegetation, a variety of plant and animal fossil remains (both cell fragments and molecules such as plant pigments), and more recently, various organic pollutants and trace elements whose biogeochemical cycles have been influenced by human activity. Over the past 30 years or so, stratigraphic analyses of long cores of lake and wetland sediments have provided important information on regional variations in past climatic conditions and watershed vegetation patterns, from which paleo-ecologists have sought to answer fundamental questions about the causes of environmental change, including questions of species extinction (e.g., Deevey, 1967). Paleolimnological studies on more recently deposited sediments have provided evidence for the timing and causes of cultural eutrophication (e.g., Brezonik and Engstrom, in press), as well as temporal patterns and geographic scales of atmospheric transport of various pollutant chemicals (e.g., Swain et al., 1992).

Lakes as Chemical Reactors

In the late 1960s, concern about the transport of chemical pollutants to lakes, their effects on biota and water quality, and their ultimate fate led to the idea that lakes can be modeled as chemical reactors. Especially important is the concept of the lake as a completely mixed flow-through reactor, often referred to as a CFSTR (continuous-flow, stirred-tank reactor; see Brezonik, 1994). Although Piontelli and Tonolli (1964) presented some initial thoughts and equations related to this idea, the CFSTR approach was developed primarily by Vollenweider (1969, 1975). He applied the principles of reactor kinetics from chemical engineering to lakes and developed a practical means to predict phosphorus concentrations in a lake from the phosphorus loading (input) rate and simple morphometric and hydrologic data. In turn, the model (and similar ones developed under this paradigm) led to the formulation of loading criteria or management guidelines to protect lakes from eutrophication problems. The reactor concept has been used subsequently to model the behavior of many other natural and anthropogenically derived chemicals in lakes, such as sulfate in acid-sensitive lakes (Baker and Brezonik, 1988; Kelly et al., 1988) and humic matter in colored lakes (Engstrom, 1987). Moreover, other concepts of reactor kinetics, such as water and substance residence times, have been used widely to provide simple quantitative measures of lake system dynamics.

Lakes as Components of Integrated Watersheds

Taken literally, the microcosm paradigm considers lakes as isolated entities and disregards the land shoreward of the water's edge, littoral wetlands, tributary streams, and the lake's terrestrial watershed. In reality, however, limnologists have recognized for many decades that lakes are not isolated from the other aquatic and terrestrial components of watersheds, even though limnologists focused on processes within lakes. As concern grew during the mid-twentieth century about the effects of excessive nutrient loadings on lake ecosystems and lake water quality, engineers and limnologists began to quantify these loadings in the form of nutrient budgets. Water budgets, of course, were found to be a necessary part of such studies, and these efforts gradually led to an increased awareness of the importance and closeness of the interactions between lakes and their surrounding landscape, including littoral or contiguous wetlands (Wetzel, 1990, 1992).

Today, policymakers and others promote the concept of integrated watershed planning and management, a "new paradigm" for water management that many hope will replace the "command-and-control" or top-down regulatory paradigm that has driven surface water management in the United States for the past 25 years or more. If this management paradigm is to be based on sound science, limnology must continue broadening and developing the paradigm of lakes as components of integrated watersheds (see Hynes, 1975, as an early example of concern about this issue). Academic programs in limnology will need to train students more broadly in fields such as hydrology, soil science, and landscape ecology. Research programs will have to recognize that lakes (and streams) mirror conditions in their watersheds and that land-water interactions are an inseparable component within the logical basic unit for limnological studies—the watershed (Cummins et al., 1995).

Other Organizing Paradigms for Lake Studies

Lewis et al. (1995) describe a variety of unifying concepts and research themes that have driven the field of limnology in recent decades. In general, these themes are not as broad as the major organizing paradigms described in previous sections, but each has been important in both a research and a management context, and several have affected the way limnology is taught in graduate programs.

Rules for predicting the interaction of predator and prey is a topic that dates back to Forbes' (1887) paper on lakes as a microcosm. In recent years, this theme has focused on such topics as size-selective predation by fish and the "trophic cascade" (Carpenter and Kitchell, 1993), that is, direct and indirect controls on lower trophic levels by predators. "Top-down control" and control by resource limitation (Tilman, 1982) are competing explanations for the control of primary production in aquatic ecosystems, and the relative importance of these two mechanisms is the subject of much current research. On the practical side, the "top-down" paradigm has led to the idea that the consequences of nutrient enrichment (i.e., high primary production) can be managed by food web management (e.g., Kitchell, 1992). This idea, sometimes called biomanipulation, was pioneered by Shapiro and his students (e.g., Shapiro et al., 1975).

The role of organic matter in natural waters has been the subject of numerous studies over the past 25 years, and substantial advances have been made in understanding the role of natural dissolved organic matter (NDOM) in controlling light transmission, in affecting the bioavailability of metals, and more recently in stimulating photolysis of synthetic organic compounds (via the production of highly reactive photointermediates such as the hydroxyl radical and singlet oxygen following the absorption of light energy by NDOM). The importance of particulate organic matter (detritus) and NDOM in energy flow within aquatic systems also has become better understood (e.g., Wetzel, 1992), but this is still an area in need of additional research.

Within the past 10 to 15 years, the principles of fluid mechanics have been applied to lakes in increasingly sophisticated ways by engineers and physical scientists, who have created the subdiscipline of *environmental fluid mechanics* (or environmental hydraulics). This work has helped limnologists understand the dynamics of water flow on a wide range of scales: from microscopic (e.g., nutrient depletion microzones surrounding plankton, and the patchy nature of the microhabitat and its role in maintaining multispecies communities in the face of competition for a common resource) to macroscopic (e.g., design of more efficient aerators and understanding the fate of silt-laden tributaries in reservoirs).

Cycling of major and minor nutrient elements (carbon, nitrogen, phosphorus, sulfur, and minor metals) within lake basins has been a topic of research in mainstream limnology for many decades. Volume 1 of Hutchinson's (1957) *Treatise on Limnology* gathered and synthesized extensive amounts of both published and unpublished data regarding these cycles in lakes and is still an important resource. Concerns about eutrophication and acid deposition stimulated much additional research on these cycles over the past few decades. Studies on elemental cycling at various scales, from ecosystems to the globe, has become a recognized field of study called biogeochemistry (e.g., Gorham, 1991a; Schlesinger, 1991). In *limnological biogeochemistry*, lakes are used as models for larger systems (e.g., the oceans) and as convenient systems for the measurement of process rates, fluxes, and storage mechanisms. Such studies are intrinsically interdisciplinary and typically involve elements of chemistry, microbiology, hydrology, and mathematical analysis (modeling).

Finally, the subject of *ecosystem energetics* has been a major organizing tool in ecology for more than 50 years, since Lindeman published his classic 1942 study of "trophic dynamics"—energy flow through the food web in Cedar Bog Lake. This study was the first important analysis of energy flow in any type of ecosystem—an example of the leadership role that inland aquatic science has played in the earth and biological sciences. Energy transfer efficiency, relationships between production and decomposition, and physical-chemical constraints on biological production have been research topics in all types of aquatic ecosystems under this general theme. The importance of linkages between wetlands and lakes and between wetlands and streams with regard to carbon and energy flows (e.g., Wetzel, 1990) ties several major research thrusts together (e.g., biogeochemical cycling, especially carbon cycling, and ecosystem energetics), and it also could serve as a unifying theme for major subfields of limnology (i.e., lake, stream and wetland ecology).

RIVERS AND STREAMS

Integrative paradigms are a much more recent development in lotic limnology[7] than in lake limnology. This reflects the less integrated nature of the disciplines contributing to stream science. Indeed, until recently, the physical, chemical, and biological components of stream limnology could be characterized as a set of independent subdisciplines that were associated more with their parent basic-science disciplines than with limnology. For example, studies on the physical aspects of water flowing in streams are associated with the fields of hydrology and hydraulics, which are taught in engineering and geoscience departments. Hydrologists have their own professional organizations or separate divisions within major engineering societies (such as the American Society of Civil Engineering and the American Geophysical Union). Similarly, the origin and development of stream channels and drainage networks is a well-developed topic within the field of fluvial geomorphology, which is a subdiscipline of geography and (to a lesser extent) geology.

Until recently, chemical studies on flowing waters have been done primarily by geochemists, environmental engineers, and lake scientists interested in quantifying fluxes of minerals and pollutants from watersheds or into standing bodies of water (lakes and oceans). That is, stream chemistry has been of interest not so much as a subject of its own but as a means to answer questions about other natural systems. To an extent, this parallels the subservient role that chemistry played in sanitary (environmental) engineering before the 1960s.

For the most part, stream limnology has been identified most strongly with stream biology, but even here, practicing stream biologists often are identified more with other "parent disciplines" such as public health biology (sanitary microbiology) and fisheries biology. Stream biology was

---

[7]Lotic refers to flowing waters (i.e., rivers and streams).

mostly descriptive and autecological through the first half of the twentieth century, focusing on the distribution, abundance, and taxonomy of stream organisms (Cummins et al., 1995). The development of stream ecology as an independent discipline is fairly recent (occurring mostly over the past 25 years) and is an outgrowth of various initiatives (described below) that began in the 1950s and 1960s. Hynes' (1970) classic book *The Ecology of Running Waters* may be regarded as the first book on stream ecology. Even today, many stream ecologists identify more with the discipline of ecology than with limnology.

Paradigms of Subdisciplines Within Stream Limnology

Organizing paradigms have been part of all the scientific disciplines that collectively constitute the field of stream limnology for many decades, as illustrated by the following examples.

The concept of the *watershed as the basic unit in hydrology* dates back at least to the 1920s (Horton, 1931; Platt, 1993), and the watershed perspective has been used both in organizing hydrologic concepts and in data collection. Classification of flowing waters according to stream order, based on their relative position in the typically dendritic (treelike) or hierarchial network in which streams are connected in a large river basin, is a long-standing organizing principle in both hydrology and geomorphology. Chaos theory and the concept of fractal dimensions have stimulated much recent research in stream geomorphology.

The modeling of *streams as plug-flow reactors* has been an organizing paradigm of sanitary and environmental engineering since the development of the first water quality model—Streeter and Phelps' 1925 model for dissolved oxygen concentrations in streams receiving point-source discharges of sewage.

The use of *benthic invertebrates as indicator organisms* for organic pollution and the *division of streams into zones of pollution and recovery* based on the presence of indicator species or groups of organisms have been major driving forces in stream biology almost since the inception of the field. Numerous classification schemes were developed under this organizing principle, starting with the European "Saprobien" system near the beginning of the twentieth century (Kolkwitz and Marsson, 1908, 1909). This paradigm stimulated much of the biological research on the structure of stream communities through the middle of this century. It also can be considered a precursor to broader indicator or classification systems such as Karr's IBI (index of biological integrity) (Karr et al., 1986, 1987) and other indices of biodiversity and biological integrity (Karr, 1991), which are currently popular topics in stream ecology.

Ecological energetics also has been an organizing concept for research in stream ecosystems for almost 40 years—since H. T. Odum's classic studies on energy flow in the Silver Springs and the Silver River (Odum, 1956, 1957). Numerous studies have measured primary production, respiration, and flow of energy and materials through stream food webs, and this work has led to more recent paradigms such as nutrient spiraling (see next section).

Just as lake scientists shifted from a descriptive or observational approach to a more experimental approach 30 to 40 years ago, stream limnologists also have adopted experimental approaches in recent decades; the experimental manipulation of a small stream with sucrose to enhance trout production (Warren et al., 1964) was the precursor of many later manipulations (e.g., see Lamberti and Steinman, 1993). Stream experiments have involved both artificial channels and real streams. Artificial channels have the advantage of allowing replication. They have been used to study the effects of chemical contaminants on stream ecosystems and to develop a mechanistic understanding of lotic processes at several locations in both this country and Europe (Lamberti and Steinman, 1993). Real streams have been used to conduct acidification and remediation manipulations; for example, lime has been added continuously to several small streams in New York and Sweden to study rates and mechanisms of biological recovery from acidic conditions. In addition, a few watershed-stream experimental manipulations have provided valuable information about linkages between terrestrial and aquatic components of watersheds (Bormann and Likens, 1967). Perhaps the best known involved a small forested watershed in the Hubbard Brook Experimental Forest (Likens et al., 1970; Vitousek et al., 1979). A dramatic increase in nitrate export from the watershed was observed when the forested catchment was clear-cut and treated with a herbicide to prevent revegetation.

The River Continuum Paradigm

Organizing paradigms that treat streams as systems and that integrate across the physical, chemical, and biological aspects of stream science are few in number and recent in origin. The River Continuum Concept (RCC) (Vannote et al., 1980) is the first and perhaps most important of these integrating paradigms. The RCC views entire fluvial systems as a continuously integrated series of physical gradients and adjustments in the associated biota (Cummins et al., 1995). Geomorphological and hydrological characteristics provide the fundamental physical template, which changes longitudinally within a drainage basin from the headwaters to the river mouth in a predictable fashion. Biological communities and associated attributes, such as the nature of the functional community groups, develop in adaptation to the fundamental physical template. The model thus has a definite watershed orientation and focuses on terrestrial and aquatic interactions. It is useful at the basin and stream scale in predicting the physical and biological characteristics of a stream; it is less useful at the scale of stream reach and also has little to say about stream chemistry.

Since its development, the RCC has been modified in many ways to accommodate a broader range of influences on stream ecology (e.g., climate, geology, tributary effects, and local geomorphology) (Minshall et al., 1983, 1985, 1992; Naiman et al., 1987; Cummins et al., 1989; Meyer, 1990). In addition, several organizing paradigms that were developed as alternatives to the RCC have been subsumed into a broader RCC. These include the *nutrient spiraling* concept, in which the unidirectional downgradient flow of streams causes nutrient cycles to be open rather than closed, so that there is a gradual downstream displacement of nutrients, the rate of which is controlled primarily by flow.

Other paradigms, such as the *patch dynamics* concept, which were viewed initially as counter to the RCC, were later shown to be compatible with a broader vision of the RCC that considers both temporal and spatial factors as influences on stream ecology. The patch dynamics concept is based on the idea that disturbance or temporal variation is the primary determinant of community organization in streams (Pringle et al., 1988). As Cummins et al. (1995) state, periods of high flow are a natural feature of most running waters that exhibit local or patch effects, even though they also show predictable longitudinal patterns. Finally, the *flood-pulse* model, an outgrowth of the patch dynamics concept, was developed to describe biological communities in rivers that regularly overtop their banks and inundate the floodplain. In such systems the cycle and extent of inundation may be the fundamental community organizer that overrides longitudinal patterns along changing stream order (Junk et al., 1989). However, Cummins et al. (1995) point out that one of the more predictable patterns of flooding along river continua is a general broadening of the floodplain along the longitudinal profile. Thus, the flood-pulse concept may be considered an extension of the RCC rather than a contradiction of it. The flood-pulse model also reminds us of the importance of stream-riparian connections; this is a topic of growing interest in stream ecology (e.g., Gregory et al., 1991).

In summary, the RCC can be described as an integrative framework for conceptualizing stream-river ecosystems. It is broad enough to accommodate many other organizing concepts to account for specific physical factors or processes that affect flowing waters. Like the microcosm concept for lakes, the RCC is likely to continue to be modified and expanded rather than supplanted.

WETLANDS

Integrated studies using modern science techniques on wetlands as ecosystems are primarily a phenomenon of the past 25 years (e.g., Mitsch and Gosselink, 1993), but the development of basic ecological concepts related to wetlands can be traced back much further. According to Gorham (1953), such studies go back at least as far as those of the mid-sixteenth century Englishman John Leland and the mid-seventeenth century Gerard Boate, "Doctor of Physick to the State of Ireland." These individuals and others in the eighteenth and early nineteenth centuries described and classified wetlands using names similar to those in use today, and they related these classes to basic hydrologic conditions. Gorham also traced the idea of wetland development and succession to an account on Irish bogs published by William King in 1685. Finally, some fundamental ideas about the chemistry of bogs were developed in the late eighteenth and early nineteenth centuries. Nonetheless, most of these pioneering studies were overlooked by twentieth century botanists and ecologists as they developed similar concepts.

Several of the major paradigms described earlier for lakes and streams also are important in wetland science:

• wetlands as microcosms (i.e., wetlands as functioning, integrated ecosystems worthy of study as entities in their own right);
• wetlands as experimental systems for scientific study and manipulable systems for remediation and rehabilitation; and
• wetlands as repositories of historical information in their sediments and, in some cases, in their vegetation (e.g., tree rings).

It is interesting to note that the use of wetland sediments for paleoecological and paleoclimatic studies actually predates the widespread use of lake sediments in this context. In addition, the trophic-dynamic concept and concepts of energy flow and nutrient cycling play the same role in wetland ecology as in lake and stream ecology.

Wetland science also has developed several organizing paradigms beyond those of the disciplines on which it is based and the general paradigms of aquatic ecology:

• wetlands as products of delicate interactions of hydrology and vegetation to produce unique, patterned landforms;
• wetlands as seres or ecotones—that is, gradient ecosystems grading between terrestrial and open-water aquatic systems; and
• wetlands as unique repositories of organic carbon (peat) with an important role in trace-gas cycling to and from the atmosphere.

In the context of global warming, there is much current interest in the role that peatlands play in fixing or releasing the "greenhouse" gases CO_2 and methane (Gorham, 1991b).

Finally, the idea that wetlands can provide service functions to humans has some unique aspects compared to related ideas about lakes and streams in serving human needs. Lakes and streams have been used for navigation, water supply, fishing, and aesthetic and recreational enjoyment for centuries. Since the "wise use" conservationists of the late nineteenth and early twentieth centuries, our stewardship responsibilities toward lakes and streams have at least been acknowledged, if occasionally ignored. In contrast, until recently, wetlands were commonly regarded as wastelands or nuisances to be "reclaimed" by draining or dredging. Their importance in regulating hydrologic processes such as floods, in providing ecological benefits such as buffering contiguous lakes and streams from impacts of human use of uplands, and in serving as habitat for wildlife has been recognized only in the past 20 years.

Ironically, as scientific and public appreciation for wetland ecosystems has grown, there also has been a trend to view them simply as natural analogues or extensions of engineered treatment systems. Thus, a paradigm has developed about *wetlands as natural treatment systems*. Several major research programs have been developed in association with academic institutions and water management agencies to study the effectiveness of wetland systems in removing nutrients and other contaminants from domestic waste effluent (e.g., Ewel and Odum, 1984) or in purifying stormwater runoff before it reaches lakes and streams (e.g., Olson and Marshall, 1992). Both natural and constructed wetlands now are being used for such purposes on a widespread scale within the United States. For some proponents of this paradigm, the quality and ecological integrity of the wetland itself appear to be less important than its ability to perform the desired function. Still, studies conducted under this paradigm have added to the heretofore meager understanding of wetland ecosystem functions and have provided some basis for preserving wetlands that otherwise might be destroyed by drainage and urban or agricultural development.

REFERENCES

Baker, L. A., and P. L. Brezonik. 1988. Dynamic model of internal alkalinity generation: Calibration and application to precipitation-dominated lakes. Water Resour. Res. 24:65–74.
Birge, E. A. 1898. Plankton studies on Lake Mendota II. The crustacea of the plankton from July, 1894, to December, 1896. Trans. Wis. Acad. Sci. Arts Lett. 11:274–448.
Birge, E. A., and C. Juday. 1911. The Inland Lakes of Wisc.: The Dissolved Gases of the Water and Their Biological Significance. Bull. Wis. Geol. Nat. Hist. Surv. 22, Sci. Ser. 7. Madison, Wisc.: Wisconsin Geological and National History Survey.
Bormann, F. H., and G. E. Likens. 1967. Nutrient cycling. Science 155:424–429.
Bradley, W. H. 1963. Paleolimnology. Pp. 621–652 in Limnology in North America, D. G. Frey, ed. Madison: University of Wisconsin Press. 734 pp.
Brezonik, P. L. 1994. Chemical Kinetics and Process Dynamics in Aquatic Systems. Boca Raton, Fla.: Lewis-CRC Press. 754 pp.
Brezonik, P. L., and D. R. Engstrom. In press. Modern and historic accumulation rates of phosphorus in Lake Okeechobee, Florida. J. Paleolimnol.

Brezonik, P. L., J. G. Eaton, T. M. Frost, P. J. Garrison, T. K. Kratz, C. E. Mach, J. H. McCormick, J. A. Perry, W. A. Rose, C. J. Sampson, B. C. L. Shelley, W. A. Swenson, and K. E. Webster. 1993. Experimental acidification of Little Rock Lake, Wisconsin: Chemical and biological changes over the pH range 6.1 to 4.7. Can. J. Fish. Aquat. Sci. 50:1101–1121.

Brooks, J. L., G. L. Clarke, A. D. Hasler, and L. E. Noland. 1951. Edward Asahel Birge. Arch. Hydrobiol. 45:235–243.

Carpenter, S. R. 1990. Large–scale perturbations: Opportunities for innovations. Ecology 71:2038–2043.

Carpenter, S. R., and J. F. Kitchell, eds. 1993. The Trophic Cascade in Lakes. New York: Cambridge University Press.

Carpenter, S. R., T. M. Frost, D. Heisey, and T. M. Kratz. 1989. Randomized intervention analysis and the interpretation of whole ecosystem experiments. Ecology 70:1142–1152.

Coffin, C. C., F. R. Hayes, L. H. Jodrey, and S. G. Whiteway. 1949. Exchange of materials in a lake as studied by the addition of radioactive phosphorus. Can. J. Res. Ser. D 27:207–222.

Cummins, K. W., M. A. Wilzbach, D. M. Gates, J. B. Perry, and W. B. Taliaferro. 1989. Shredders and riparian vegetation. BioScience 39:24–30.

Cummins, K. W., C. E. Cushing, and G. W. Minshall. 1995. Introduction: An overview of stream ecosystems. In River and Stream Ecosystems, Ecosystems of the World, vol. 22, C. E. Cushing, K. W. Cummins, and G. W. Minshall, eds. New York: Elsevier.

Deevey, E. S., Jr. 1967. Introduction. Pp. 63–72 in Pleistocene Extinctions: The Search for a Cause, P. S. Martin and H. E. Wright, Jr., eds. New Haven, Conn.: Yale University Press.

Edmondson, W. T. 1994. What is limnology? In Limnology Now: A Paradigm of Planetary Problems, R. Margalef, ed. New York: Elsevier.

Einsele, W. 1941. Die Umsetzung von zugeführtem anorganischen Phosphat in eutrophen See und ihre Rückwirkung auf seinen Gesamthaushalt. Z. Fisch. 39:407–488.

Emerson, S. 1975. Chemically enhanced CO_2 gas exchange in a eutrophic lake: A general model. Limnol. Oceanogr. 20:743–753.

Engstrom, D. R. 1987. Influence of vegetation and hydrology on the humus budgets of Labrador lakes. Can. J. Fish. Aquat. Sci. 44:1306–1314.

Ewel, K. C., and H. T. Odum, eds. 1984. Cypress Swamps. Gainesville: University of Florida Press. 472 pp.

Forbes, S. A. 1887. The lake as a microcosm. Bull. Peoria Sci. Assoc. Reprinted in Bull. Ill. Nat. Hist. Surv. 15:537–550.

Forel, F. A. 1892, 1895, 1904. Le Leman: Monographie Limnologique, vols. 1–3. Lausanne: F. Rouge.

Frey, D. G. 1963. Wisconsin: The Birge and Juday years. Chapter 1 in Limnology in North America, D. G. Frey, ed. Madison: University of Wisconsin Press. 734 pp.

Gilbert, G. K. 1891. Lake Bonneville. U.S. Geol. Surv. Monogr. I. 438 pp.

Gorham, E. 1953. Some early ideas concerning the nature, origin and development of peat lands. J. Ecol. 41:257–274.

Gorham, E. 1991a. Biogeochemistry: Its origins and development. Biogeochemistry 13:199–239.

Gorham, E. 1991b. Northern peatlands: Role in the carbon cycle and probable responses to climatic warming. Ecol. Appl. 1:182–195.

Gorham, E. 1992. Atmospheric deposition to lakes and its ecological effects: A retrospective and prospective view of research. Jpn. J. Limnol. 53:231–248.

Gregory, S. V., F. J. Swanson, W. A. McKee, and K. W. Cummins. 1991. An ecosystem perspective of riparian zones. BioScience 41:540–551.

Hasler, A. D. 1964. Experimental limnology. BioScience 14(7):36–38.

Horton, R. E. 1931. The field, scope, and status of the science of hydrology. Pp. 189–202 in Reports and Papers, Hydrology, Trans. AGU. Washington, D.C.: National Research Council.

Hurlbert, S. J. 1984. Pseudoreplication and the design of ecological field experiments. Ecol. Monog. 54:187–211.

Hutchinson, G. E. 1957. A Treatise on Limnology, Vol. I. New York: John Wiley & Sons.

Hutchinson, G. E., and V. T. Bowen. 1947. A direct demonstration of the phosphorus cycle in a small lake. Proc. Natl. Acad. Sci. USA 33:148–153.

Hynes, H. B. N. 1970. The Ecology of Running Waters. Toronto: University of Toronto Press.

Hynes, H. B. N. 1975. The stream and its valley. Verh. Int. Verein. Limnol. 19:1–15.

Johnson, W. E., and A. D. Hasler. 1954. Rainbow trout production in dystrophic lakes. J. Wildl. Manage. 18:113–134.

Juday, C., and C. L. Schloemer. 1938. Effects of fertilizers on plankton production and on fish growth in a Wisconsin lake. Progr. Fish-Cult. 40:24–27.

Junk, W. J., P. B. Bayley, and R. E. Sparks. 1989. The flood pulse concept in river floodplain systems. Pp. 110–127 in Proceedings of the International Large River Symposium, D. P. Dodge, ed. Special Publication in Fisheries and Aquatic Sciences 106. Ottawa: Fisheries and Oceans Canada.

Karr, J. R. 1991. Biological integrity: A long-neglected aspect of water resource management. Ecol. Appl. 1:66–84.

Karr, J. R., K. D. Fausch, P. L. Angermeier, P. R. Yant, and I. J. Schlosser. 1986. Assessing biological integrity in running waters: a method and its rationale. Ill. Nat. Hist. Surv. Spec. Publ. No. 5, Champaign, Illinois Natural History Survey.

Karr, J. R., P. R. Yant, K. D. Fausch, and I. J. Schlosser. 1987. Spatial and temporal variability of the index of biotic integrity in three midwestern streams. Trans. Am. Fish. Soc. 116:1–11.

Kelly, C. A., J. W. M. Rudd, R. H. Hesslein, D. W. Schindler, P. J. Dillon, C. T. Driscoll, S. A. Gherini, and R. E. Hecky. 1988. Prediction of biological acid neutralization in acid-sensitive lakes. Biogeochemistry 3:129–140.

Kitchell, J. F., ed. 1992. Food Web Management: A Case Study of Lake Mendota. New York: Springer–Verlag.

Kolkwitz, R., and M. Marsson. 1908. Ökologie der pflanzlichen Saprobien. Ber. Dtsch. Bot. Ges. 26a:505–551.

Kolkwitz, R., and M. Marsson. 1909. Ökologie der tierischen Saprobien. Int. Rev. Ges. Hydrobiol. Hydrol. 2:126–152.

Lamberti, G. A., and A. D. Steinman, eds. 1993. Research in artificial streams: Application, uses, and abuses. J. N. Am. Benthol. Soc. 12:313–384.

Lewis, W. M., Jr., S. Chisholm, C. D'Elia, E. Fee, N. G. Hairston, Jr., J. Hobbie, G. Likens, S. Threlkeld, and R. G. Wetzel. 1995. Challenges for limnology in North America: an assessment of the discipline in the 1990s. Am. Soc. Limnol. Oceanogr. Bull. 4(2):1–20.

Likens, G. E., and A. D. Hasler. 1960. Movement of radiosodium in a chemically stratified lake. Science 131:1676–1677.

Likens, G. E., F. H. Bormann, N. M. Johnson, D. W. Fisher, and R. S. Pierce. 1970. Effects of forest cutting and herbicide treatment on nutrient budgets in the Hubbard Brook watershed-ecosystem. Ecol. Monogr. 40:23–47.

Lindeman, R. L. 1942. The trophic-dynamic aspect of ecology. Ecology 23:399–418.

Meyer, J. 1990. A blackwater perspective on riverine ecosystems. BioScience 40:643–651.

Minshall, G. W., R. C. Petersen, K. W. Cummins, T. L. Bott, J. R. Sedell, C. E. Cushing, and R. L. Vannote. 1983. Interbiome comparison of stream dynamics. Ecol. Monogr. 53:1–25.

Minshall, G. W., K. W. Cummins, R. C. Petersen, C. E. Cushing, D. A. Bruns, J. R. Sedell, and R. L. Vannote. 1985. Developments in stream ecosystem theory. Can. J. Fish. Aquat. Sci. 42:1045–1055.

Minshall, G. W., R. C. Petersen, T. L. Bott, C. E. Cushing, K. W. Cummins, R. L. Vannote, and J. R. Sedell. 1992. Stream ecosystem dynamics of the Salmon River, Idaho: An 8th order system. J. N. Am. Benthol. Soc. 11:111–137.

Mitsch, W., and J. G. Gosselink. 1993. Wetlands, 2nd ed. New York: Van Nostrand Reinhold. 722 pp.

Mortimer, C. H. 1956. E. A. Birge: An explorer of lakes. Pp. 165–211 in E. A. Birge: A Memoir. Madison: University of Wisconsin Press.

Naiman, R. J., J. M. Melillo, M. A. Lock, T. E. Ford, and S. R. Reice. 1987. Longitudinal patterns of ecosystems processes and community structure in a subarctic river continuum. Ecology 1139–1156.

Nipkow, F. 1920. Vorläufige Mitteilungen über Untersuchungen des Schlammabsatzes im Zürichsee Rev. Hydrol. 1:100–122.

Odum, H. T. 1956. Primary production in flowing waters. Limnol. Oceanogr. 1:102–117.

Odum, H. T. 1957. Trophic structure and productivity of Silver Springs, Florida. Ecol. Monogr. 27:55–112.

Olson, R. K., and K. Marshall, eds. 1992. The role of created and natural wetlands in controlling nonpoint source pollution. Ecol. Eng. 1:1–170.

Piontelli, R., and V. Tonolli. 1964. Il tempo di residenza della acque lacustri in relazione ai fenomeni di arrichimento in sostanze immesse, con particolare riguardo al Lago Maggiore. Mem. Ist. Ital. Idrolbiol. 17:247–266.

Platt, R. H. 1993. Geographers and water resource policy. Pp. 36–54 in Water Resource Adminstration in the United States, M. Reuss, ed. East Lansing: Michigan State University Press.

Pringle, C. M., R. J. Naiman, G. Bretschko, J. R. Karr, M. W. Oswood, J. R. Webster, R. L. Welcomme, and M. J. Winterbourn. 1988. Patch dynamics in lotic systmes: the stream as a mosaic. J. N. Am. Benthol. Soc. 7:503–524.

Russell, I. C. 1885. Lake Lahontan. U.S. Geological Survey, Monogr. 11. Reston, Va.: U.S. Geological Survey.

Schindler, D. W. 1974. Eutrophication and recovery in the experimental lakes: Implications for lake management. Science 184:897–898.

Schindler, D. W. 1990. Experimental perturbations of whole lakes as tests of hypotheses concerning ecosystem structure and function. Oikos 57:25–41.

Schindler, D. W., K. H. Mills, D. F. Malley, D. L. Findley, J. A. Shearer, I. J. Davies, M. A. Turner, G. A. Lindsey, and D. R. Cruikshank. 1985. Long-term ecosystem stress: The effects of years of experimental acidification on a small lake. Science 228:1395–1401.

Schindler, D. W., T. M. Frost, K. H. Mills, P. S. S. Chang, I. J. Davies, L. Findlay, D. F. Malley, J. A. Shearer, M. A. Turner, P. J. Garrison, C. J. Watras, K. E. Webster, J. M. Gunn, P. L. Brezonik, and W. A. Swenson. 1992. Comparisons between experimentally- and atmospherically-acidified lakes during stress and recovery. Proc. R. Soc. Edinburgh 97B:193–226.

Schlesinger, W. H. 1991. Biogeochemistry: An Analysis of Global Change. Orlando, Fla.: Academic Press.

Schmitz, W. R., and A. D. Hasler. 1958. Artificial induced circulation of lakes by means of compressed air. Science 128:1088–1089.

Shapiro, J., V. Lamarra, and M. Lynch. 1975. Biomanipulation: An ecosystem approach to lake restoration. Pp. 85–96 in Water Quality Management Through Biological Control: Proceedings of a Symposium, P. L. Brezonik, and J. L. Fox, eds. Gainesville: University of Florida.

Stewart-Oaten, A., W. W. Murdoch, and K. R. Parker. 1986. Environmental impact assessment: "Pseudoreplications" in time? Ecology 67:929–940.

Stewart-Oaten, A., J. B. Bence, and C. W. Osenberg. 1992. Assessing effects of unreplicated perturbations: No simple solutions. Ecology 73:1396–1404.

Streeter, H. W., and E. B. Phelps. 1925. A study of the pollution and natural purification of the Ohio River. III. Factors concerned in the phenomena of oxidation and reaeration. U.S. Public Health Serv. Public Health Bull. 146.

Stross, R. G., and A. D. Hasler. 1960. Some lime-induced changes in lake metabolism. Limnol. Oceanogr. 5:265–272.

Swain, E. B., D. R. Engstrom, M. E. Brigham, T. A. Henning, and P. L. Brezonik. 1992. Increasing rates of atmospheric mercury deposition in midcontinental North America. Science 257:784–787.

Tansley, A. G. 1935. The use and abuse of vegetational concepts and terms. Ecology 16:284–307.

Tilman, D. 1982. Resource Competition and Community Structure. Princeton, N.J.: Princeton University Press.

Vannote, R. L., G. W. Minshall, K. W. Cummins, J. R. Sedell, and C. E. Cushing. 1980. The river continuum concept. Can. J. Fish. Aquat. Sci. 37:130–137.

Vitousek, P. M., J. R. Gosz, C. C. Grier, J. M. Melillo, W. A. Reiner, and R. L. Todd. 1979. Nitrate losses from disturbed ecosystems. Science 204:469–474.

Vollenweider, R. A. 1969. Möglichkeiten und Grenzen elementarer Modelle der Stoffbilanz von Seen. Arch. Hydrobiol. 66:1–36.

Vollenweider, R. A. 1975. Input-output models with special reference to the phosphorus loading concept in limnolgy. Schweiz. Z. Hyrol. 37:53–84.

Warren, C. E., J. H. Wales, G. E. Davis, and P. Doudoroff. 1964. Trout production in an experimental stream enriched with sucrose. J. Wild. Manage. 28:617–660.

Wetzel, R. G. 1990. Land-water interfaces: Metabolic and limnological regulators. Verh. Int. Verein. Limnol. 24:6–24.

Wetzel, R. G. 1992. Gradient-dominated ecosystems: Sources and regulatory functions of dissolved organic matter in freshwater ecosystems. Hydrobiologia 229:181–189.

Zicker, E. L. 1955. The Release of Phosphorus from Bottom Muds and Light Penetration in Northern Wisconsin Bog Lake Waters as Influenced by Various Chemicals. Ph.D. dissertation. University of Wisconsin, Madison.

Linkages Among Diverse Aquatic Ecosystems: A Neglected Field of Study

Eville Gorham
Department of Ecology, Evolution, and Behavior
University of Minnesota, St. Paul
St. Paul, Minnesota

SUMMARY

This paper discusses the scarcity of studies of linkages (functional couplings) among wetlands, streams, rivers, and small and large lakes. It examines difficulties in the pursuit of such studies and the drawbacks of restricting studies to single ecosystems. Several examples of ecosystem linkages are described, along with examples of important linkages deserving greater consideration in the future. The paper suggests mechanisms for fostering closer ties among students of different kinds of ecosystems in both research and teaching.

INTRODUCTION

Aquatic scientists commonly combine a variety of scientific approaches—physical, chemical, biological and ecological—in the study of their particular ecosystem, whether it be stream, lake, wetland, or ocean. It has been far less common, even unusual, for these same aquatic scientists to investigate in any detailed way the interactions of the type of ecosystem in which they specialize with other types of aquatic ecosystems, despite the fact that all are linked through the hydrological cycle and in many other ways. Indeed, it is more common for such ecologists to examine interactions with upland ecosystems, for instance, between streams and their valleys (Hynes, 1975; Vannote et al., 1980; Gregory et al., 1991), or with the atmosphere, as in studies of methane emissions from peatlands (Bartlett and Harriss, 1993; Harriss et al., 1993), than to focus on the linkages among streams, wetlands, lakes, and oceans.

Figure 1 illustrates a way of categorizing research studies by both research field and ecosystem type (the atmosphere is added as a source of inputs). A reading of the technical literature indicates that most ecosystem studies proceed horizontally in Figure 1. One important exception is the study of anadromous fish that migrate between the ocean and fresh waters. Another is the study of nutrient cycling, in which atmospheric inputs are followed through the gravitational phase of the hydrological cycle (i.e., vertically in Figure 1) to catchments, streams, lakes, and wetlands (e.g., Bormann and Likens, 1979; Urban and Eisenreich, 1988). Nutrient inputs from rivers to estuaries and oceans have been followed similarly. Given the great success and significance of such nutrient-cycling studies, which are very widely cited in the literature of ecosystem ecology and biogeochemistry, a strong argument can be made for broadening greatly such vertically oriented research on the linkages among aquatic ecosystems to include other aspects of ecosystem function, diverse examples of which are given in the following pages. There are, however, impediments to the pursuit of such studies.

DIFFICULTIES IN THE PURSUIT OF INTERECOSYSTEM STUDIES

It is difficult for any one person to acquire knowledge of more than one type of ecosystem. Few have expertise in both streams and lakes, or

Ecosystems	Major Fields of Research					
	Physics	Chemistry	Organismal Biology	Community and Ecosystem Ecology	Landscape Ecology	Global Studies
Atmosphere (deposition)						
Upland catchments						
Wetlands						
Streams and rivers						
Small lakes						
Large lakes						
Oceans						

FIGURE 1   A matrix illustrating scientific approaches and the different sorts of ecosystems to which they are applied.

combine one of those two with an interest in wetland ecosystems or in oceans. Despite the need for them, generalists in ecosystem studies are rare, just as are field botanists who can identify more than one group among the vascular plants, bryophytes, lichens, and algae. Given the rarity of generalists, team approaches are useful, although these have their own problems of interpersonal relations, differences in the scientific cultures (with their unfortunate "pecking orders") from which team members spring, leadership, funding, etc.

One result of the lack of generalists is the infrequent occurrence of studies focusing on linkages among aquatic ecosystems. An examination of titles (and, in a few cases where the title was insufficient, abstracts) in recent issues of specialist journals revealed the following ratios of linkage-oriented papers to the total number: *Limnology and Oceanography* (1 to 88), *Journal of the North American Benthological Society* (0 to 70), *Wetlands* (5 to 83), *Journal of Marine Research* (0 to 60), and *Canadian Journal of Fisheries and Aquatic Sciences* (4 to 99). In total, only 2.5 percent of the papers discussed interecosystem studies. This lack of examples means that the topic is seldom brought to the attention of either practicing scientists or students beginning to read technical literature. Indeed, the need for studies of functional couplings among streams, lakes, and wetlands is mentioned seldom, and without detailed consideration, in recent documents discussing the future of freshwater studies (Lewis et al., 1995; Naiman et al., 1995), although Lewis et al. do deplore the fragmentation of freshwater scientists into separate societies dealing with streams (North American Benthological Society), lakes and oceans (American Society of Limnology and Oceanography), and wetlands (Society of Wetland Scientists).

Because interecosystem studies are unusual, it may be difficult to attract research funds for them, as is sometimes the case with interdisciplinary studies. Members of review panels usually specialize in one of the major types of aquatic ecosystem, and some of them—consciously or unconsciously—may not be sympathetic to proposals to examine linkages among the various types. The reverse may of course be true of other reviewers; more needs to be known about such reviewer bias in order to find ways of countering it.

EXAMPLES OF INTERDISCIPLINARY AND INTERECOSYSTEM STUDIES

The nature and importance of interdisciplinary and interecosystem studies can be illustrated most clearly by a series of examples that demonstrate, first, the linkages among disciplines and, second, the functional couplings among different types of inland aquatic ecosystems.

Linkages Among Major Fields of Research

Examples of the employment of linked physical, chemical, and biological approaches to the study of aquatic ecosystems are numerous and readily found in any modern textbook on lakes (Wetzel, 1983), streams (Hynes, 1970; Allan, 1995), wetlands (Mitsch and Gosselink, 1993), and oceans (Duxbury and Duxbury, 1994), or in books on individual ecosystems such as those of the Hubbard Brook Experimental Forest (Likens et al., 1977; Bormann and Likens, 1979). The following few examples illustrate their nature and their relevance to the solution of important environmental problems.

Example 1. It has long been known (Mortimer, 1956) that physical stratification of shallow, highly productive lakes in summer, by isolating their deeper waters from contact with the atmosphere, is responsible for the severe depletion of oxygen that makes those deeper lakes uninhabitable by many organisms, including a number of fish species and their prey. Likewise, the freezing phase of the annual temperature cycle in northern latitudes may lead to oxygen depletion throughout shallow lakes and therefore to the winterkill that often decimates sport fisheries.

Example 2. Cultural eutrophication, involving nuisance algal blooms caused by nutrient enrichment from fertilizers and sewage, is a serious environmental problem. Paleoecological reconstruction of its history usually employs radioisotope dating of sediment cores, combined with the measurement of sedimentation rates for nutrients such as phosphorus and for the plant and animal microfossils and fossilized plant pigments that are indicators of aquatic productivity (Brugam, 1978; Engstrom et al., 1985). Conversely, studies of bryophyte fossils in radiocarbon-dated peat cores (Janssens, 1983) have allowed the construction of profiles of past levels of acidity and water tables in peatlands (Gorham and Janssens, 1992; Janssens et al., 1992) so that we can see if future human activities causing acid deposition from the atmosphere or climatic warming lead to changes that transcend those that have occurred in past decades, centuries, or millennia.

Example 3. In studies of the toxic effects of acid deposition from the atmosphere on stream and lake biota—with special reference to sport and commercial fisheries—it has been necessary to consider patterns of air mass movement, amounts of precipitation, the timing of snowmelt, and other physical factors that control the deposition and pathways of strong acids resulting from urban-industrial air pollution. Also involved are studies of geology, soil chemistry, alkalinity generation by biological processes, physiological analysis of gill function in response to high aluminum concentrations, examination of various alterations in biotic communities and their food chains, and research on a host of other biological responses to interacting physical and chemical properties of the environment (Freedman, 1989a; Anonymous, 1990).

Example 4. Recent studies in northern Minnesota (Siegel, 1983, 1988; Glaser et al., 1990) have shown that upwelling of circumneutral ground water rich in calcium bicarbonate is vital to the development of patterned peatlands in northern landscapes. This upwelling water forms distinct fen water tracks that become dominated by sedges and flow through the peatlands in a way that separates them sharply from the ovoid, acid sphagnum bogs that are the other distinctive feature of such landscapes. Because peatlands are an important source and sink for greenhouse gases, and fens and bogs differ in their production and consumption of such gases, it is important to understand the hydrology that controls their development.

Fens and bogs also differ greatly in their acidity, and hence in their export of acidity to streams and lakes, which in turn influences plants and animals downstream from these peatlands (see next section).

Linkages Among Different Types of Aquatic Ecosystem

Studies that link different aquatic ecosystems, although by no means lacking, are much more difficult to find in the literature, as shown previously. The following are several examples of such studies.

Example 1. Among the earliest studies to link peatlands and lakes were those that described the temporal linkages involved in the pattern of aquatic succession leading—on a time scale of decades to millennia—from lake to sedge meadow (DeLuc, 1810) and even to raised bog (Aiton, 1811; see also Gorham, 1953). Raymond Lindeman (1941) was one of many later investigators of such phenomena. Such successional studies can be very important in devising programs of ecosystem rehabilitation, which are best carried out from detailed knowledge of how a given ecosystem reached its present state.

Example 2. Chemical budgets for lakes, reservoirs, and wetlands necessarily involve measurement of varying flow rates and associated chemical fluxes through inflow and outflow streams (Schindler et al., 1976; Likens et al, 1977; Urban and Eisenreich, 1989; Urban et al., 1990). Such budgets are of vital importance to understanding whether a given aquatic ecosystem can store nutrients or toxins to a significant degree and thus prevent runoff to downstream ecosystems. Such storage may also be a source of later problems if disturbance—for instance, drawdown of a wetland—releases such materials downstream (Bayley et al., 1992).

Example 3. Many scientists and engineers concerned with problems of cultural eutrophication have examined the hypothesis that wetlands can act as traps for nutrients such as phosphorus and nitrogen in runoff from adjacent upland soils and thus mitigate nuisance algal blooms in adjacent lakes (Johnston, 1991; Richardson and Craft, 1993). Ecological limitations on such uses of wetlands have been discussed by Guntenspergen and Stearns (1981).

Example 4. Analysis of carbon isotopes has shown that particulate organic detritus eroded from tundra peatlands is a significant source of nutrition for food chains leading to fish and ducks in nearby ponds and lakes (Schell, 1983). This sort of information provides a better understanding of food chain dynamics in freshwater habitats, a matter of particular concern for indigenous peoples and sport fishermen.

Example 5. Geochemical study of iron and aluminum in acid Nova Scotian lakes (Urban et al., 1990) has shown that both elements, weathered from upland catchments, are combined with dissolved organic matter (DOM), supplied from nearby peatlands, to form metal humates. Precipitation of such humates from solution may regulate the concentrations of both iron and aluminum, and to a lesser extent DOM, in these acid lakes. DOM concentrations in the lake water are also related to the proportion of the catchment that is covered by peat (Gorham et al., 1986; Kortelainen, 1993a). The importance of such studies is twofold. Dissolved organic matter in these waters consists largely of colored organic (humic) acids that can, like acid rain, acidify fresh waters. Unlike acid rain, however, which releases aluminum from soils and sediments in forms toxic to fish, humic acids form nontoxic complexes with aluminum.

Example 6. Studies of diverse catchments in northwestern Ontario (St. Louis et al., 1994) have revealed that wetlands are important sources of methylmercury to downstream aquatic ecosystems. In its methylated form, mercury can bioaccumulate to toxic levels in aquatic food chains, leading ultimately to humans through fish (Minnesota Fish Consumption Advisory, 1994).

Example 7. Studies of northern pike, an important North American sport fish, show that it depends on wetlands flooded in spring for spawning and nursery areas. This is undoubtedly true for other organisms.

Example 8. The influence of beaver on small streams, ponds, and lakes is profound (Johnston and Naiman, 1990a,b) and can be understood only at the landscape level because these furbearing mammals range over all three types of water body, as well as adjacent uplands. All of these habitats are affected substantially by beaver use, which has an influence that may last as a legacy on the landscape for decades or more after they have left the area.

Example 9. Wetlands can have an appreciable influence on the water budgets of streams and rivers. An extreme example is the White Nile, which flows through a vast wetland, the Sudd. As much as half of the water flowing into the Sudd is lost by evapotranspiration from the wetland (Melack, 1992), which must greatly affect all aspects of the river downstream.

DRAWBACKS TO FRAGMENTATION BY ECOSYSTEM

The previous examples illustrate clearly the importance of studying functional couplings among aquatic ecosystems as well as to the surrounding uplands; without such studies, our understanding would be seriously inadequate. The nature of various problems of environmental degradation, many of which affect all upland and aquatic ecosystems in one way or another, provides a further and extremely cogent reason for such interecosystem studies, as shown by the following examples:

Example 1. If, as expected, climate warming becomes a serious environmental problem in the next century (Intergovernmental Panel on Climate Change, Working Group 1, 1992), the Great Plains of North America will likely see a sharp decline in the number of prairie potholes combining wetlands suitable for the breeding of ducks with open water suitable for their feeding. Interpreting changes in these prairie potholes and in their associated biota as they dry out owing to greatly increased drought frequency will require a careful assessment of the linkages among lakes, ponds, and wetlands as they are affected in differing degrees by alterations in ground water hydrology. The prospect for devastating effects on duck populations in the Great Plains—already threatened by human overexploitation—is alarming.

Example 2. Studies of lake acidification ignored for a long time the possibility that upstream and marginal peatlands can be a potent source of acid inputs; these studies ascribed lake acidification solely to atmospheric deposition of sulfuric and nitric acids derived from air pollutants. Organic acids from peat bogs (Gorham et al., 1985) have, however, proved to be significant and, in some cases, predominant contributors to lake acidification in areas of Nova Scotia (Gorham et al., 1986; Kerekes et al., 1986; Gorham et al., in review), Norway (Brakke et al., 1987), and Finland (Kortelainen, 1993b). They must, therefore, be taken into account in much of the north temperate zone, where bogs are common features of the landscape. On the one hand, ignoring such natural acid inputs may overemphasize the effects of acid rain caused by human activity, but on the other hand, natural acidification of this kind may predispose some aquatic ecosystems to damage by relatively small inputs of acid rain.[1]

Example 3. Peats have a great capacity to bind toxaphene and DDT (Rapaport and Eisenreich, 1986), as well as other organochlorine micropollutants. The possibility exists, therefore, that wetlands—by their effective trapping of such materials—are a significant factor in lessening the

---

[1]It should be noted that some clear water lakes on rocks that are not readily weathered can become quite acid through a variety of natural processes (Ford, 1990), though not to the degree brought about by acidic deposition from the atmosphere (Renberg, 1990) or substantial inputs from peat bogs in the catchment (Gorham et al., 1986).

buildup of such toxic pollutants in downstream ecosystems. If severe climate warming comes to pass and leads to drawdown of peatland water tables, it may well be that subsequent peat aeration and oxidation will release these toxins—largely concentrated in the surface peats—downstream. The same may be true of nitrogen and sulfur, stored in these peatlands in large amounts (Gorham, 1994) and liable to be released in acid form by drought and fire (Bayley et al., 1992). Bog mosses and lichens also have a strong affinity for radioactive fallout (Gorham, 1958, 1959; Miettinen, 1969), which could be released similarly. In all three cases there is the potential for ecological damage, but at present we lack sufficient information to evaluate its likely significance.

Example 4. Oil pollution is a largely local problem in all sorts of ecosystems from the uplands to the ocean. Many different organisms from diverse habitats are capable of degrading different petroleum compounds (Freedman, 1989b). The search for the most efficient among them for use in bioremediation should encompass the full range of those habitats.

In all of these examples, restriction of studies to a single type of ecosystem would also restrict, to a very significant extent, both our thinking about and our understanding of serious environmental problems.

IMPORTANT QUESTIONS FOR FUTURE RESEARCH ON THE LINKAGES AMONG AQUATIC ECOSYSTEMS

More attention to interecosystem linkages can be justified only if truly important research questions, both fundamental and practical, can be asked about them. The following are a few examples suggested from experience in studying lakes, streams, and wetlands. Individual scientists will be able to add their own examples, and environmental surprises in the years to come will reveal many more cases in which a better understanding of interecosystem linkages would have served us well.

Example 1. How does the influence of organic carbon compounds (both particulate and dissolved) on (1) aquatic productivity; (2) acid-base balance; (3) oxidation-reduction potential; (4) trace-metal transport; and (5) emissions of volatile carbon, nitrogen, and sulfur compounds change from the time of its deposition as dead organic matter (forest litter and aquatic plant remains) to the time of its ultimate arrival in the open ocean, having passed through streams, wetlands, small and large lakes, large rivers, and estuaries? All of these items are important elements in the metabolism of aquatic ecosystems, on which sport and commercial fisheries depend, but although we have fitted some parts of the metabolic puzzle together we do not yet have anything close to a complete picture spanning the full range of aquatic ecosystems.

Example 2. Do chemical inputs of nutrients and toxins from wetlands have an influence on the biodiversity of receiving streams and lakes? It is all but certain that they do, but we have far too little information about such matters, which may affect the overall functioning of these aquatic ecosystems and also be important to the success or failure of threatened and endangered species.

Example 3. In a broader way, we may ask how riparian wetlands, which are extensions of the littoral zones of streams and rivers, influence the functioning of the actual littoral zones and their adjacent water bodies. Wetzel (1990, 1992) has provided a general account of effects on carbon cycling, but much remains to be done in terms of the cycles of nutrients and toxins and their effects upon species composition, productivity, food chains, etc. The invasion of North American wetlands by the Eurasian exotic purple loosestrife (Thompson et al., 1987), as it affects the functioning of adjacent streams and lakes, is a specific case worthy of further investigation in this context. Also relevant are the movements of organisms such as insects, amphibians, and furbearing mammals to and from uplands, wetlands, and adjacent littoral areas of streams and lakes.

Example 4. In a related question, what is the influence of streams and lakes on the plant and animal communities of adjacent wetland and upland ecosystems? This is an important matter that concerns the design of buffer zones around lakes and the management of riparian zones.

Example 5. How will effects of global warming on wetlands influence ecosystem structure and function in the streams and lakes that receive inputs from them (for instance, of nutrients such as nitrogen and phosphorus, acidifying agents such as organic (humic) acids and sulfuric acid, or stored metal and organic pollutants)? Meyer and Pulliam (1992) provide some interesting insights that suggest future research directions.

Example 6. An extremely important problem in landscape ecology (Risser, 1987), as well as in regional and global studies, is that of scaling up from purely local studies of individual ecosystems to develop understanding at the landscape or higher level. It is here that interactions among aquatic (and upland) ecosystems are of major importance, because they must be taken into account at each successive step in the scaling process if we are to understand adequately, for instance, nutrient cycling and the processing of toxic contaminants, as well as their influence on species composition, productivity, carbon storage, etc.—all vital components in the functioning and management of the biosphere.

In many cases, processes of ecosystem interaction will be nonlinear and may involve threshold and stochastic phenomena, making it necessary to study them over long periods for adequate understanding (Likens, 1989). For instance, consider the influence of peatlands on small lakes in the same catchment and receiving drainage from them. The studies of Gorham et al. (1986) indicate that the influence of such peatlands on the concentrations of colored organic acids in the lakes increases in nonlinear fashion with the ratio of peatland area to lake area. Even a very small proportion of peatland in a catchment leads to a rapid increase in lake acidity, but the rate of increase slows steadily as the proportion of peatlands in the catchment rises. As an example of a threshold phenomenon, peatlands may shift rather suddenly—over a period of decades to centuries and after millennia of relative stability—from circumneutral sedge meadows to strongly acid moss bogs as the level of upward peat accumulation reaches a threshold that shuts off supplies of bases from the mineral soil, so that acid-loving *Sphagnum* mosses can invade the sedge communities (Gorham and Janssens, 1992). The consequences for the acidity of downstream waters, and hence for their plants and animals, can be considerable.

Elements of the answers to all these questions undoubtedly exist in technical literature, but more active research on interecosystem processes would help bring them into focus, filling many gaps in our present information.

MECHANISMS FOR FOSTERING CLOSER TIES AMONG STUDENTS OF DIFFERENT AQUATIC ECOSYSTEMS

If scientists are to be encouraged to study functional couplings among the diverse types of aquatic ecosystems, increased attention to both mechanisms and incentives for bringing them together will be required. That attention must focus not only on research itself, but also on professional recognition of the importance of generalist ecosystem studies and on the teaching of future generations of aquatic scientists.

Research

The best way to encourage research on ecosystem linkages is to provide research and training grants explicitly for that purpose. Foundations may perhaps be more willing to encourage innovative ideas for such studies, but many federal agencies are engaged in the study of environmental problems that, as noted earlier, involve a variety of interecosystem processes. The National Science Foundation (NSF), without such a problem-oriented mandate, might also be approached to begin a modest pioneer program on ecosystem linkages in the same way that it began the Long-Term Ecological Research Program under the Ecosystem Studies Program. Alternatively, it could be suggested for inclusion in the new NSF-Environmental Protection Agency partnership for environmental research.

Another way to foster research is to hold workshops and symposia devoted to bringing together the scientists who study different kinds of aquatic ecosystems. Such meetings may be sponsored by scientific societies, government agencies, or private foundations; scientific journals—societal or private—frequently assist the process by publishing their proceedings. Gordon Conferences, the Dahlem Conferences, and North Atlantic Treaty Organization Advanced Institutes are excellent examples. Scientific societies can also hold joint meetings to bring their members together, as happened at Edmonton recently with the American Society of Limnology and Oceanography and the Society of Wetland Scientists. It is not enough, however, merely to bring scientists from different backgrounds together; they must also be brought to focus directly in individual sessions on the functional couplings among wetlands, streams, lakes, and oceans rather than simply to compare ecosystem processes such as net primary production, food chain dynamics, cycling of nutrients and toxins, and emissions of trace gases, valuable as such environmental and biotic comparisons may be.

Professional recognition for ecosystem generalists who study functional couplings is also important. If they are regarded merely as dilettantes, their careers are likely to suffer in departments staffed largely by specialists.

Teaching

Students can be recruited more effectively for graduate studies that deal with linkages among aquatic ecosystems (and with upland ecosystems and the atmosphere) if they learn about them as undergraduates. Such teaching should not be left too late or students will already have become restricted in their studies, and barriers will have arisen to an interecosystem approach. Under what rubric such an approach can be taught most effectively remains to be seen; if a few examples are presented in courses strongly focused primarily on one of the above-mentioned disciplines, they are unlikely to achieve their purpose. Courses on ecosystem ecology and biogeochemistry may be the best places to proselytize for interecosystem studies, especially if they are team-taught by specialists in the different types of ecosystems who are committed to a focus on linkages in both teaching and research. Support for curriculum development could be sought from education-oriented private foundations and from the recently expanded Education Program of the National Science Foundation.

Organizational support for teaching the functional couplings among ecosystems (and with the atmosphere) might come most readily from the linkage of specialist departments into interdisciplinary programs, centers, institutes, etc. The Watershed Science Program of Utah State University is an example, linking groups in fisheries and wildlife, forest resources, geography and earth resources, and range science. Such programs, however, may not always have the assured staffing and line-item budgets required for long-term commitment, and the allegiance of program faculty (and often their graduate students) to their home departments may make such programs less than ideal. Departments that are organized specifically for interdisciplinary studies of the environment—as, for instance, at the University of Virginia where the Department of Environmental Sciences encompasses four tracks (ecology, geology, hydrology, and atmospheric sciences)—probably provide the optimal model, with their own line-item and supply budgets.

CONCLUSION

Given the fundamental scientific and practical economic importance of linkages among aquatic (and other) ecosystems and the present difficulties of studying them, a concerted effort to foster research on them is badly needed. Initiatives should therefore be developed at this time, in both teaching and research, to attract scientists to their study. A broader graduate training, developing thorough familiarity with more than one type of ecosystem, would be greatly conducive to thinking about the functional couplings among ecosystems.

ACKNOWLEDGMENTS

I thank Pat Brezonik for assigning me this topic. Jackie MacDonald and my fellow committee members provided many helpful comments and suggestions, as did Jesse Ford and Susan Galatowitsch.

REFERENCES

Aiton, W. 1811. Treatise on the Origin, Qualities, and Cultivation of Moss-Earth, with Direction for Converting It into Manure. Ayr, Scotland: Wilson and Paul.

Allan, J. D. 1995. Stream Ecology: Structure and Function of Running Waters. London: Chapman and Hall.

Anonymous. 1990. Acid Deposition: State of Science and Technology. Washington, D.C.: National Acid Precipitation Assessment Program.

Bartlett, K. B., and R. C. Harriss. 1993. Review and assessment of methane emissions from wetlands. Chemosphere 26:261–320.

Bayley, S. E., D. W. Schindler, B. R. Parker, M. P. Stainton, and K. G. Beaty. 1992. Effects of forest fire and drought on acidity of a base-poor boreal forest stream: Similarities between climatic warming and acidic precipitation. Biogeochemistry 17:191–204.

Bormann, F. H., and G. E. Likens. 1979. Pattern and Process in a Forested Ecosystem. Berlin: Springer-Verlag.

Brakke, D. F., A. Henriksen, and S. A. Norton. 1987. The relative importance of acidity sources for humic lakes in Norway. Nature 329:432–434.

Brugam, R. B. 1978. Human disturbance and the historical development of Linsley Pond. Ecology 59:19–36.

DeLuc, J. A. 1810. Geological Travels, Vol. 1. London: Rivington.

Duxbury, A. C., and A. B. Duxbury. 1994. An Introduction to the World's Oceans, 4th ed. Dubuque, Ia.: W. C. Brown.

Engstrom, D. R., E. B. Swain, and J. C. Kingston. 1985. A paleolimnological record of human disturbance from Harvey's Lake, Vermont: Geochemistry, oscillaxanthin, and diatoms. Freshwater Biol. 15:261–288.

Ford, M. S. 1990. A 10,000 year history of natural ecosystem acidification. Ecol. Monogr. 60:57–89.

Freedman, B. 1989a. Acidification. Pp. 81–123 in Environmental Ecology. San Diego, Calif.: Academic Press.

Freedman, B. 1989b. Oil Pollution. Pp. 138–158 in Environmental Ecology. San Diego, Calif.: Academic Press.

Glaser, P. H., J. A. Janssens, and D. I. Siegel. 1990. The response of vegetation to chemical and hydrological gradients in the Lost River Peatland, northern Minnesota. J. Ecol. 78:1021–1048.

Gorham, E. 1953. Some early ideas concerning the nature, origin and development of peat lands. J. Ecol. 41:257–274.

Gorham, E. 1958. Accumulation of radioactive fallout by plants in the English Lake District. Nature 181:152–154.

Gorham, E. 1959. A comparison of lower and higher plants as accumulators of radioactive fallout. Can. J. Bot. 37:327–329.

Gorham, E. 1994. The biogeochemistry of northern peatlands and its possible responses to global warming. In Biotic Feedbacks in the Global Climate System, G. M. Woodwell, ed. Oxford, England: Oxford University Press.

Gorham, E., and J. A. Janssens. 1992. The paleorecord of geochemistry and hydrology in northern peatlands and its relation to global change. Suo 43:9–19.

Gorham, E., S. J. Eisenreich, J. Ford, and M. V. Santelmann. 1985. The chemistry of bog waters. Pp. 339–362 in Chemical Processes in Lakes, W. Stumm, ed. New York: John Wiley & Sons.

Gorham, E., J. K. Underwood, F. B. Martin, and J. G. Ogden III. 1986. Natural and anthropogenic causes of lake acidification in Nova Scotia. Nature 324:451-453.

Gorham, E., J. K. Underwood, J. A. Janssens, B. Freedman, W. Maass, D. H. Waller, and J. G. Ogden III. In review. The chemistry of streams in southwestern and central Nova Scotia, with particular reference to the influence of dissolved organic carbon from wetlands.

Gregory, S. V., F. J. Swanson, W. A. McKee, and K. W. Cummins. 1991. An ecosystem perpective of riparian zones. BioScience 41:540–551.

Guntenspergen, G., and F. Stearns. 1981. Ecological limitations on wetland use for wastewater treatment. Pp. 273–284 in Wetland Values and Management, B. Richardson, ed. St. Paul: Minnesota Water Planning Board.

Harriss, R., K. Bartlett, S. Frolking, and P. Crill. 1993. Methane emission from high-latitude wetlands. Pp. 449–486 in Biogeochemistry of Global Change: Selected Papers from the Tenth International Symposium on Environmental Biogeochemistry, R. S. Oremland, ed., San Francisco, 1991. New York: Chapman and Hall.

Hynes, H. B. N. 1970. The Ecology of Running Waters. Toronto: University of Toronto Press. 555 pp.

Hynes, H. B. N. 1975. The stream and its valley. Proc. Int. Assoc. Theor. Appl. Limnol. 19:1–15.

Intergovernmental Panel on Climate Change, Working Group 1. 1992. Supplement to Scientific Assessment of Climate Change. Geneva: World Meteorological Organization and United Nations Environmental Program.

Janssens, J. A. 1983. A quantitative method for stratigraphical analysis of bryophytes in peat. J. Ecol. 71:189–196.

Janssens, J. A., B. C. S. Hansen, P. H. Glaser, and C. Whitlock. 1992. Development of a raised-bog complex in Northern Minnesota. Pp. 189–221 in Patterned Peatlands of Northern Minnesota, H. E. Wright, Jr., B. Coffin, and N. Aaseng, eds. Minneapolis: University of Minnesota Press.

Johnston, C. A. 1991. Sediment and nutrient retention by freshwater wetlands: Effects on surface water quality. Crit. Rev. Environ. Control 21:491–565.

Johnston, C. A., and R. J. Naiman. 1990a. The use of a geographic information system to analyze long-term landscape alteration by beaver. Landscape Ecol. 4:5–19.

Johnston, C. A., and R. J. Naiman. 1990b. Aquatic patch creation in relation to beaver population trends. Ecology 71:1617–1621.

Kerekes, J., S. Beauchamp, R. Tordon, and T. Pollock. 1986. Sources of sulphate and acidity in wetlands and lakes in Nova Scotia. Water Air Soil Pollut. 31:207–214.

Kortelainen, P. 1993a. Content of total organic carbon in Finnish lakes and its relationship to catchment characteristics. Can. J. Fish. Aquat. Sci. 50:1477–1483.

Kortelainen, P. 1993b. Contribution of Organic Acids to the Acidity of Finnish Lakes. Publications of the Water and Environment Research Institute, No. 13. Helsinki: National Board of Waters and the Environment, Finland, No. 13.

Lewis, W. M., Jr., S. Chisholm, C. D'Elia, E. Fee, N. G. Hairston, J. Hobbie, G. E. Likens, S. Threlkeld, and R. G. Wetzel. 1995. Challenges for limnology in North America: an assessment of the discipline in the 1990s. Am. Soc. Limnol. Oceanogr. Bull. 4(2):1–20.

Likens, G. E., ed. 1989. Long-Term Studies in Ecology, Approaches and Alternatives. New York: Springer Verlag.

Likens, G. E., F. H. Bormann, R. S. Pierce, J. S. Eaton, and N. M. Johnson. 1977. Biogeochemistry of a Forested Ecosystem. New York: Springer Verlag. 146 pp.

Lindeman, R. L. 1941. The developmental history of Cedar Bog Lake, Minnesota. Am. Midl. Nat. 25:101–112.

Melack, J. M. 1992. Reciprocal interactions among lakes, large rivers, and climate. Pp. 68–87 in Global Climate Change and Freshwater Ecosystems, P. Firth and S. G. Fisher, eds., New York: Springer Verlag.

Meyer, J. L., and W. M. Pulliam. 1992. Modification of terrestrial-aquatic interactions by a changing climate. Pp. 177–191 in Global Climate Change and Freshwater Ecosystems, P. Firth and S.G. Fisher, eds. New York: Springer Verlag.

Miettinen, J. 1969. Enrichment of radioactivity by Arctic ecosystems in Finnish Lappland. Pp. 23–31 in Symposium on Radioecology, D.J. Nelson and R.C. Evans, eds. Clearinghouse for Federal Scientific and Technical Information, CONF-670503, Biology and Medicine (TID-4500) Springfield, Va.: National Bureau of Standards.

Minnesota Fish Consumption Advisory. 1994. Minneapolis: Minnesota Department of Health.

Mitsch, W. J., and J. G. Gosselink. 1993. Wetlands, 2nd ed. New York: Van Nostrand Reinhold. 722 pp.

Mortimer, C. H. 1956. E. A. Birge: An explorer of lakes. Pp. 165–211 in E. A. Birge: A Memoir, by G. C. Sellery. Madison: University of Wisconsin Press.

Naiman, R. M., J. J. Magnuson, D. M. McKnight, and J. A. Stanford. 1995. The Freshwater Imperative: A Research Agenda. Washington, D.C.: Island Press.

Rapaport, R., and S. J. Eisenreich. 1986. Atmospheric deposition of toxaphene to eastern North America derived from peat accumulation. Atmos. Environ. 20:2367–2379.

Renberg, I. 1990. A 12,600 year perspective on the acidification of Lilla Öresjön, southwest Sweden. Philos. Trans. R. Soc. Lond. B 327:357–361.

Richardson, C. J., and C. B. Craft. 1993. Effective phosphorus retention in wetlands: Fact or fiction? Pp. 271–282 in Constructed Wetlands for Water Quality Improvement, G. A. Moshiri, ed. Boca Raton, Fla.: Lewis Publishers.

Risser, P. G. 1987. Landscape ecology: State of the art. Pp. 3–14 in Landscape Heterogeneity and Disturbance, M. G. Turner, ed. New York: Springer Verlag.

Schell, D. M. 1983. Carbon-13 and carbon-14 abundances in Alaskan aquatic organisms: Delayed production from peat in Arctic food webs. Science 219:1068–1071.

Schindler, D. W., R. W. Newbury, K. G. Beaty, and P. Campbell. 1976. Natural water and chemical budgets for a small Precambrian lake basin in central Canada. J. Fish. Res. Bd. Can. 33:2526–2543.

Siegel, D. I. 1983. Groundwater and evolution of patterned mires, Glacial Lake Agassiz Peatlands, northern Minnesota. J. Ecol. 71:913–922.

Sieges, D. I. 1988. Evaluating cumulative effects of disturbance on the hydrologic function of bogs, fens, and mires. Environ. Manage. 12:621–626.

St. Louis, V. L., J. W. M. Rudd, C. A. Kelly, K. G. Beaty, N. S. Bloom, and R. J. Flett. 1994. Importance of wetlands as sources of methyl mercury to boreal forest ecosystems. Can. J. Fish. Aquat. Sci. 51:1065–1076.

Thompson, D. Q., R. L. Stuckey, and E. B. Thompson. 1987. Spread, Impact, and Control of Purple Loosestrife (*Lythrum salicaria*) in North American Wetlands. Fish Wild. Res., No. 2. Washington, D.C.: U.S. Fish and Wildlife Service.

Urban, N. R., and S. J. Eisenreich. 1988. Nitrogen cycling in a forested Minnesota bog. Can. J. Bot. 66:435–449.

Urban, N. R., E. Gorham, J. K. Underwood, F. B. Martin, and J. G. Ogden III. 1990. Geochemical processes controlling concentrations of Al, Fe, and Mn in Nova Scotia lakes. Limnol. Oceanogr. 35:1516–1534.

Vannote, R. L., G. W. Minshall, K. W. Cummins, J. R. Sedell, and C. E. Cushing. 1980. The river continuum concept. Can. J. Fish. Aquat. Sci. 37:130–137.

Wetzel, R. G. 1983. Limnology, 2nd ed. Philadelphia: Saunders College Publishing.

Wetzel, R. G. 1990. Land-water interfaces: Metabolic and limnological regulators. Verh. Int. Verein. Theoret. Angew. Limnol. 24:6–24.

Wetzel, R. G. 1992. Gradient-dominated ecosystems: Sources and regulatory functions of dissolved organic matter in freshwater ecosystems. Hydrobiologia 229:181–198.

Training of Aquatic Ecosystem Scientists

Robert G. Wetzel
Department of Biological Sciences
University of Alabama
Tuscaloosa, Alabama

SUMMARY

Limnology is a distinct professional discipline that examines the structure, functions, and management of inland, primarily freshwater, aquatic ecosystems. Limnological expertise is urgently required to understand the mechanisms regulating the operation of lake, reservoir, river, and wetland ecosystems. Only with such understanding can aquatic ecosystems be managed effectively.

Undergraduate programs in the United States usually do not specifically train students in the integrated discipline of freshwater ecology and aquatic environmental problem solving. Even graduates in limnology are frequently inadequately trained and are inexperienced in management of environmental problems. Improvements in instruction and research training in limnology are needed, particularly with regard to interdisciplinary breadth, ecosystem integration, and practical experience in problem solving.

This paper proposes combined programs of rigorous, science-based undergraduate and graduate training at professional schools of limnology, in which practitioners and researchers are trained in inland aquatic ecology with true ecosystem perspectives by means of an integrated problem-solving environment. Programs should be coordinated to ensure minimal professional standards of limnological education. Advanced programs at regional schools of limnology are designed to augment and strengthen existing traditional programs, not supplant them, and to provide guidance to existing instructional programs for improving training in the discipline.

INTRODUCTION

Limnology is an integrative discipline of inland waters. The subject clearly should address the coupled spatial and temporal variations in physical, chemical, and biological properties; the way in which these properties influence aquatic biota and their growth, dynamics, and productivities; and how the community biological metabolism affects geochemical properties. The aquatic components are integrated in an interactive ecosystem that extends considerably beyond the traditional shoreline boundary of the lake, reservoir, wetland, or stream (Likens, 1984; Wetzel, 1990a; Wetzel and Ward, 1992). Physical, geological, hydrological, chemical, and biological characteristics and processes are examined along a large range of scales, for example, from individual chemical reactions to chemical fluxes within entire ecosystems. A fundamental aspect of aquatic ecosystems that is overlooked frequently, however, is that they are biogeochemical systems; biological processes are essential components of all qualitative and many quantitative aspects of inland aquatic ecosystems.

DEVELOPMENT OF THE ECOSYSTEM PERSPECTIVE

Past approaches to the study of inland waters were initially descriptive, which was common to many disciplines in the past century. Massive comparative analyses of physical, chemical, and biological properties of inland waters evolved rapidly from 1920 to 1950, particularly in Europe. Greater emphasis was placed on analytical evaluations of intercoupled relationships, particularly among nutrient-phytoplankton interactions, in the subsequent period (1950s–1970s). Simultaneously, great emphasis on feeding relationships emerged in 1970–1980 in concert with marked advances in predator-prey relationships. During the last and present decades, limnologists have begun to recognize the inadequacy of examining pelagic communities independently from the littoral-wetland and land-water interface regions of most lake and river ecosystems. Couplings of all components of the ecosystem, the drainage basin, land-water interface communities, and open-water communities, are critical to both the qualitative and the quantitative understanding of lake and river ecosystems. This essential ecosystem perspective is now being incorporated gradually into the management of inland waters. For a number of complex reasons discussed below, however, the ecosystem perspective is not being incorporated effectively into the undergraduate and graduate training of students in aquatic ecology.

UNDERPINNINGS OF THE ECOSYSTEM PERSPECTIVE
IN EDUCATION

There are two major needs for educating limnologists according to a broad, ecosystem perspective. Foremost, there is a need to promulgate the values of freshwater ecosystems properly in economic terms in our teachings at all levels of education. The importance of the availability of high-quality freshwater resources, including ground water, is almost universally underappreciated as essential to the economic viability of society. Responses to reductions and localized crises of water supply or quality are generally evasive, involving cosmetic, temporary technological responses that do not correct the problems (Francko and Wetzel, 1983; Rogers, 1993). Many developed countries, such as the United States, have made concerted and partially successful efforts toward regulated treatment and release of used water. However, with exponentially increasing demands from increasing populations, it will be difficult to maintain present standards of water quality in both developed and emerging countries (Wetzel, 1992).

A second major need is imparting to students that the most effective and economical management of aquatic ecosystems results from an understanding of the mechanisms governing the integrated hydrology, chemistry, and biology of these ecosystems. The correct diagnoses of freshwater problems and their corrective management are most effective when the dynamics of controlling processes are quantified. The management to control and correct eutrophication in fresh waters is an excellent example. Corrective techniques (e.g., Chapra and Reckhow, 1983; Cooke et al., 1993) have been possible because of the basic understanding that existed in plant and algal physiology and ecology. Einar Naumann recognized in the 1920s that inorganic nutrients, particularly nitrogen and phosphorus, likely influenced the growth of phytoplankton. These conclusions were inferred from agricultural studies before nutrients could be measured effectively in water. Similarly, Birge and Juday (1934) inferred the importance of hydrological renewal rates to material loadings and lake productivity. Subsequent extensive evaluations of nutrient regulation of algal growth and productivity, biogeochemistry of nutrients, and hydraulic characteristics of lakes permitted the evolution of effective models with some empirical predictive capabilities. These couplings between phosphorus loading and retention as developed by Piontelli and Tonolli (1964) and Vollenweider (1969), for example, were possible only because of extensive scientific underpinnings. That understanding allows the development of modeling and remedial or restorative programs to cope with individual characteristics of specific lakes and rivers.

Many decisions concerning the management and use of freshwater resources, however, are presently based on trial and error or correlative methods that may have little application to real-world conditions. Many management decisions are made by default because of lack of real information, based largely on assumed physical processes and chemical reactions from models or pure solution chemistry. The pivotal roles played by organisms and biological metabolism in these physical (e.g., altered heating and stratification cycles) and particularly chemical processes are still often held subservient or discarded as irrelevant in freshwater supply and quality management. Biogeochemical regulation of the metabolism, energy fluxes, and productivity of inland waters, all at the ecosystem level of integration, is relevant to all managerial procedures. Limnologists must be educated to understand and manage inland waters as multidimensional ecosystems, which can be done more effectively than in the past through enhanced foundational understanding from research into causes and control relationships and by application of this understanding to management of the integrated ecosystems. Both are essential, and limnologists should be versed in the basics of both.

FRAGMENTATION OF DISCIPLINES

The ecosystem perspective is uniformly recognized and espoused as essential to both research and training in aquatic ecology, but because of specialized training among most faculty, this approach requires integrated, interdisciplinary instruction. In a few cases, integrated formal educational programs are appropriate, particularly at the undergraduate levels to prepare students adequately for advanced training in aquatic sciences. The opposite conditions prevail, however, in nearly all American universities. Some claim to have comprehensive and integrated programs in limnology, but most exist largely on paper.

Based on a review of limnological programs at a variety of U.S. universities, I conclude that faculty in needed component subdisciplines (e.g., hydrology, aquatic chemistry, applied health, aquatic law, or engineering facets) may exist on campus, but most do not participate in training aquatic ecologists. Stated specialty courses, if taught, are disparate, offered infrequently, and almost always optional to aquatic ecology students as electives. Essential courses, such as hydrology for nonengineering majors or limnology for nonscience majors, are rarely offered; extant courses often have excessive requirements for an ecosystem-oriented program. Such specialized courses are necessary for advanced training, but they can inhibit interdisciplinary integration if no alternative courses at the general level are offered. True instructional interaction among faculty of different departments rarely exists in reality. Often members of the aquatic faculty of the same university never meet or interact because of ideological differences and the general time constraints in a very demanding profession.

RESEARCH SYNERGISM WITH EDUCATION

Strong educational programs for aquatic ecologists are most frequently associated with strong research programs, where students have opportunities for direct field and experimental involvement in problem solving.

Ecosystem-oriented programs are essential, but such instructional programs are rare owing to a number of synergistic factors.

Government support of faculty-student research programs in limnology is inadequate, given the value of freshwater resources and the crucial importance of basic and applied research and education to effective management of these resources (Lewis et al., 1995). The National Science Foundation (NSF) is the only significant agency that supports fundamental limnological research in academic institutions. With extremely severe competition for these limited funds, the probability of long-term support, which is required for ecosystem research, is low. Alternative funding sources without specific mission commitments are very few. The weak governmental support and budgetary anonymity of limnology have contributed to a decline in the independence and recognition of the discipline (Lewis et al., 1995). This support structure has also contributed to fragmentation of the discipline into specialized fields of inquiry and weakening of their interconnections. As a result, university researchers tend to conduct research in small, specialized areas in which specific results can be obtained relatively rapidly. Extreme competition for scarce resources also promotes isolation among faculty and researchers. Instruction by faculty tends to become specialized and insular, with minimal interdisciplinary collaboration.

In addition to the inadequate federal funding of limnological programs and the shift of fiscal responsibilities for higher education to state and internal sources, universities commonly encourage popular and relatively well-funded subdisciplines, such as molecular biology or medicine. A number of particularly strong research and instructional programs in aquatic ecology in major universities (Yale University, Indiana University, University of Washington, and others) have been terminated in the past decade. This decline is in sharp contrast to the marked increase in aquatic ecosystem programs in many other industrialized countries where the critical importance of research foundations to effective management of fresh waters is recognized. Strong instructional and research programs in limnology have emerged in Denmark, Sweden, and Norway, where the research and instructional liaisons between universities and environmental agencies are particularly vigorous and there is recognition of the importance of understanding the quantitative dynamics of controlling factors for effective management and restoration of freshwater ecosystems.

In the United States, several alternatives have emerged by default. Limnology has often languished in departments of biological sciences, which are too narrowly based, while emerging in departments of fisheries and schools of natural resources, where stream ecology and wetlands are more relevant. Aquatic chemistry and environmental hydraulics programs related to lakes and rivers have developed in engineering departments as natural extensions of those fields, but with minimal training in the importance of *bio*geochemistry to ecosystem functioning. Environmental resource programs have proliferated in geography and resource groups with minimal science underpinnings, but effective solutions of water resource problems require an understanding of metabolic constraints within ecosystems and the biogeochemical dependencies of ecosystem functioning. Such fragmentation frustrates students who wish to obtain essential interdisciplinary training and faculty who wish to communicate and collaborate in both instruction and research. Many small aquatic foci within a university also compete less effectively with larger, departmentally oriented programs for funding, positions, and program development.

Another development is an increase in the programs in ecology and environmental sciences in non-research-oriented colleges and universities. Often these programs develop in response to perceived needs for broadly trained individuals in environmental sciences, deficiencies of those programs in larger universities, and dedicated individual faculty members. Students of the better programs are involved in research projects, and dedicated instruction is common. Many of these programs, however, lack the necessary physical, chemical, and biological expertise. Often a single committed individual has developed an admirable but modest program without the programmatic resources required. A number of exceptional programs exist at undergraduate schools; these should be promoted and enhanced. In times of limited resources, dilution of resources at the expense of quality is unwise. (See discussion of a national initiative in general education below.)

OPTIMAL CRITERIA FOR LIMNOLOGICAL EDUCATION

Effective management of freshwater resources ultimately must be based on an in-depth understanding of the structure and physical, chemical, and biological mechanisms governing biotic development within lake, river, and wetland ecosystems. This understanding must be sufficiently detailed to encompass both the individualities of the ecosystems and the functional commonalities that prevail among them.

Limnological education should strive to train limnologists (1) with the critical scientific underpinnings required for understanding integrative ecosystem processes and (2) with sufficient understanding of ecosystem components to make effective managerial and regulatory decisions. These objectives are rarely accomplished in training programs.

Limnology students frequently are trained in general biology or environmental engineering, with specialized exposure to a course in general limnology and one or more courses in the biology of aquatic organisms (e.g., algae, aquatic insects). Limnology is usually taught as a brief lecture course, with no exposure to field conditions. Rarely are students more than superficially versed in ecology, quantitative statistics, the conditions of natural communities (particularly, for example, under ice cover or during high river flows), dominating irregular nonequilibrium conditions, growth and reproductive characteristics, environmental heterogeneity, etc. Dissertational research in graduate school, although often of excellent quality, is frequently narrow and laboratory oriented. Recently, a few vocal schools have advocated empirical correlational modeling in limnology with no appreciable understanding of causality or controlling variables.

A strong bias exists toward zoological aspects of limnology. Deeply rooted in historical and in some cases religious foundations,[1] limnology has been, and still is, taught primarily by biologists with zoological training and interests. The importance of consumers in determining the biomass, species composition, and production of prey is paramount among the principles governing aquatic food web structure. Size-selective predation by fish on zooplankton is among the most predictable community phenomena. Yet generally less than 10 to 20 percent of aquatic ecosystem energetics and regulation is associated with animals (Wetzel, 1995). The pivotal importance of organic matter produced by photosynthetic organisms both within the lake or river and within the drainage basin and imported to the water body, and of degradation, biogeochemical cycling, and energy fluxes, is markedly understudied and poorly taught. It is important that the enormous existing zoological information be integrated correctly into educational and research evaluations of ecosystem operations and regulation.

Integration at the ecosystem level is required of studies and teaching of system components. Limnology is a composite of physical, chemical, geological, and biological topics, and an integration among these subdisciplines is essential for the interdependent ecosystem perspective and effective management of inland aquatic ecosystems. Coupled research and teaching are essential to achieve this training.

---

[1]The premier position that animals have assumed in biological study, ecological research, and conceptual developments of ecology cannot be questioned. Historical roots of zoological dominance in aquatic ecological study and conceptual developments are varied and include the food and economic importance of fish and aquatic insects, the early relative ease of sampling and examination of population and community interactions of larger organisms, and—in part—the religion-inspired omnipotence of humans and other animals over other organisms, particularly plants and microbes. The idea of humans as supreme over the environment has prevailed in recent history, particularly in the schools of Goethe, Spencer, and Darwin. The behavioral characteristics of animals, fish as a protein source, and human biology related to medicine have contributed further to a greater emphasis on animals than on plants or microbiota and to weakened understanding of the couplings and interactive regulations at the ecosystem levels (cf. review of Wetzel, 1995).

A National Initiative: General Education

Educational programs in limnology should be redesigned and strengthened to achieve the breadth of the ecosystem perspective and to couple that perspective with prudent uses and management of freshwater resources. The public must be kept informed about the importance of ecosystem-oriented limnology to the wise management of inland waters, the essential characteristics of inland waters, and the value of these resources. Instruction in general limnology or aquatic ecology (not just biology) should be conducted at *every* institution of higher education, preferably by faculty with some interest and training in limnology. Training for professional limnologists (inland aquatic ecologists) obviously must be much more intensive and interdisciplinary. Most institutions, however, are not committed to the development of a program in limnology.

A National Initiative: Coordinated Schools of Limnology

There is an urgent need to properly train limnologists in the United States at both a research level and a practicing level. There are many viewpoints about how limnological training should be carried out. Present education in aquatic ecology suffers from inertia, and laissez-faire attitudes among faculty are common. It is my thesis that programs must be structured more rigorously than has been the case in the past and that research and practical training are best done simultaneously.

Several universities in the United States should make coordinated commitments, rigorously screened by a panel organized by the National Academy of Sciences with national scientific societies in aquatic ecology and supported by the federal government, to develop regional university-based schools of limnology (Wetzel, 1991). These schools would train both limnological practitioners and researchers from the undergraduate through the doctoral and postdoctoral levels. Excellence in the medical profession emanates from medical schools that both train practicing physicians and conduct basic research. Similarly, schools of limnology should train limnologists to function as effective diagnosticians and problem solvers and should also train professional researchers to conduct active research on the fundamentals of aquatic ecosystems. Just as in the medical profession, professional researchers and faculty would be very few in relation to the practitioners that are applying the results of research to practical problems. Professional researchers must demonstrate the capacity for continuing innovative contributions to the discipline.

Improved programs of instruction and training must be phased into the existing spectrum of largely biology and engineering programs. The proposed schools of limnology are designed to augment existing programs, not supplant them (Figure 1). Most of the existing educational routes, largely through departments of biological sciences, would continue their traditional programs in aquatic biology, water resources, fisheries management, etc. Freshwater resources are of such value to the economy and health of the country (Francko and Wetzel, 1983; Benke, 1990; Thornton et al., 1990; van der Leeden et al., 1990; Wetzel, 1992; Callow and Petts, 1993; Gleick, 1993; Rogers, 1993) that expanded training of limnological leaders to enhance the understanding and invigorate the management of fresh waters is greatly needed.

The interdisciplinary nature of limnology mandates that programs or schools of limnology consist of integrated instruction from disciplines not normally aggregated into a single department or even division. Rigorous

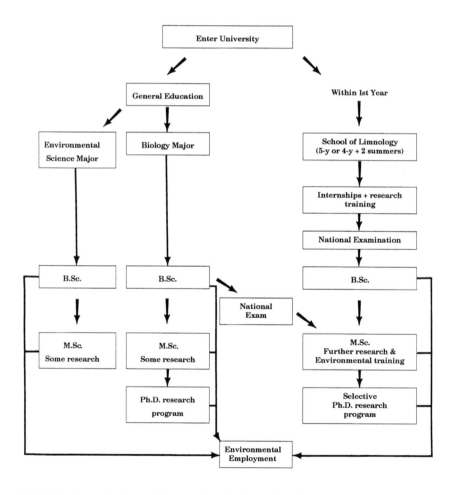

FIGURE 1   Example of possible educational tracks in limnology.

instructional programs are needed. Just as chemists must be versed in physical, inorganic, organic, and other facets of chemistry before specialization, limnologists should be required to know the basics of geomorphology, hydrology, aquatic inorganic and organic chemistry, biochemistry, biology from bacteria to fish, biostatistics, and other facets of limnology. Ideally, students would commit early to limnological training. A basic two-year curriculum in mathematics and science should be followed by upper-level courses that maximize understanding of inland aquatic ecosystems, their biota, biogeochemical cycling, and management (Box 1).

A rigorous program of instruction of this depth and thoroughness will require discipline and perseverance. Time is inadequate in a traditional four-year curriculum to include the necessary training and practical experience. Electives in liberal education are limited to the early phases of the curriculum, as is the case in nearly every structured professional program (e.g., engineering, medicine, nursing, business). Claims that limiting liberal arts electives would produce narrow-minded graduates are not substantiated. In contrast, there is abundant experience that graduates poorly trained in aquatic ecology are often functionally disadvantaged and require many years of expensive and inefficient on-the-job training before becoming moderately productive. That internal training commonly comes from individuals who received their training 15 to 25 years earlier, thus frequently perpetuating antiquated methods and understanding.

Options include a four-year program in which two or more full summers are devoted to internships with government environmental agencies, consulting firms, university research projects, and other training programs or a five-year curriculum similar to professional nursing programs. The fifth year would be relegated to "practicals," in which participants analyze problem ecological situations at an integrated ecosystem level. Limnological conditions and problems would be diagnosed, evaluated, and actually corrected where possible. Students would participate in field courses and be encouraged to gain experience in ongoing field experiments and analytical programs. In the final or fifth year, students would also be expected to interact extensively with graduate students and their research programs. Most universities in Europe require a mandatory examination, consisting of oral and written components, in order to receive the B.Sc. degree. Professional examinations (boards) are required on completion of the B.Sc. degree in many disciplines (e.g., nursing) in the United States in order to practice. Similar minimal national examination standards are essential if limnology is ever to develop into the rigorous professional discipline that freshwater resources deserve.

Graduate Programs

Most graduates of the schools of limnology at the B.Sc. and M.Sc. levels would become practitioners who apply their training in diagnostic and

BOX 1   OUTLINE OF REQUIREMENTS FOR A RIGOROUS UNIVERSITY LIMNOLOGY PROGRAM

I. Undergraduate

A. Mandatory subjects (one semester or equivalent unless noted)
   1. Mathematics through basic calculus and differential equations
   2. Statistics (see biology) (two semesters minimum)
   3. Physics (thermodynamics, heat, light)
   4. Chemistry
      a. Inorganic (two semesters)
      b. Organic
      c. Biochemistry
      d. Aquatic chemistry
   5. Geology/geomorphology
   6. Hydrology, with emphasis on ground water
   7. Biology
      a. General biology for majors (two semesters)
      b. Invertebrate biology, including aquatic invertebrates
      c. Algae
      d. Aquatic plants
      e. Genetics and Molecular biology
      f. Biostatistics (two semesters minimum)
         i. Parametric/nonparametric
         ii. Analyses of variance; problem applications
      g. Limnology (four-semester sequence)
         i. Physical/chemical limnology
         ii. Biological limnology
         iii. Lake/reservoir limnology practicum
         iv. River limnology practicum
      h. Microbiology
      i. Microbial ecology
      j. Ichthyology/fisheries management
      k. Limnological analyses (problem-solving practicum, two semesters)
      l. Ecology
      m. Ecosystem ecology

B. Highly recommended subjects
   1. Physical chemistry
   2. Advanced hydrology
   3. Landform analyses/geographic information systems
   4. Aquatic biotic productivity
   5. Wetland ecology
   6. Environmental law
   7. Public speaking

II. Graduate (M.Sc. and Ph.D.)

A. Course studies for the M.Sc. and 48 semester credits for the Ph.D. Minimum of 24 semester credits of advanced studies in addition to those cited in the Undergraduate curriculum (above).[1] Courses should emphasize advanced integrated treatment of the ecology, biogeochemistry, evolution, and systematics of freshwater ecosystems.

B. Dissertational research and thesis composition, in which independent thought and research capacity is evidenced. Research must be publishable in refereed scientific literature. Research for the M.Sc. degree should serve as an evaluation period to determine reasonable qualifications for Ph.D. studies. Often the research will serve as pilot studies for research of greater depth in the Ph.D. program.

---

[1] Bypassing the M.Sc. with direct entrance into the Ph.D. program is not recommended. The M.Sc. degree should serve as an evaluation period in advanced study and research abilities, from which only the best qualified are encouraged to obtain the Ph.D.

advisory capacities. Only a small percentage would enter the advanced research training stages toward higher degrees. Entrance into the advanced degree program in a school of limnology also must be rigorous. For example, students with a basic undergraduate degree from another university would be required to fulfill the minimal standards of the undergraduate program in the school of limnology before being able to participate in the graduate program within that school (see Figure 1). There should be a screening entrance examination in ecology, as well as demonstration of acceptable Graduate Record Examination grades, prior to admission.

The master's degree should serve several purposes simultaneously. A primary objective is to gain additional experience in limnology concerned with the regulation of biotic growth and productivity, biogeochemical dynamics, and other community processes in inland waters. Small independent research projects are mandatory to gain experience in the design and execution of research and the communication of results to scientific peers. The fundamental final step of research is written communication of those results to scientific peers. Although research conducted at the M.Sc. level is often preliminary and designed largely for training purposes, summaries of the results must be published, at least at the regional level. Research conducted at the Ph.D. level must provide advances to fundamental knowledge of the discipline. Results that are not publishable in refereed journals are not satisfactory for the Ph.D. degree.

In both the master's and the doctoral degree programs, a number of advanced speciality courses should be mandatory to give greater depth and experience to the students. Students could select a number of different areas of specialization. Particularly to be encouraged are interdisciplinary areas such as wetland ecology, aquatic environmental law, and ground water pollution. In schools of limnology, however, it is essential that a uniform and rigorous undergraduate training be acquired. The weaknesses and wide disparities in undergraduate training place such a burden on graduate programs that corrections are not made well and the quality of graduate training and research is often compromised. Field experience is not only desirable at the undergraduate level, but should be mandatory. The most effective research programs couple in situ analyses with the rigor of controlled experimentation in both the laboratory and the field. In all cases, undergraduate students should be incorporated into research programs wherever possible, both to give them experience and to gain their fresh insights on problems, environmental circumstances, and management.

Graduate instruction and an active research program, preferably at the interdisciplinary ecosystem level, are critical to effective professional instruction. The faculty and associated research scientists provide the essential personnel and milieu for dynamic teaching at all levels. Conversely, the fresh insights and perceptions of students are critical to advances in basic research. The collective integration of teaching and research at all levels, undergraduate through postdoctoral, is the most effective means of increasing fundamental understanding of aquatic ecosystems. Several named schools of limnology should be developed nationally to the minimal standards suggested here, and preferably some or most of them should be in the nonglaciated regions of the United States, where most of the population resides but which are least understood limnologically. Limnology is a profession, and professionally trained practitioners are needed in nearly every county of every state. Several enlightened European countries have professionally trained limnologists in every county assisting with resource decisions. In Sweden, for example, nearly every county has at least one Ph.D.-level limnologist who works with resource planners and management specialists to assist in science-based decisions. Progress in management depends upon acquiring fundamental understanding of aquatic ecosystems. Research advancement and its application are interdependent and self-reinforcing.

Coordination

Some coordination of limnological programs would be desirable to ensure minimal standards. Standards should be initiated and administered through an independent overseeing group, perhaps coordinated by limnological sections of the American Society of Limnology and Oceanography, the Ecological Society of America, the North American Benthological Society (containing most stream or river limnologists), the North American Lake Management Society, and possibly certain specialty groups such as the Society of Wetland Scientists.

The freshwater resources of the United States are of such major value to the economy and human health of the country that it is simple economic prudence to collectively support training of effective diagnosticians and problem-solving managers of these critical resources. The initial phases of the development of regional schools of limnology in rigorously selected universities should be subsidized by the federal government, probably through the auspices of the National Science Foundation. Once several programs are in place and operating, they should become self-sustaining by means of direct support and other subventions.

At least four regional schools of limnology should be established initially. Instructional programs should be unified at functional levels. The current clustering of limnological training in glaciated lakes regions, however, is a historical artifact that should be addressed, since there are great needs in nonglaciated regions. Further neglect is not only unrealistic but unwise economically. Training should include river and reservoir characteristics and problems of the Southeast and Southwest; surface and ground water resources of the Great Plains; and the traditional lake analyses of the alpine and northern regions of the United States.

The universities that develop such schools of limnology must emphasize instructional objectives in both undergraduate and graduate programs and must recognize that teaching and research are synergistic and self-reinforcing. In some cases, specific courses must be designed for the program. For example, basic training in hydrology is essential for all limnologists. Extant courses in hydrology in many universities are taught in upper levels of engineering programs and have several years of advanced mathematics as prerequisites, thus potentially excluding biological limnologists. Yet the necessary basics of hydrology can and should be taught for the program students who are not specializing in hydrology. Other interdisciplinary courses requiring careful development include biogeochemistry, aquatic chemistry, wetland ecology, and possibly environmental economics and law. Such courses will require directed efforts and certain nontraditional faculty coordinations. In some cases, new faculty may be required to accommodate increased instructional loads. Name changes of traditional courses simply will not satisfy the demands of an innovative interdisciplinary program.

The importance of problem-solving and research components within the training program and the feedback mechanisms between practitioners and researchers cannot be overemphasized. An analogous situation often occurs in the best of medical and dental schools as the problems of practitioners and the developmental advances of researchers are exchanged, applied, and used to rapidly advance the disciplines.

Recruiting and Placement Programs

The need is acute to inform potential students about the importance of freshwater sciences. Designated limnologists should interact with potential candidates at the high school, college, and university levels to explain freshwater resources, their management, and the scientific understanding that is required to correctly shepherd fresh waters and guide their sustainable use. Public relations efforts concerning limnology as a profession are necessary. Although faculty need not necessarily perform these liaisons with potential students, their counsel to the messages being related by point persons is essential. Students of all backgrounds, including women and minorities, will probably be more attracted to limnology if they are contacted as undergraduates, when they can still meet rigorous programmatic needs. These problems must be addressed as an essential aspect of program development.

Graduates at all levels from these innovative and directed programs should have no problems finding jobs. The excellence of their training and the importance of that training for subsequent employment should be used in recruitment. Concerted advertising efforts are necessary, however, for schools of limnology to promote their graduates at all levels (B.Sc., M.Sc., and Ph.D.) as among the best-trained and qualified for positions in local, state, and federal agencies; consulting companies; industry; and even within large municipalities. An active placement office should evaluate hiring practices of agencies, private and government laboratories, and universities and should inform potential employers of graduates' directed training.

ACKNOWLEDGMENTS

The author acknowledges with appreciation the extensive discussions of this subject with many colleagues, in particular Gordon L. Godshalk, Gene E. Likens, Peter H. Rich, and Amelia K. Ward. Most helpful constructive comments on an earlier draft of this manuscript were given by Patrick L. Brezonik, Eville Gorham, Jacqueline A. MacDonald, G. W. Minshall, and Dean B. Premo.

REFERENCES

Benke, A. C. 1990. A perspective on America's vanishing streams. J. N. Am. Benthol. Soc. 9:77–88.

Birge, E. A., and C. Juday. 1934. Particulate and dissolved organic matter in inland lakes. Ecol. Monogr. 4:440–474.

Callow, P., and G. E. Petts, eds. 1993. Rivers Handbook, Vol. 2. Oxford: Blackwell Scientific Publications.

Chapra, S. C., and K. H. Reckhow. 1983. Engineering Approaches for Lake Management, Vol. 2: Mechanistic Modeling. Woburn, Mass.: Butterworth.

Cooke, G. D., E. B. Welch, S. A. Peterson, and P. R. Newroth. 1993. Restoration and Management of Lakes and Reservoirs, 2nd ed. Boca Raton, Fla. : Lewis Publication. 548 pp.

Francko, D. A., and R. G. Wetzel. 1983. To Quench Our Thirst: Present and Future Freshwater Resources of the United States. Ann Arbor: University of Michigan Press. 148 pp.

Gleick, P. H., ed. 1993. Water in Crisis: A Guide to the World's Freshwater Resources. Oxford: Oxford University Press.

Lewis, W. M., Jr., S. Chisholm, C. D'Elia, E. Fee, N. G. Hairston, Jr., J. Hobbie, G. E. Likens, S. Threlkeld, and R. G. Wetzel. 1995. Challenges for limnology in North America: An assessment of the discipline in the 1990s. Bull. Am. Soc. Limnol. Oceanogr. 4(2):1–20.

Likens, G. E. 1984. Beyond the shoreline: A watershed-ecosystem approach. Verh. Internat. Verein. Limnol. 22:1–22.

Piontelli, R., and V. Tonolli. 1964. Il tempo di residenza delle acque lacustri in relazione ai fenomeni di arricchimento in sostanze immesse, con particolare riguardo al Lago Maggiore. Mem. Ist. Ital. Idrobiol. 17:247–266.

Rogers, P. 1993. America's Water. Cambridge, Mass.: MIT Press.

Thornton, K. W., B. L. Kimmel, and F. E. Payne, eds. 1990. Reservoir Limnology: Ecological Perspectives. New York: John Wiley & Sons.

van der Leeden, F., F. L. Troise, and D. K. Todd. 1990. The Water Encyclopedia. 2nd ed. Chelsea, Mich.: Lewis Publication.

Vollenweider, R. A. 1969. Möglichkeiten und Grenzen elementarer Modelle der Stoffbilanz von Seen. Arch. Hydrobiol. 66:1–36.

Wetzel, R. G. 1990a. Land-water interfaces: Metabolic and limnological regulators. Verh. Int. Verein. Limnol. 24:6–24.

Wetzel, R. G. 1990b. Reservoir ecosystems: Conclusions and speculations. Pp. 227–238 in Reservoir Limnology: Ecological Perspectives, K. W. Thornton, B. L. Kimmel, and F. E. Payne, eds. New York: John Wiley & Sons.

Wetzel, R. G. 1991. On the teaching of limnology: Need for a national initiative. Limnol. Oceanogr. 36:213–215.

Wetzel, R. G. 1992. Clean water: A fading resource. Hydrobiologia 243/244:21–30.

Wetzel, R. G. 1995. Death, detritus, and energy flow in aquatic ecosystems. Freshwater Biol. 33:83–89.

Wetzel, R. G., and A. K. Ward. 1992. Primary production. Pp. 354–369 in Rivers Handbook, Vol. 1, P. Calow, and G. E. Petts, eds. Oxford: Blackwell Scientific Publications.

Wetlands: An Essential Component of Curricula in Limnology

Eville Gorham
Department of Ecology, Evolution, and Behavior
University of Minnesota, St. Paul
St. Paul, Minnesota

SUMMARY

Wetlands have long received little attention in traditional limnology courses. Yet they are a critical resource, providing habitat for important species, significant links in the cycling of nutrients and the global storage of carbon, buffering against pollutants, and other services. Limnology courses should, therefore, be broadened to cover wetlands more thoroughly. Students should be taught how wetlands are defined, categorized, and distributed locally and globally; their patterns of development; their ecological and biogeochemical functions; their values to human society; the causes of wetland degradation and destruction; concepts and techniques in wetland restoration and creation; and issues in wetland management. They should also learn about key research areas in wetland science, dealing in particular with concerns about their future in the face of increasing human disturbance.

INTRODUCTION

Wetlands are waterlogged landscapes that cover approximately 8.6×10^6 km², which amounts to 6.4 percent of the world's land surface (Mitsch and Gosselink, 1993). They cover, therefore, a much greater area than the 1.2×10^6 km² of freshwater lakes and the 0.8×10^6 km² of saline lakes (Shiklomanov, 1993). A little less than half of wetland area is peatland, with deposits of organic detritus more than 30 cm deep. About 2 percent of the global wetland area is located in polar regions, 30 percent in the boreal zone, 12 percent in the subboreal, 25 percent in the subtropics, and 31 percent in the tropics. In the conterminous United States, wetlands cover about 5 percent of land area.

Wetlands may range in size from wet hollows a few meters square to the vast peatlands of the west Siberian Plain (Neishstadt, 1977; Walter, 1977) and the Hudson/James Bay Lowland (Wickware et al., 1980; Pala and Weischet, 1982), which between them cover more than 10^6 km^2. They occur in a great diversity of types, which can be grouped into the seven major categories listed in Table 1 (Mitsch and Gosselink, 1993). Each category can be divided into two or more subcategories. Northern peatlands, for instance, are usually divided into circumneutral fens and strongly acid bogs (Gorham and Janssens, 1992a). In fens, the bicarbonate buffer system regulates pH, whereas in bogs, colored organic acids buffer the aqueous system at a pH of about 4.

Despite the significance of wetland ecosystems, which cover 26 percent more area in the conterminous United States than do fresh and saline lakes, reservoirs, rivers, and bays (Frayer, 1991), they seldom receive much attention in curricula designed for limnologists. In six familiar limnology textbooks (Table 2), wetlands account for a little more than 1 percent of the textual material.

This paper summarizes material that might be included in a limnology course encompassing wetland ecosystems (see also Gore, 1983; Mitsch and Gosselink, 1993).

DESCRIPTION OF WETLANDS

Wetlands are waterlogged landscape features and often develop at the margins of rivers and lakes. In the latter case the lakes may eventually—through the deposition of silt and peat—become converted wholly to wetland. The largest wetlands, however, generally form on very flat terrain in which damp mineral soils are invaded in their wettest parts by peat-forming vegetation. Peat impedes and dams up the natural drainage and brings about an expansion of the waterlogged area, eventually swamping large areas of upland forest. Very large peatlands often develop intricate and beautiful landscape patterns, which represent perhaps the most delicate mutual interaction between hydrology and vegetation on the surface of the earth (Sjörs, 1961; Heinselman, 1963; Wright et al., 1992). Ground water upwelling is a major factor in the development of these

TABLE 1   Major Types of Wetlands

Inland	Coastal
Freshwater	Tidal salt marshes
Riparian wetlands	Tidal freshwater marshes
Southern deepwater swamps	Mangrove swamps
Northern peatlands	

TABLE 2   Coverage of Streams and Wetlands in Six Familiar
Limnology Texts

Author	Date	Title	Total Text	Streams	Wetlands
Welch	1935	*Limnology*	394	30	4
Hutchinson	1957	*A Treatise on Limnology, V.1.*	902	0	0
Ruttner	1963	*Fundamentals of Limnology, 3rd ed.*	249	22	6
Cole	1983	*Textbook of Limnology, 3rd ed.*	344	22	0
Wetzel	1983	*Limnology, 2nd ed.*	753	0	8
Horne and Goldman	1994	*Limnology, 2nd ed.*	520	51	24
Total			3,162	125	41
Percentage of total			100	4.0	1.3

NOTE: The table excludes occasional brief mentions of streams and wetlands in material on lakes.

patterns (Siegel, 1983, 1988; Glaser et al., 1990) and is likely to be influenced strongly by global warming. The importance of hydrology in wetland studies has long been understood (Ingram, 1983), and the physiographic and climatic factors that influence it are becoming increasingly clear (Winter, 1992).

Wetlands have their own distinct flora and fauna and are associated with about one-third of the rare, threatened, and endangered species in the United States (Niering, 1988). Among the showy or unusual plants of the wetlands are diverse carnivorous plants (sundews, pitcher plants, Venus flytraps, and bladderworts), numerous kinds of orchids, and the bog moss *Sphagnum*. Animals include a variety of invertebrates, among the most noticeable being the butterflies, and vertebrates as diverse as lemmings, moose, and alligators. Many invertebrate groups are poorly known, particularly among the insects (Danks and Rosenberg 1987), and await detailed taxonomic study.

PATTERNS OF DEVELOPMENT

Peat deposits provide a long-term archive of landscape dynamics, which can be complex and exhibit a great variety of pathways (e.g., Tallis, 1983; Janssens et al., 1992). Fossil assemblages, as well as physical and

chemical data, analyzed in dated peat cores allow reconstruction of historical records of the successional development of the deposit, as well as of
past levels of acidity and watertable depth, that can serve as baselines
against which to assess recent and current impacts of human activity
(Gorham and Janssens, 1992b). Pathways of development are driven both
by external environmental factors and by the internal dynamics of the
system itself.

ECOLOGICAL AND BIOGEOCHEMICAL FUNCTIONS

Wetland productivity ranges from a few hundred grams of dry matter
per square meter per year in many peatlands of the northern boreal zone
to thousands of grams in coastal salt marshes and inland freshwater
marshes (Mitsch and Gosselink, 1993). In northern peatlands, about 8
percent of the 296 grams of carbon fixed annually from the atmosphere
is preserved as peat, which forms a carbon pool of 412×10^{15} g (Woodwell
et al., 1995). The average annual accumulation rate over the postglacial
period is currently thought to be 96×10^{12} g per year, or 230 kg per
hectare per year (Gorham, 1995). About 7 percent of carbon in northern
peatlands is exported in streams as dissolved organic carbon, with the
potential to acidify downstream waters (Gorham et al., 1986; Kerekes et
al., 1986), and about 1.4 percent is emitted to the atmosphere as methane
(Gorham, 1995). The rest, 84 percent, returns to the atmosphere as carbon dioxide.

Wetlands are important in the cycling of nitrogen and sulfur to the
atmosphere because their anoxic soils and peats are habitats well suited
to microbes capable of reducing nitrate and sulfate. Denitrification occurs
readily in circumneutral waterlogged soils (Kaplan et al., 1979) but less
so in acid oligotrophic peatlands (Gorham, 1995). Nitrogen fixation takes
place in a variety of ways, more particularly in circumneutral wetlands.
Shrubs such as *Alnus* and *Myrica* have actinomycete nodules on their
roots, and cyanobacteria (blue-green algae) are often important in marshes
with standing water (Dickinson, 1984; Mitsch and Gosselink, 1993). Sulfate
reduction in waterlogged soils results in the emission of volatile sulfur
gases, but relatively few measurements have been made (Castro and
Dierberg, 1987; Faulkner and Richardson, 1989). Northern peatlands can
be important sinks for nitrogen and sulfur, with a nitrogen stock of about
16×10^{15} g and a sulfur stock of about 1.3×10^{15} g (Gorham, 1991). Their
average accumulation rates during the postglacial period are estimated
at 10 and 0.83 kg per hectare per year, respectively.

Wetlands are extremely efficient sinks for a diverse array of chlorinated
hydrocarbons, including DDT, toxaphene, and PCBs (polychlorinated
biphenyls; Rapaport and Eisenreich, 1988). Mercury, on the other hand,

is more mobile. St. Louis et al. (1994) report that despite relatively low total yields of mercury in waters flowing from catchments dominated by wetlands, yields of methylmercury were relatively high compared to yields from upland catchments. A purely upland catchment appeared to retain or demethylate methylmercury, whereas catchments containing wetlands were net producers of methylmercury. Rudd (1995) has reviewed the sources of methylmercury to freshwater ecosystems.

HUMAN VALUES

In recent times wetlands—whether pristine or not—have been seen to have a broad array of values to human society (reviewed by Greeson et al., 1979, and Sharitz and Gibbons, 1989; see also Richardson, 1994). Among the physical values are such properties as shoreline stabilization, flood-peak reduction, and ground water recharge. On the other hand, peatlands can also serve as physical barriers to human movement, as they do in northern Canada and Russia and as they did for Napoleon in his Russian campaign of 1812 (Chandler, 1966). Environmental problems of transport, pipeline building, and waste disposal have been described by Radforth and Brawner (1977).

Wetlands function chemically to improve water quality as filters, transformers, and sinks for materials delivered to them by human activities. For instance, they can filter 60-90 percent of suspended solids from added wastewater and as much as 80 percent of sediment in runoff from agricultural fields (Richardson, 1989). Wetlands to which nitrogen fertilizer is added can transform it by microbial denitrification to gaseous molecular nitrogen and nitrous oxide at rates substantially higher than those in similar unfertilized wetlands (Martikainen et al., 1992). As mentioned earlier, wetlands are sinks for carbon, sulfur, and nitrogen, and they retain in their peat deposits or volatilize a good deal of these materials when added from human by-products such as acid deposition (Bayley et al., 1987), agricultural runoff, and sewage wastewater (Richardson, 1989).

Wetlands also provide a variety of biological benefits to human society. In the natural state, these include substantial forest resources and minor agricultural resources such as marsh hay, wild rice, and cranberries. After drainage, both forestry and agriculture are possible on a broader scale. Waterfowl and furbearers such as muskrat, beaver, and mink inhabit many natural wetlands, and many riparian wetlands serve as protective nurseries and food sources for young fish prior to moving out into open water.

Aesthetic values of wetlands include unique plants, animals, and scenery. The patterned nature of many large northern peatlands as seen from the air also provides an aesthetically as well as scientifically valuable experience (see, for example, the illustrations in Wright et al., 1992). In addition, wetlands are often the least disturbed among urban ecosystems and provide locally valuable aesthetic and educational experiences for city dwellers. The values of wetlands to scientists interested particularly in ecology and biogeochemistry should not be ignored. As noted earlier in connection with ecological functions, wetlands are important on both local and global scales in the cycles of carbon, nitrogen, and sulfur (Deevey, 1970, 1973), and their patterned development—often driven by upwelling ground water (Siegel, 1983, 1988; Glaser et al., 1990)—is a matter of intense interest to peatland ecologists (Janssens et al., 1992).

DEGRADATION AND DESTRUCTION OF WETLANDS

Until recently, wetlands have been regarded generally as wastelands fit for nothing until "reclaimed" by human manipulation (Patrick, 1994). The largest human uses of wetlands are for forestry and agriculture (Kivinen, 1980). For northern peatlands the percentage of use varies from as high as 97 percent in Denmark and 94 percent in Poland to as low as 1 percent in Canada and 9 percent in the former Soviet Union, the two countries possessing the largest areas of peatland in the world.

Wetland protection is very limited. Usually, at most, a few percent of wetlands are protected, New Zealand being a striking exception with 33 percent (Kivinen and Pakarinen, 1980). In the contiguous United States, 54 percent of the original wetlands had been lost by the 1970s (Frayer, 1991), primarily to agriculture. In some states, such as California, Ohio, and Iowa, less than 10 percent of the original wetlands remains, with consequent losses of wildfowl, furbearers, and fish. In the decade from the 1970s to the 1980s, the United States lost 2.6 percent of its wetlands (Frayer, 1991).

Fuel mining is an important use for peat in northern countries, particularly in Russia and to a lesser degree in Ireland and Finland (Kivinen, 1980). Peat has scarcely been exploited in North America, although after the first "oil shock," consideration was given to the possibilities of converting peat in northern Minnesota to natural gas. Some peat has also been used as a chemical feedstock (Botch and Masing, 1984).

Urban development has led to the dredging and filling of many wetlands in urban areas. Others have been flooded or drained by highway construction with inadequate provision for culvert drainage. Stormwater and wastewater diversion to wetlands (Environmental Protection Agency, 1984) has added both nutrients and toxins, the former especially leading to major alterations of flora and fauna. Air pollution has also affected many wetlands, in particular through acidification by sulfuric acid and the addition of nitrate as a significant plant nutrient (Gorham et al., 1984). The subject of wetland ecotoxicology has been reviewed by Catallo (1993).

Biochemical indicators of environmental stress in wetland plants, whether natural or anthropogenic, have been investigated recently by Mendelssohn and McKee (1992).

Human activities have led to the decline or extinction of certain wetland plants and animals; as mentioned earlier, wetlands are habitats for a disproportionate share of rare, threatened, or endangered species. Humans have also facilitated the spread of exotic species into wetlands, a notable current example being the invasion of many North American wetlands by Eurasian purple loosestrife (*Lythrum salicaria*) with consequent loss of wildlife habitat (Thompson et al., 1987).

WETLAND RESTORATION

According to the National Research Council (NRC, 1992), restoration should involve the return of a given ecosystem to a state approximating that in which it existed prior to disturbances; the NRC states that, "the goal is to emulate a natural, functioning, self-regulating system that is integrated with the ecological landscape in which it occurs."

As presently practiced, wetland restoration is a very imperfect science (Mitsch and Gosselink, 1993). The conservation reserve program in the United States has encouraged farmers to restore wetlands previously drained for crop production, but the degree to which they approximate their former condition is largely unknown. Restoration of forested wetlands is difficult, given the long lives of trees. Early successional wetlands (e.g., cattail marshes) are the best candidates for restoration, whereas restoration of the large patterned peatlands of the boreal zone—developed over millennia of interactions of local and regional hydrology (Siegel, 1988)—must be regarded as impossible once they are seriously degraded. In the intermediate case of natural prairie wetlands, Galatowitsch and van der Valk (in press) have observed that the efficient-community hypothesis—"all plant species that can become established and survive under the environmental conditions found at a site will eventually be found growing there and/or will be found in its seed bank"—cannot be accepted completely as a basis for restoration, particularly for the species of sedge meadows.

WETLAND CREATION

New wetlands are often constructed in the United States to replace nearby wetlands destroyed by development, following rules laid down in Section 404 of the Clean Water Act (Kusler and Kentula, 1989). The process is described as mitigation and is intended to replace lost wetland functions in the landscape. Unfortunately, there is often no record of what the destroyed wetland was like, and there is seldom any follow-up evaluation of created wetlands, so that the success of the mitigation effort cannot be evaluated (Erwin, 1989; K. Bettendorf, personal communication, 1993). Erwin (1991), in a rare follow-up study, reports a very limited record of success for wetland mitigation in South Florida. According to Bedford (in press), reviews of hundreds of mitigation projects have demonstrated that hydrology has not been given adequate attention, nor has the reestablishment of specific types of plant communities. According to D'Avanzo (1989), efforts at wetland creation by the U.S. Army Corps of Engineers have been among the most successful.

The current U.S. policy of "no net loss" of wetlands has stimulated their creation as an effort at mitigation where wetlands have been destroyed. As discussed above, this can be successful where early successional wetlands are replaced. However, replacing mature wetlands that have taken centuries or millennia to develop with cattails goes against the no-net-loss concept because it results in a major shift in biodiversity and ecological function. As argued cogently by Bedford (1996), mitigation should be evaluated on the basis of hydrologic equivalence on a landscape scale.

WETLAND MANAGEMENT

Management of wetlands (Mitsch and Gosselink, 1993) may involve a wide range of practices, including draining and converting them to agriculture or forestry, using them for wastewater processing, "improving" them for enhancement of wildlife for hunters and trappers, or maintaining them in a relatively undisturbed state. For all these purposes, but particularly for the last, a major legal objective is to delineate exactly the boundaries of wetlands. Guidelines to assist such delineation have been prepared by federal agencies; they include characteristics of wetland hydrology, vegetation, and soils, along with a variety of field indicators. Short courses on wetland delineation are now widely available and are used in particular by consulting firms engaged in the process.

Ecosystem management is, of course, closely tied to assessment of ecological risk and to the consequent determination of public policy (Pastor, 1995). It also requires, for its basis, a sound program of basic research; see, for example, Mendelssohn and McKee (1989). Whether such management is sustainable in the long run remains uncertain (Stanley, 1995).

MAJOR QUESTIONS CONCERNING WETLANDS

For all wetlands, major questions include the following:

1. What have been their patterns of development in time and space?
2. To what degree has that development been controlled by environmental factors, hydrology in particular (Siegel, 1988; Winter, 1992), and

to what degree has it been a consequence of autogenic processes such as peat accumulation?

3. Do changes in the environment generally cause slow and steady changes in wetlands, or do threshold phenomena dictate relatively sudden responses to environmental stresses that build up over time (see Gorham and Janssens, 1992b)?

For areas of extensive peatland, which constitute a global carbon pool of about 412×10^{15} g as compared to about 694×10^{15} g in plants, 1,600 $\times 10^{15}$ g in soils (including peat), and more than 700×10^{15} g in the atmosphere (Woodwell et al., 1995), the following are major questions:

1. How has this pool accumulated over time and space?
2. What environmental factors have controlled its accumulation?
3. What is the role of these peatlands in the global carbon cycle, particularly with regard to the greenhouse gases carbon dioxide and methane (Billings, 1987; Gorham, 1995)?

For areas of extensive peatland, future concerns center around what may happen to the very large pool of carbon locked up in peat as global warming occurs:

1. Will falling water tables lead to oxidation of surface peats, large emissions of carbon dioxide to the atmosphere, and a strong positive feedback to global warming—particularly in the southern boreal zone where fire frequency will increase considerably and peatland fires may smolder for years in remote areas?

2. Will such emissions of carbon dioxide be offset (or more than offset) by a shutdown of methane emissions from previously waterlogged anoxic peats?

3. How will the melting of permafrost in the north temperate zone affect the carbon cycle in its peatlands (i.e., how much of the landscape will be flooded and how much will be dried out by runoff to the ocean)?

4. Can northward migration of peat-forming plant communities into the Arctic, as global warming provides suitable conditions there, take place as rapidly as peatland degradation along the southern boundary of the boreal zone (Gorham, 1991, 1994, 1995)?

For smaller wetlands, major questions for the future include the following:

1. What will be the effects of global warming (Poiani and Johnson, 1991), acid rain (Gorham et al., 1984), and nutrient enrichment by atmospheric nitrogen deposition upon them.

2. Can the pace of wetland drainage for human activities (agriculture, forestry, urban development, etc.) be slowed sufficiently to allow a reasonable degree of wetland protection and preservation?

3. Can techniques of mitigation, restoration, and wetland creation (Kusler and Kentula, 1989) be placed on a firm scientific footing to prevent further degradation, rehabilitate damaged wetlands, and replace those that must inevitably be destroyed?

ACKNOWLEDGMENT

This review owes much to the textbook of Mitsch and Gosselink (1993), which should be consulted for further details on most of these topics. I thank also the many colleagues who have sent me, for the past 45 years, reprints of their articles on wetland ecology and biogeochemisty.

REFERENCES

Bayley, S. E., D. H. Vitt, R. W. Newbury, K. G. Beaty, R. Behr, and C. Miller. 1987. Experimental acidification of a *Sphagnum*-dominated peatland: First year results. Can. J. Fish. Aquat. Sci. 44(Supplement 1):194–205.

Bedford, B. L. 1996. The need to define hydrologic equivalence at the landscape scale for freshwater wetland mitigation. Ecol. Appl. 6:57–68.

Billings, W. D. 1987. Carbon balance of Alaskan tundra and taiga ecosystems: Past, present, and future. Quaternary Sci. Rev. 6:165–177.

Botch, M. S., and V. V. Masing. 1984. Mire ecosystems in the U.S.S.R. Pp. 95–152 in Ecosystems of the World, 4B, Mires: Swamp, Bog, Fen and Moor, Regional Studies, A. J. P. Gore, ed. Amsterdam: Elsevier.

Castro, M. S., and F. E. Dierberg. 1987. Biogenic hydrogen sulfide emissions from selected Florida wetlands. Water Air Soil Pollut. 33:1–3.

Catallo, W. J. 1993. Ecotoxicology and wetland ecosystems: current understanding and future needs. Environ. Toxicol. Chem. 12:2209–2224.

Chandler, D. G. 1966. The campaigns of Napoleon. New York: Macmillan.

Cole, G. A. 1983. Textbook of Limnology, Third Edition. St. Louis: Mosby.

Danks, H. V., and D. M. Rosenberg. 1987. Aquatic insects of peatlands and marshes in Canada: Synthesis of information and identification of needs for research. Mem. Entomol. Soc. Can. 140:163–174.

D'Avanzo, C. 1989. Long-term evaluation of wetland creation projects. Pp. 75–84 in Wetland Creation and Restoration: The Status of the Science, vol. 2, J. A. Kusler and M. E. Kentula, eds. Report 600/3-89/038. Corvallis, Ore.: Environmental Protection Agency, Environmental Research Laboratory.

Deevey, E. S. 1970. In defense of mud. Bull. Ecol. Soc. Am. 51:5–8.

Deevey, E. S. 1973. Sulfur, nitrogen and carbon in the biosphere. Pp. 182–190 in Carbon and the Biosphere, G. M. Woodwell and E. V. Pecan, eds. CONF-720510. Washington, D.C.: U.S. Atomic Energy Commission, Technical Information Center, Office of Information Services.

Dickinson, C. H. 1984. Micro-organisms in peatlands. Pp. 225–245 in Ecosystems of the World, 4A, Mires: Swamp, Bog, Fen and Moor, Regional Studies, A. J. P. Gore, ed. Amsterdam: Elsevier.

Environmental Protection Agency (EPA). 1984. The Ecological Impacts of Wastewater on Wetlands: An Annoted Bibliography. Report 905/3-84-002. Chicago: EPA.

Erwin, K. L. 1991. An Evaluation of Wetland Mitigation in the South Florida Management District, Vol. 1. West Palm Beach: South Florida Management District.

Faulkner, S. P., and C. J. Richardson. 1989. Physical and chemical characteristics of freshwater wetland soils. Pp. 41–72 in Constructed Wetlands for Wastewater Treatment, D. A. Hammer, ed. Chelsea, Mich.: Lewis Publishers.

Frayer, W. E. 1991. Status and Trends of Wetlands and Deepwater Habitats in the Contermi- nous United States, 1970s to 1980s. Houghton, Mich.: Michigan Technological Univer- sity.

Galatowitsch, S. M., and A. G. van der Valk. 1996. The vegetation of restored and natural prairie wetlands. Ecol. Appl 6:102–112.

Glaser, P. H., J. A. Janssens, and D. I. Siegel. 1990. The response of vegetation to chemical and hydrological gradients in the Lost River Peatland, northern Minnesota. J. Ecol. 78:1021–1048.

Gore, A. J. P., ed. 1983. Ecosystems of the World, 4A, Mires: Swamp, Bog, Fen and Moor. Amsterdam: Elsevier.

Gorham, E. 1991. Human influences on the health of northern peatlands. Trans. R. Soc. Can. Ser. VI 2:199–208.

Gorham, E. 1994. The future of research in Canadian peatlands: A brief survey with particular reference to global change. Wetlands 14:206–215.

Gorham, E. 1995. The biogeochemistry of northern peatlands and its possible responses to global warming. Pp. 169–187 in Biotic Feedbacks in the Global Climate System, G. M. Woodwell and F. T. Mackenzie, eds. New York: Oxford University Press.

Gorham, E., and J. A. Janssens. 1992a. Concepts of fen and bog re-examined in relation to bryophyte cover and the acidity of surface waters. Acta Soc. Bot. Pol. 61:7–20.

Gorham, E., and J. A. Janssens. 1992b. The paleorecord of geochemistry and hydrology in northern peatlands and its relation to global change. Suo 43:9–19.

Gorham, E., S. E. Bayley, and D. W. Schindler. 1984. Ecological effects of acid deposition upon peatlands: a neglected field in "acid-rain" research. Can. J. Fish. Aquat. Sci. 41:1256–1258.

Gorham, E., J. K. Underwood, F. B. Martin, and J. G. Ogden III. 1986. Natural and anthropo- genic causes of lake acidification in Nova Scotia. Nature 324:451–453.

Greeson, P. E., J. R. Clark, and J. E. Clark, eds. 1979. Wetland Values and Functions: The State of Our Understanding. Technical Publication TPA 79-2. Minneapolis, Minn.: American Water Resources Association.

Heinselman, M. L. 1963. Forest sites, bog processes, and peatland types in the glacial Lake Agassiz region, northern Minnesota. Ecol. Monogr. 33:327–374.

Hutchinson, G. E. 1957. A Treatise on Limnology, Vol. 1: Geography, Physics, and Chemistry. New York: John Wiley & Sons.

Ingram, H. A. P. 1983. Hydrology. Pp. 67–158 in Ecosystems of the World, 4A, Mires: Swamp, Bog, Fen and Moor, A. J. P. Gore, ed. Amsterdam: Elsevier.

Janssens, J. A., B. C. S. Hansen, P. H. Glaser, and C. Whitlock. 1992. Development of a raised-bog complex in Northern Minnesota. Pp. 189–221 in Patterned Peatlands of Northern Minnesota, H. E. Wright, Jr., B. Coffin, and N. Aaseng, eds. Minneapolis: University of Minnesota Press.

Kaplan, W., I. Valiela, and J. M. Teal. 1979. Denitrification in a salt marsh ecosystem. Limnol. Oceanogr. 24:726–734.

Kerekes, J., S. Beauchamp, R. Tordon, and T. Pollock. 1986. Sources of sulphate and acidity in wetlands and lakes in Nova Scotia. Pp. 207–214 in Proceedings of the International Symposium on Acidic Precipitation, Vol. 2, Muskoka, Ontario, Canada, H. C. Martin, ed. Dordrecht, The Netherlands: Reidel.

Kivinen, E. 1980. New statistics on the utilization of peatlands in different countries. Pp. 48–51 in Proceedings of the Sixth International Peat Congress, Duluth, Minn. Jyska, Finland: International Peat Society.

Kivinen, E., and P. Pakarinen. 1980. Peatland areas and the proportion of virgin peatland in different countries. Pp. 52–54 in Proceedings of the Sixth International Peat Congress, Duluth, Minn. Jyska, Finland: International Peat Society.

Kusler, J. A., and M. E. Kentula. 1989. Wetland Creation and Restoration: The Status of the Science. Vol. 1: Regional Reviews, Vol. 2: Perspectives. Report 600/3-89/038. Corvallis, Ore.: Environmental Protection Agency, Environmental Research Laboratory.

Martikainen, P. J., H. Nykänen, and J. Silvola. 1992. Emissions of methane and nitrous oxide from peat ecosystems. Pp. 199–204 in The Finnish Research Programme on Climate Change: Progress Report. Helsinki: VAPK.

Mendelssohn, I. A., and K. L. McKee. 1989. The use of basic research in wetland management decisions. Pp. 354–364 in Marsh Management in Coastal Louisiana: Effects and Issues, W. G. Duffy and D. Clark, eds. Biol. Rep. 89(22). Washington, D.C.: U.S. Fish and Wildlife Service.

Mendelssohn, I. A., and K. L. McKee. 1992. Indicators of environmental stress in wetland plants. Pp. 603–624 in Ecological Indicators, Vol. 1, D. H. McKenzie, D. E. Hyatt, and V. J. McDonald, eds. New York: Elsevier.

Mitsch, W. J., and J. G. Gosselink. 1993. Wetlands, 2nd ed. New York: Van Nostrand.

National Research Council (NRC). 1992. Restoration of Aquatic Ecosystems. Washington, D.C: National Academy Press.

Neishstadt, M. I. 1977. The world's largest peat basin, its commercial potentialities and protection. Int. Peat Soc. Bull. 8:37–43.

Niering, W. A. 1988. Endangered, threatened and rare wetland plants and animals of the continental United States. Pp. 227–238 in The Ecology and Management of Wetlands, Vol. 1, D. D. Hook, ed. Portland, Ore.: Timber Press.

Pala, S., and W. Weischet. 1982. Toward a physiographic analysis of the Hudson-James Bay Lowland. Nat. Can. 109:639–651.

Pastor, J. 1995. Ecosystem management, ecological risk, and public policy. BioScience 45:286–288.

Patrick, W. H., Jr. 1994. From wasteland to wetlands. J. Environ. Qual. 23:892–896.

Poiani, K. A., and W. C. Johnson. 1991. Global warming and prairie wetlands. BioScience 41:611–618.

Radforth, N. W., and C. O. Brawner, eds. 1977. Muskeg and the Northern Environment in Canada. Toronto: University of Toronto Press.

Rapaport, R. A., and S. J. Eisenreich. 1988. Historical atmospheric inputs of high molecular weight chlorinated hydrocarbons to eastern North America. Environ. Sci. Technol. 22:931–941.

Richardson, C. J. 1989. Freshwater wetlands: Transformers, filters, or sinks? Pp. 25–46 in Freshwater Wetlands and Wildlife, R. R. Sharitz and J. W. Gibbons, eds. CONF-8603101. Washington, D.C.: Department of Energy, Office of Health and Environmental Research.

Richardson, C. J. 1994. Ecological functions and human values in wetlands: A framework for assessing forestry impacts. Wetlands 14:1–9.

Rudd, J. W. M. 1995. Sources of methyl mercury to freshwater ecosystems: A review. Water Air Soil Pollut. 80:697–713.

Ruttner, F. 1963. Fundamentals of Limnology, 3rd ed. (translated by D. G. Frey and F. E. J. Fry). Toronto: University of Toronto Press.

Sharitz, R. R., and J. W. Gibbons, eds. 1989. Freshwater Wetlands and Wildlife. Symposium Series No. 61 (CONF-8630101). Oak Ridge, Tenn.: Department of Energy.

Shiklomanov, I. A. 1993. World fresh water resources. Pp. 13–24 in Water in Crisis, P. H. Gleick, ed. New York: Oxford University Press.

Siegel, D. I. 1983. Ground water and evolution of patterned mires, Glacial Lake Agassiz peatlands, northern Minnesota. J. Ecol. 71:913–922.

Siegel, D. I. 1988. Evaluating cumulative effects of disturbance on the hydrologic function of bogs, fens, and mires. Environ. Manage. 12:621–626.

Sjörs, H. 1961. Surface patterns in boreal peatlands. Endeavour 20:217–224.

Stanley, T. R. Jr. 1995. Ecosystem management and the arrogance of humanism. Conserv. Biol. 9:255–262.

St. Louis, V. L., J. W. M. Rudd, C. A. Kelly, K. G. Beaty, N. S. Bloom, and R. J. Flett. 1994. Importance of wetlands as sources of methylmercury to boreal forest ecosystems. Can. J. Fish. Aquat. Sci. 51:1065–1076.

Thompson, D. Q., R. L. Stuckey, and E. B. Thompson. 1987. Spread, Impact, and Control of Purple Loosestrife (Lythrum salicaria) in North American Wetlands. Fish and Wildlife Research, No. 2. Washington, D.C.: U.S. Fish and Wildlife Service.

Tallis, J. 1983. Changes in wetland communities. Pp. 311–347 in Ecosystems of the World 4A, Mires: Swamp, Bog, Fen and Moor, General Studies, A. J. P. Gore, ed. Amsterdam: Elsevier.

Walter, H. 1977. The oligotrophic peatlands of western Siberia—the largest peinohelobiome in the world. Vegetatio 34(3):167–178.

Welch, P. S. 1935. Limnology. New York: McGraw Hill.

Wetzel, R. G. 1983. Limnology, 2nd ed. Orlando, Fla.: Harcourt Brace Jovanovich, Inc., Saunders College Publishing.

Wickware, G. M., D. W. Cowell, and R. A. Sims. 1980. Peat resources of the Hudson Bay Lowland coastal zone. Pp. 138–143 in Proceedings of the Sixth International Peat Congress, Duluth, Minn. Jyska, Finland: International Peat Society.

Winter, T. C. 1992. A physiographic and climatic framework for hydrologic studies of wetlands. Pp. 127–148 in Aquatic Ecosystems in Semi-Arid Regions: Implications for Resource Management, R. D. Robarts and M. L. Bothwell, eds. N.H.R.I. Symposium Series 7. Saskatoon: Environment Canada.

Woodwell, G. M., F. T. Mackenzie, R. A. Houghton, M. J. Apps, E. Gorham, and E. A. Davidson. 1995. Will the warming speed the warming? Pp. 393–411 in Biotic Feedbacks in the Global Climate System, G. M. Woodwell and F. T. Mackenzie, eds. New York: Oxford University Press.

Wright, H. E., Jr., B. Coffin, and N. Aaseng, eds. 1992. Patterned Peatlands of Northern Minnesota. Minneapolis: University of Minnesota Press.

Applied Aquatic Ecosystem Science

Dean B. Premo
White Water Associates
Amasa, Michigan

Douglas R. Knauer
Wisconsin Department of Natural Resources
Monona, Wisconsin

SUMMARY

The need for practical application of aquatic ecosystem science to water quality and ecosystem health challenges has never been more pressing. Human activities and support systems are inextricably woven together with freshwater and aquatic ecosystems, yet human activities continue to degrade the quality of both. This paper discusses human impacts on aquatic ecosystems of the United States by reviewing the status of rivers, lakes, and wetlands and summarizing the most important sources of degradation and loss. It outlines areas where improved knowledge can help avoid and mitigate water resource problems; urges a commitment to interdisciplinary, interagency research on the part of governments, agencies, and scientists; cites two successful models of such commitment; and champions the role of monitoring in improving conservation technology and guiding future research.

INTRODUCTION

One measure of the relevance of a natural resource science is how much it contributes to management of an environment stressed by innumerable human uses and activities. The paradoxical predicament of humans, who simultaneously require fresh water and intact aquatic ecosystems yet diminish the quality and amount of these resources, can be remedied only by concerted application of aquatic ecosystem science integrated with the efforts of many other disciplines and interests. This paper addresses several issues that compel and frame applied research in aquatic ecosystem science.

Four topics related to applied research are discussed. The section on human impacts on aquatic ecosystems outlines examples of the major societal concerns regarding degradation of aquatic ecosystems and characterizes the scope and severity of the problems. These issues are presented to illustrate the diversity of challenges that face today's aquatic ecosystem scientists. Often societal and scientific opinions differ as to the relative importance of these issues. The section on technology and research needs frames the diversity of practical knowledge required to address today's aquatic resource problems. The section on scientific infrastructure discusses the academic, governmental, private, and societal foundation necessary to advance aquatic resource conservation. The final section on monitoring management strategies, details the importance of careful review of management activities as the best way to improve conservation technology and to guide future research. It promotes monitoring as a legitimate research activity equally deserving of funds.

HUMAN IMPACTS ON AQUATIC ECOSYSTEMS

Human-caused impacts on aquatic ecosystems are inevitable. As aquatic ecosystems evolve, society's course of action should be to understand the rate of change that is occurring and to decide whether society can, or should, intercede to slow the processes that control change. These decisions require social, economic, and scientific considerations. There are numerous human-caused environmental impacts on rivers, lakes, ground water, and wetlands. Some of these impacts are highly visible, such as draining of wetlands and flooding of large tracts of land to create reservoirs. Other impacts are not so easily observed, such as the bioaccumulation of mercury in top predators and the loss of biological diversity in lakes sensitive to acid deposition. Finally, there are undoubtedly impacts unknown to the current generation of scientists.

Under Section 305(b) of the federal Clean Water Act, states must report the status of water quality assessments to the Environmental Protection Agency (EPA). According to these state assessments, improvements in wastewater treatment have led to enhanced stream water quality in the 20 years since the Clean Water Act was first passed, but nonpoint sources of water pollution and toxic substances remain serious problems (EPA, 1994). The EPA recommends that states assess water quality based on the following individual beneficial uses:

* *Aquatic life support:* The water body provides suitable habitat for survival and reproduction of desirable fish, shellfish, and other aquatic organisms.
* *Fish consumption:* The water body supports a population of fish free from contamination that could pose a human health risk to consumers.

• *Shellfish harvest:* The water body supports a population of shellfish free from toxicants and pathogens that could pose a human health risk to consumers.

• *Drinking water supply:* The water body can supply safe drinking water with conventional treatment.

• *Primary contact recreation or swimming:* People can swim in the water body without risk of adverse human health effects.

• *Secondary contact recreation:* People can perform activities on the water (such as canoeing) without risking adverse human health effects from occasional contact with the water.

• *Agriculture:* The water quality is suitable for irrigating fields or watering livestock.

Regarding these beneficial uses, there are five levels of use support recognized by the EPA:

1. *Fully supporting overall use:* All designated beneficial uses are supported fully.

2. *Threatened overall use:* One or more of the designated beneficial uses are threatened and the remaining uses are supported fully.

3. *Partially supporting overall use:* One or more of the designated beneficial uses are partially supported and the remaining uses are supported fully.

4. *Not supporting overall use:* One or more designated beneficial uses are not supported.

5. *Not attainable:* The state has performed a use-attainability study and documented that support of one or more designated beneficial uses is not achievable.

The EPA defines *impaired waters* as the sum of water bodies partially supporting uses and not supporting uses.

The data on rivers, lakes, and wetlands below are from 1992 state assessments under section 305(b) of the Clean Water Act as summarized by EPA (1994) for use in the Congress.

Rivers

The United States has approximately 3.5 million miles of rivers and streams. Of the 642,881 miles of rivers assessed by the states in 1992, more than one-third did not fully support designated uses. The five leading sources of river water quality impairment listed by the states were (1) agriculture, (2) municipal point sources, (3) urban runoff or storm sewers, (4) resource extraction, and (5) organic enrichment or low dissolved oxygen. Forty-five states identified almost 160,000 river miles impaired by agricultural sources.

Lakes

The United States has about 39.9 million acres of lakes (excluding the Great Lakes). Of the 18 million acres assessed by the states in 1992, 35 percent failed to meet designated use criteria at times, 9 percent frequently failed to meet designated use criteria, and less than 1 percent of the lakes could not be used at all due to irreversible natural conditions or human activity. Metals and nutrients are the most common causes of nonsupport. An example of metal problems is the accumulation of excessive amounts of mercury in top predator game fish in lakes throughout the Northern Hemisphere. More than 33 states have issued fish consumption advisories because of elevated concentrations of mercury in game fish.

Nutrient problems were reported widely by states. Forty-five states reported that agricultural runoff is the leading source of pollutants, impairing more lake acres than any other source. Agricultural runoff includes nutrients, organics, and pesticides. Thirty states reported that siltation impaired lakes and reservoirs. Priority organic chemicals, such as PCBs (polychlorinated biphenyls), were reported as significant in the number of lake acres they impaired (ranking eighth).

Another indicator of water quality problems in lakes is the growth of organizations such as the North American Lake Management Society (NALMS) that resonate a strong societal concern. By 1994, NALMS had 20 state chapters comprised of local citizens interested in solving or preventing lake water quality problems. NALMS has actively promoted the involvement of citizens in lake sampling under the umbrella of government environmental agencies. The number of volunteers trained in lake sampling is impressive: 600 in Wisconsin, 3,000 in Texas, and 1,500 in Florida.

Wetlands

When European settlers first arrived in America, about 89 million hectares of wetlands existed in what would eventually become the conterminous states. Today, more than half of the original wetlands have been destroyed by filling, draining, polluting, channelizing, clearing, and other modifications resulting from human activities. In their water quality assessments for the Clean Water Act, 27 states listed agriculture and commercial development as the leading cause of wetland loss. Of 14 states that identified sources of pollutants that degrade wetlands, 11 ranked agriculture as the number one source. Agricultural runoff includes excessive levels of nutrients, organic matter, and pesticides. A U.S. Fish and Wildlife Service study of wetland loss found that 1.1 million hectares of wetlands were lost over a nine-year period (mid-1970s to mid-1980s), or about 117,000 hectares per year. Although this is a seemingly unacceptable rate of loss, it is an improvement compared with the 1950s to 1970s, when wetlands were lost at a rate of 185,000 hectares per year. This improvement, however, cannot be considered satisfactory, because the United States continues to lose wetlands at a rate of at least 1 percent per year in spite of mitigation and restoration attempts. The poor-quality wetlands that result from many mitigation and restoration efforts are no substitute for the lost habitats, processes, and functions of natural wetlands. Serious consequences have resulted nationwide from the loss and degradation of wetlands, including species decline and extinction, water quality deterioration, and increased incidence of flooding.

TECHNOLOGY AND RESEARCH NEEDS

The need for enhanced interaction among the subdisciplines of aquatic science is nowhere more evident than in the area of applying this science to management and conservation of freshwater resources. In fact, knowledge from other disciplines (including social science and education) is necessary to improve the ability to address challenges of management, mitigation, and perpetuation of aquatic resources. The fact that aquatic science favors an ecosystem approach demands this interdisciplinary perspective—one that includes humans as an ecosystem component to be considered in nearly every applied strategy.

Specific aquatic science technology and research needs are too many to list but include such disparate subjects as on-site waste disposal systems, the relationship between macrophyte beds and fish production, the role of nitrogen in macrophyte growth, bioaccumulation, wetland functions, transport and deposition of toxics, waterborne pathogens, buffer zones, best-management practices, control of exotic species, water quality reference sites, and ecological risk assessments. Organizing these and many other needs, prioritizing them, and then tackling them with well-designed research is a daunting responsibility, but it is crucial to the perpetuation of aquatic resources.

The field of conservation biology has promulgated the construct of *coarse filter* and *fine filter* in order to advance biodiversity conservation (Noss, 1987; Hunter, 1990). This approach is used by the Nature Conservancy to address management issues ranging from endangered species to ecosystem perpetuation (Hudson, 1991). Hunter (1990) provided a lucid description of this approach. The coarse-filter approach to perpetuating biodiversity involves maintaining a variety of ecosystems, and assuming that this will include the majority of species in a region. The fine-filter approach is used in the case of individual species known to be very rare and vulnerable to "passing through" the coarse filter. A perpetual dilemma in applied aquatic ecosystem science is the need to proceed with management activities despite the lack of complete knowledge required to solve a particular resource problem. Given this reality, the fine- and coarse-filter construct provides a feasible tactic for applying aquatic ecosystem science to conservation and management.

A hypothetical example of the fine-filter approach would be the perpetuation of a natural breeding population of brook trout (*Salvelinus fontinalis*) in a cold-water stream. Meeting this objective would involve identifying healthy populations and appropriate spawning habitat and using this and other specific information to design suitable management strategies to maintain conditions. In this case, the management focus is on a single species. Control of cyanobacteria in a lake is also an application of a fine-filter approach.

An example of the coarse-filter approach would be to plan and implement a variety of watershed management practices that perpetuate habitat for a natural diversity of warm- and cold-water fish species in various parts of the watershed. In this approach, neither brook trout nor cyanobacteria is the primary focus, but both are affected by the prescribed watershed practices. A coarse-filter approach allows the management net to be cast wide, even in the absence of complete information, and increases the likelihood that desirable outcomes will result. The fine-filter approach is a desirable complementary tactic, allowing treatment of issues not satisfactorily addressed by a broader suite of management practices.

In the context of management of aquatic resources, a coarse-filter tactic can equate to a landscape or watershed approach. Integrating the aquatic and terrestrial systems, including the riparian ecotones (transition zones), is a necessary strategy to management of most aquatic ecosystems. The watershed perspective represents one of the identified practical approaches to defining ecological management landscapes around which conservation participants can rally and cooperate (Rogers and Premo, 1994). This perspective provides the appropriate ecological context for applying aquatic ecosystem science to conservation and management.

There are research needs in the social sciences that will advance application of aquatic ecosystem science to resource problems. Many difficult jurisdictional and socioeconomic questions are often unresolved. Who conducts the management? Who is responsible for monitoring? Who funds the research? Who regulates compliance? If a landscape or watershed approach is adopted, answers to these questions are even more complex. In the United States, the "command-and-control" approach to "end-of-pipe" pollution has been successful. What remains, however, are the onerous problems of nonpoint sources of water quality degradation and the challenges of inducing landscape owners and managers to cooperate in management and land use so that nonpoint-source pollution is ameliorated. These social science challenges are not unique to aquatic ecosystem management. They represent a research need for most fields of resource management (Gerlach and Bengston, 1994; Great Lakes Natural Resource Center, 1994; Rogers and Premo, 1994).

A challenge of the landscape approach is developing cooperation among resource agencies charged with various aspects of maintaining the integrity of aquatic ecosystems or the quality of drinking water. Too often policies and activities directly conflict or are poorly coordinated. Research that provides guidance for cross-agency cooperation would be beneficial to applied aquatic ecosystem science.

There is also a need for improved education. Policymakers and managers have an ongoing need for information. Education must also extend to interest groups, including environmentalists, industry representatives, school children, and other members of the public.

Good technology and well-intentioned goals will contribute little to conservation of intact and functioning aquatic ecosystems if they cannot be translated into practical action in specific watersheds. Effective restoration, conservation, and perpetuation of water resources must integrate sound science with practical approaches that can be understood and accepted by the public.

SCIENTIFIC INFRASTRUCTURE

The most difficult part of implementing the interdisciplinary research model is that most government agencies, universities, and other public and private institutions have not demonstrated a willingness to relinquish their territorial claims on individual parts within the ecosystem. Within government environmental agencies, a department of fisheries often has a much different mandate and environmental perspective than a department of water resources management. For example, during a Clean Lakes project in Kentucky, watershed nutrient reduction plans were being discussed with the Division of Water for a eutrophic 21-hectare lake even as state fisheries biologists added nutrients directly to the lake as part of a standard practice to maintain a productive ecosystem. From a fisheries perspective, a very nutrient-rich lake resulted in a better fishery. From a water quality perspective, the overabundant plant biomass and lack of dissolved oxygen in summer throughout much of the water column was a concern. Separate administrative units within state government needed to integrate their efforts to sustain a lake ecosystem that could support a sport fishery, assimilate the watershed nutrient loads, and improve water quality for other recreational endeavors. Similar circumstances exist for universities. The college of agriculture may focus its attention on promoting a very high soil phosphorus buildup without considering the water quality concerns expressed by the college of natural resources.

The solution requires a policy and financial commitment by governments to make applied interdisciplinary research a priority. Equally important is progress by the scientific community to overcome territorial problems associated with individual recognition versus team effort.

There are several models that have been successful at integrating across institutional and environmental disciplines. Two examples are the Experimental Lakes Area (ELA) and the Long-Term Ecological Research (LTER) program. Both examples are based on a commitment of federal dollars for support.

The mandate for ELA was to conduct applied research to solve priority environmental problems (e.g., eutrophication and acid rain). To accomplish that mandate, the federal government of Canada financed the site operation, assigned federal scientists to the location, and encouraged interdisciplinary research. The commitment of federal funding encouraged the development of on-site laboratories and logistical facilities and the continuous collection of basic water quality information. The success of the ELA approach is not measured in the number of publications, although there are many (e.g., Schindler, 1974, 1988; Fee, 1976; Schindler and Turner, 1982; Schindler et al., 1985, 1986; St. Louis et al., 1994), but in the value of attracting interdisciplinary teams of scientists with individuals representing universities as well as federal and provincial governments to conduct holistic research on real and pressing environmental problems.

The LTER model involves federal funding to establish environmental monitoring and research sites throughout the United States. In the case of LTER's aquatic program for temperate lakes, the funding is directed to the University of Wisconsin to establish a suite of lakes and their respective watersheds in northern Wisconsin to act as sentinels for long-term regional changes in the environment. Federal funds help support on-site laboratories and logistics at the Trout Lake university facilities and a continuous monitoring of a variety of lake and watershed parameters. As in the ELA model, the advantage of supporting an established site to conduct holistic environmental research is that it attracts interdisciplinary studies to the area. For example, an interdisciplinary, multi-institutional study on the biogeochemical fate of mercury in lakes was attracted to the Trout Lake university facilities because of the existing laboratory and scientific support. The $5 million project involves scientists from three universities, two federal agencies, three private contractors, and state government, with combined expertise in analytical chemistry, atmospheric chemistry and transport, aquatic microbiology, ground water hydrology and geochemistry, fisheries, biology, physicochemical and biological limnology, and environmental modeling. Many valuable contributions resulted from this effort (e.g., Bloom, 1989; Fitzgerald and Watras, 1989; Wiener et al., 1990; Bloom et al., 1991; Krabbenhoft and Babiarz, 1992; Watras and Bloom, 1992; Porcella, 1994; Watras et al., 1994).

Both of these programs attracted scientists from a variety of disciplines and institutions who worked together to solve aquatic resource problems needing attention by decisionmakers. Essentially, if resources are allocated to building an infrastructure that is conducive to multi-institutional, inter-disciplinary cooperative research, top-quality scientists will become involved.

MONITORING MANAGEMENT STRATEGIES

It is clear that society cannot wait until all scientific questions are answered before acting to perpetuate healthy aquatic ecosystems. Research, education, and management must proceed simultaneously, but always with a feedback mechanism through which each can be assessed and altered if insufficient or undesirable outcomes are detected. Management activities themselves, sometimes large-system manipulations, can be viewed as primary tools for experimentation, fueling a process that is often called "adaptive management" (Walters, 1986). The adaptive management approach is an ongoing effort. New scientific knowledge is integrated continually into practical and appropriate management strategies. Techniques may require frequent refinement and alteration. Society's goals for the desired state of its resources in the future require consideration and update.

For this process to continue successfully, effective long- and short-term monitoring of the integrity of existing aquatic ecosystems or changes resulting from management regimes is crucial. Nevertheless, the preponderance of public and private research dollars is funneled toward projects that test theories or design mathematical models. The assessment of effects of management strategies is often not perceived as being scientifically rigorous enough to deserve a share of limited funding (Noss, 1990). Yet monitoring the efficacy of science applied to the solution of problems of aquatic ecosystems is a critical step toward improvement. In addition, such research often results in fundamental scientific findings.

REFERENCES

Bloom, N. S. 1989. Determination of picogram levels of methyl mercury by aqueous phase ethylation, followed by cryogenic gas chromatography with cold vapor atomic fluorescence detection. Can. J. Fish. Aquat. Sci. 46:1131–1140.

Bloom, N. S., C. J. Watras, and J. P. Hurley. 1991. Impact of acidification on the methyl mercury cycle of remote seepage lakes. Water Air Soil Pollut. 56:477–492.

Environmental Protection Agency (EPA). 1994. National Water Quality Inventory: 1992 Report to Congress. EPA 841-R-94-001. Washington, D.C.: EPA.

Fee, E. J. 1976. The vertical and seasonal distribution of chlorophyll in lakes of the Experimental Lakes Area, NW Ontario: Implication for primary production estimates. Limnol. Oceanogr. 21:767–783.

Fitzgerald, W. F., and C. J. Watras. 1989. Mercury in the surficial waters of rural Wisconsin lakes. Sci. Tot. Environ. 87/88:223–232.

Gerlach, L. P., and D. N. Bengston. 1994. If ecosystem management is the solution, what's the problem? Eleven challenges for ecosystem management. J. For. 92(8):18–21.

Great Lakes Natural Resource Center. 1994. Seeing the Forest Through the Trees: A Model Biodiversity Collaboration Strategy for the Lake Superior Basin. Ann Arbor, Mich.: National Wildlife Federation.

Hudson, W. E. 1991. Landscape Linkages and Biodiversity/Defenders of Wildlife. Washington, D.C.: Island Press. 196 pp.

Hunter, M. L., Jr. 1990. Wildlife, Forests, and Forestry: Principles of Managing Forests for Biological Diversity. Englewood Cliffs, N. J.: Prentice Hall. 370 pp.

Krabbenhoft, D. P., and C. L. Babiarz. 1992. The role of groundwater transport in aquatic mercury cycling. Water Resour. Res. 28:3119–3128.

Noss, R. F. 1987. From plant communities to landscapes in conservative inventories: A look at the Nature Conservancy (USA). Biol. Conserv. 41:11–37.

Noss, R. F. 1990. Indicators for monitoring biodiversity: A hierarchical approach. Conserv. Biol. 4:355–364.

Porcella, D. B. 1994. Mercury in the environment: Biogeochemistry. Pp. 3–19 in Mercury Pollution: Integration and Synthesis, C. Watras and J. Huckabee, eds. Boca Raton, Fla.: Lewis Publishers.

Rogers, E. I., and D. B. Premo. 1994. Model biodiversity management plan—Scientific section In Seeing the Forest Through the Trees: A Model Biodiversity Collaboration Strategy for the Lake Superior Basin. Great Lakes Natural Resource Center. Ann Arbor, Mich.: National Wildlife Federation.

Schindler, D. W. 1974. Eutrophication and recovery in experimental lakes: Implication for lake management. Science 184:897–899.

Schindler, D. W. 1988. Effects of acid rain on freshwater ecosystems. Science 239:149–157.

Schindler, D. W., and M. A. Turner. 1982. Biological, chemical and physical responses of lakes to experimental acidification. Water Air Soil Pollut. 18:259–271.

Schindler, D. W., K. H. Mills, D. F. Malley, D. L. Findlay, J. A. Shearer, I. J. Davies, M. A. Turner, G. A. Linsey, and D. R. Cruikshank. 1985. Long-term ecosystem stress: The effects of years of experimental acidification on a small lake. Science 228:1395–1401.

Schindler, D. W., M. A. Turner, M. P. Stainton, and G. A. Linsey. 1986. Natural sources of acid neutralizing capacity in low alkalinity lakes of the Precambrian Shield. Science 232:844–847.

St. Louis, V. L., J. W. M. Rudd, C. A. Kelly, K. G. Beaty, N. S. Bloom, and R. J. Flett. 1994. Importance of wetlands as sources of methyl mercury to boreal forest ecosystems. Can. J. Fish. Aquat. Sci. 51:1065–1076.

Walters, C. 1986. Objectives, constraints, and problem bounding. Chapter 2 in Adaptive Management of Renewable Resources, W. M. Getz, ed. New York: Macmillan.

Watras, C. J., and N. S. Bloom. 1992. Mercury and methylmercury in individual zooplankton: Implications for bioaccumulation. Limnol. Oceanogr. 37:1313–1318.

Watras, C. J., N. S. Bloom, R. J. M. Hudson, S. Gherini, R. Munson, S. A. Claas, K. A. Morrison, J. Hurley, J. H. Wiener, W. F. Fitzgerald, R. Mason, G. Vandal, D. Powell, R. Rada, L. Rislov, M. Winfrey, J. D. Krabbenhoft, A. W. Andren, C. Babiarz, D. B. Porcella, and J. W. Huckabee. 1994. Sources and fates of mercury and methylmercury in Wisconsin lakes. Pp. 153–177 in Mercury Pollution: Integration and Synthesis, C. Watras and J. Huckabee, eds. Boca Raton, Fla.: Lewis Publishers.

Wiener, J. G., W. F. Fitzgerald, C. J. Watras, and R. G. Rada. 1990. Partitioning and bioavailability of mercury in an experimentally acidified Wisconsin lake. Environ. Toxicol. Chem. 9:909–918.

Fundamental Research Questions in Inland Aquatic Ecosystem Science

Diane M. McKnight
Water Resources Division
U.S. Geological Survey
Boulder, Colorado

Elizabeth Reid Blood
J. W. Jones Ecological Research Center
Newton, Georgia

Charles R. O'Melia
Department of Geography and Environmental Engineering
The Johns Hopkins University
Baltimore, Maryland

SUMMARY

This background paper presents one possible set of fundamental questions for inland aquatic ecosystems. There are other questions that may be more fundamental or more pressing for meeting the needs of society in managing water resources. The general view we would like to convey is that although there is a need for additional monitoring, better analytical techniques, and more extensive systematics and classification, aquatic scientists are constrained most by incomplete understanding of the aquatic ecosystems themselves. Scientific progress and improved management of aquatic resources will be achieved by improving understanding of aquatic systems and by applying and communicating that understanding.

INTRODUCTION

Limnology is an integrative science, combining a knowledge of physical, chemical, and biological processes to promote understanding of aquatic systems. For any lake, stream or wetland, there are innumerable unanswered questions; some are obvious, others are more subtle. Some key questions become apparent only after detailed long-term study. Through observation of aquatic ecosystems, patterns or relationships are discovered that cannot be readily explained. Even the seemingly simple determination of whether an observed pattern is caused by hydrologic, chemical, or biological processes may be difficult. Incorrect assumptions about unexplained patterns or observations can lead to confusion.

An inherent challenge in limnology is "scaling up" from controlled laboratory experiments to real aquatic ecosystems. In studies of the biodegradation of organic contaminants or of many physical processes, for example, it may be true that a beaker is equivalent to a lake or an experimental flume is equivalent to a stream. Conversely, field experiments, such as ecosystem manipulations, are difficult to replicate and may be influenced by uncontrolled variables.

Within this context of limnology as an integrative science addressing aquatic systems, it is a challenge to distinguish which questions are fundamental and which are merely intriguing. A fundamental question may lead to a domino effect, where one answer illuminates answers to many other questions. Further, a fundamental question may or may not be a "big" question, that is, an area in which we recognize a gaping void in understanding or knowledge. For instance, the role of peatlands in the global carbon cycle is a fundamental question, but a big question that is not fundamental in the same sense is just how large the peat reservoir is, particularly in Russia. We know that Russia has enormous deposits of peat, especially in the west Siberia Plain, but the numbers on both peatland area and mass of peat are of questionable accuracy.

INLAND AQUATIC ECOSYSTEMS

For the purposes of providing a background for this study, we have chosen to pose questions at a level relevant to the full range of inland aquatic ecosystems (lakes, wetlands, streams, big rivers, ponds, etc.). An ecosystem is defined by its boundaries, and an inland aquatic ecosystem can range in size from a small transient desert pool that forms after a storm to the watershed of the Colorado River. Boundaries can be set based on their utility for studying a given process or interaction. For example, defining an aquatic ecosystem as an entire watershed may be useful for studying responses to acidic deposition, but it may not be useful for studying the daily migration of zooplankton between depths in a lake within that watershed.

Several approaches can be used to classify inland waters. To illustrate their diversity and distribution, we have broadly classified inland waters based on the physical or chemical characteristics that exert major influences on ecosystem properties (see Table 1). We have also indicated the general occurrence of such ecosystems in different regions of North America (Table 1 and Figure 1). The three main classes (streams or rivers, lakes, and wetlands) represent a gradient in hydrologic residence time from rapid to slow. It is important to remember, however, that there are not always clear distinctions between these ecosystems because of their interconnections.

General types of inland waters are identified in the next level of classification. Saline lakes and blackwater rivers have chemical characteristics with a pervasive influence on ecosystem properties. For example, the limited transmission of light in blackwater systems restricts photosynthesis. Saline lakes are closed-basin lakes, with high concentrations of dissolved solids. Prairie wetlands and intermittent streams are distinguished because the climatic sensitivity of their hydrologic regimes can cause

TABLE 1   Occurrence of Various Classes of Inland Waters in Regions of North America

Classification	Region 1	2	3	4	5	6	7	8
Streams/rivers								
Blackwater rivers	—	—	—	—	X	—	—	—
Alpine streams	—	X	X	—	—	X	—	—
Intermittent streams	—	—	X	—	—	X	X	X
Perennial streams	X	X	X	X	X	X	X	X
Lakes								
Large lakes	X	X	X	—	X	X	—	—
Saline lakes	—	X	—	—	X	X	X	X
Small lakes/ponds	X	X	X	X	X	X	X	X
Reservoirs	X	—	X	X	X	X	X	X
Wetlands								
Large riparian wetlands	X	X	X	X	X	X	X	—
Large swamps	—	—	—	—	X	—	—	—
Prarie wetlands	—	—	—	—	—	X	X	—
Small bogs/fens	X	X	X	X	X	X	—	—
Large peatlands	X	X	—	—	—	—	—	—
Coastal zone areas	X	X	—	X	X	X	—	X

NOTE: Numbers correspond to the following regions:
1 Laurentian Great Lakes and Precambrian Shield of the United States and Canada.
2 Arctic and subarctic areas of Canada.
3 Rocky Mountains in the United States and Canada.
4 Mid-Atlantic and New England area of the United States.
5 Southeastern United States and coastal Mexico.
6 Pacific coast mountains and western Great Basin.
7 Great Plains of the United States and Canada.
8 Basin and range regions and adjacent arid and semiarid regions of the United States and Mexico.

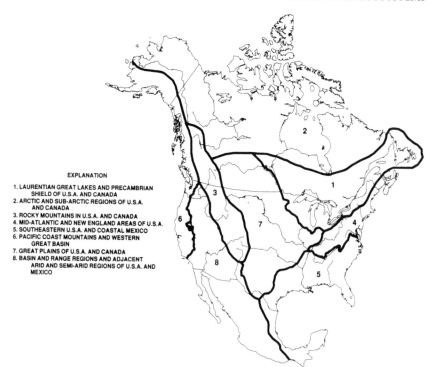

EXPLANATION
1. LAURENTIAN GREAT LAKES AND PRECAMBRIAN
   SHIELD OF U.S.A. AND CANADA
2. ARCTIC AND SUB-ARCTIC REGIONS OF U.S.A.
   AND CANADA
3. ROCKY MOUNTAINS IN U.S.A. AND CANADA
4. MID-ATLANTIC AND NEW ENGLAND AREAS OF U.S.A.
5. SOUTHEASTERN U.S.A. AND COASTAL MEXICO
6. PACIFIC COAST MOUNTAINS AND WESTERN
   GREAT BASIN
7. GREAT PLAINS OF U.S.A. AND CANADA
8. BASIN AND RANGE REGIONS AND ADJACENT
   ARID AND SEMI-ARID REGIONS OF U.S.A. AND
   MEXICO

FIGURE 1   Regions of North America corresponding to the designations in Table 1.
SOURCE: American Society of Limnology and Oceanography 1994). © by the American
Society of Limnology and Oceanography.

profound annual variations in these ecosystems. Fresh deep lakes show
physical properties that are unique because the temperature of maximum
density (which does not exist in salt water) is dependent on pressure and
salinity. Thus, it appears that the circulation of deep water in deep lakes
(e.g., Lake Baikal in Russia) is a physical mystery that may be explained
by the theory of neutral surfaces and by the phenomena of caballing
and thermobaricity (D. Imboden, Swiss Federal Institute of Technology,
personal communication, 1994).

Despite its simplicity, this classification illustrates the similarities and
differences in the occurrence of inland aquatic ecosystems at the regional
scale for the North American continent. Three types of inland waters are
commonly found throughout the continent: (1) small lakes and ponds,
(2) large rivers, and (3) perennial streams. On the other hand, large peatlands, large swamps, prairie wetlands, very large lakes, and blackwater
rivers are less widely distributed in North America.

Insight into one type of aquatic ecosystem can be gained from study of another. Similarly, study of processes occurring in lakes can provide insight into processes occurring in oceans. Two examples of processes that are important in marine systems and can be studied readily in lakes are turbulent mixing at boundaries and sediment-water interactions, including the microstructure of the sediment-water interface and the resulting microchemical environments. Other examples of transferability of knowledge between aquatic ecosystems are complementary studies of the hyporheic zone (the area under and adjacent to the open channel), which have been conducted in both small first-order mountain streams and large rivers.

FUNDAMENTAL QUESTIONS

To provide a structure for this chapter, we have grouped the fundamental questions identified into three broad categories relating to (1) the temporal and spatial dynamics of inland aquatic ecosystems (understanding time and space), (2) aquatic organisms occurring in inland aquatic ecosystems (the ecological significance of the species), and (3) interactions involving both major and trace chemical constituents of natural waters (natural water—a chemical world). Within the three categories, questions related to several topics are raised.

Understanding Time and Space

For terrestrial ecosystems, the land surface can be considered stationary during the life span of most of the organisms within the ecosystem. However, for aquatic ecosystems, climatically driven water (the milieu within which the organisms live) is moving through the ecosystem, as driven by the hydrologic cycle. Aquatic ecosystems are therefore highly variable over the full range of temporal and spatial scales. In this section, questions going from long-term to daily time scales and then questions going from very small scales to regional scales are considered.

Paleolimnology

A historical perspective can provide insight into contemporary phenomena. In limnology, history is archived in lake or floodplain sediments and in peatlands. Environmental change, such as change in nutrient concentration, climatic conditions, or hydrologic regime, has the potential to affect chemical, physical, and biological processes of lakes, streams, and wetlands. For example, materials entering a lake from inflowing streams, atmospheric deposition, or biomass produced within the lake can settle to the lake bottom. When changes in inflow or in lake productivity occur, evidence of the change may become preserved in the sediments. Paleolimnologists can examine an integrated record of lake and watershed history, often in relatively undisturbed sediments. The complexities of a three-dimensional, dynamic aquatic system are compressed into layers of mud. The challenge in paleolimnology, therefore, is in interpretation. The primary goal is often to reconstruct the physical or chemical record from the biological record, such as diatom frustules, algal pigments, or ostracod fossils. This primary goal of detecting a signal can be obtained by a variety of tools; the more tools used, the more robust is the interpretation. Multiple lines of evidence can be used to detect often subtle environmental change.

Question: Are all important limnological characteristics associated with an event or a gradual shift recorded in sediments? How can interpretations of the sediment record be strengthened?

The role that time, location, stability, and complexity play in creating an entire system all enter into consideration. Further, temperate lakes move through large fluctuations during a year (ice cover, spring overturn, summer stratification, and fall overturn). The more subtle year-to-year fluctuations are superimposed upon the signal left by this intra-annual cycle. The impact and detectability of an event are dependent on the time scale over which the event occurs. Some events are abrupt or even catastrophic, such as those associated with major floods or volcanic eruptions. Other events are more gradual or prolonged, such as droughts or annual variations driven by ENSO (El Niño Southern Oscillation) events. The stability of a given parameter in relation to change can vary. For example, arctic diatom species may respond to small changes in nutrient concentrations but be insensitive to temperature changes of several degrees. Complex systems can respond to change in unexpected ways, and deciphering the complexity is the major task ahead.

Question: Is a climatic signal local to one particular lake, or does it operate over a regional scale?

The location of a lake is important in determining the wider implications of its sedimentary record. One successful strategy is to study sediment cores from many lakes in a region and look for correlations in the records. Careful study of just one core using a variety of methods is a very labor-intensive task; studying numerous cores from a region requires a monumental effort. Advances in analytical methods and data analysis may make the application of paleolimnological techniques more practical for regional issues.

Importance of Winter

As we recognize that human activities have the potential to change global climate, there is an interest in understanding the responses of freshwater ecosystems to all types of temperature ranges and climate scenarios.

Question: What are the critical events and conditions that control autotrophic and heterotrophic processes during winter?

Field studies of inland aquatic ecosystems are not distributed evenly among the seasons of the year because the "field season" for limnologists is generally late spring and summer. For many ecosystems, the dominant processes operative in summer have already been studied in detail. For example, the dynamic succession of phytoplankton from spring through summer into fall has been studied in numerous temperate lakes. In contrast, much less is known about the dynamics of inland aquatic ecosystems during winter because it is clearly more difficult to conduct research under winter conditions. For the northern areas of North America, winter is a time when many lakes, streams, and rivers are covered by ice and snow, and wetlands are frozen and snow covered. In lakes, for example, light transmission through the ice cover can vary during the winter with the accumulation and dispersal of snow. Therefore, photosynthesis by phytoplankton under the ice may be temporally dynamic, and the sequence of phytoplankton species dominance may vary from one year to the next. Even though there is an awareness that aquatic ecosystems are not in suspended animation under the ice and snow, these logistical difficulties combined with the constraints of academic calendars tend to discourage studies of winter dynamics.

Question: What critical processes may occur in winter that control the behavior of an ecosystem in the subsequent summer and spring?

Primary production refers to the rate at which algae and aquatic plants assimilate carbon dioxide and produce biomass through photosynthesis, whereas heterotrophic degradation refers to the rate at which algal and plant biomass is consumed by nonphotosynthetic organisms, from microorganisms to fish, producing carbon dioxide. Because of decreased light intensities associated with snow and ice cover, the ratio of primary production to heterotrophic degradation undoubtedly decreases greatly in many aquatic ecosystems in the transition from fall to winter. In temperate streams, for example, the extent of heterotrophic degradation during winter of fallen leaves and other (nonliving) organic material from the watershed may depend on hydrologic and climatic conditions, such as the timing and duration of ice cover. The amount and quality of this detrital material available to benthic invertebrates at the beginning of the summer thus may vary between years. For wetlands, the timing of the freeze and thaw of the wetland surface may control the extent of heterotrophic degradation and annual production as well.

Morning, Noon, and Night

The water that surrounds an aquatic organism is continually changing, in both physical and chemical properties. For this reason, the solar cycle has a pervasive influence on aquatic ecosystems, raising a fundamental issue.

Question: How do the diverse 24-hour (diel) cycles driven by the solar cycle fit together to structure the daily pulse of an aquatic ecosystem?

Photosynthesis and respiration obviously vary during the day, but these are not the only processes driven by the solar cycle. Examples of other biological processes include nutrient uptake and cell division by algae, both of which are controlled by circadian clocks synchronized to the solar cycle. Processes controlling the chemical speciation of biologically reactive constituents are also driven by sunlight or by changes in pH and oxygen concentration associated with photosynthesis. For example, photoreduction of particulate iron and manganese oxides in the water column or in the streambed can release sorbed phosphate, which is then assimilated by algae. Cascading chemical reactions involving photoproducts (such as hydroxyl radicals) and other metals or organic compounds can also occur.

It is important not only to become aware of these diel variations, but also to understand in more detail how the "clockwork" controls the ecosystem and drives changes that occur over weeks and seasons. We should expect that aquatic ecosystems would have evolved a temporal "fine structure," and it could be that this fine structure is where the system is most vulnerable to ecological insult from anthropogenic contaminants.

Living at the Interface

The hydrologic sciences have developed with an emphasis on understanding hydrologic processes at the macroscopic scale. This is the scale most useful for understanding the development of stream networks, for predicting flood responses and bed load transport, and for determining the transport of pollutant discharges. Averaged parameters, such as a roughness coefficient and hydraulic conductivity, are effective in describing physical interactions between the streambed and the flowing water and ground water in aquifers. Key features of the habitat may not be reflected in the lumped or averaged parameters developed to answer larger-scale questions.

Question: How do fluid mechanics in fluvial systems determine the habitat as experienced by the organism?

Many aquatic organisms live on a surface exposed to moving water. Examples range from a bacterial cell adhered to the surface of a suspended particle to a mayfly grazing on an algal-covered rock. A quantitative understanding of the physical behavior of this habitat requires an approach to environmental fluid mechanics at a smaller scale. A sound physical understanding is a necessary basis for the study of many chemical and biological processes, such as microbial degradation of contaminants sorbed on particulate or sediment surfaces and particle feeding by caddis and blackflies.

Regional Scales, Spatial Heterogeneity and Ecosystem Linkages

Large discrepancies exist between the scale of information needed to manage aquatic resources effectively and the scale at which current research and management occur (Repetto, 1986).

Question: How can we develop an understanding of inland aquatic ecosystems at regional scales?

Most current ecological studies are conducted at local to watershed scales. Resource management activities often operate independently of an ecological perspective or other concurrent management activities. Current resource management activities are carried out at the "end-of-the-pipe" to subwatershed scale. In contrast, interactions among human, ecological, and atmospheric systems largely occur at the regional scale (areas larger than ecosystems or landscapes but subcontinental), and these interactions drive local and regional structure. Environmental policy, economic incentives, and return rates are governed by political units that correspond to the regional scale. Cultural and social traditions that control the ethics of land use also occur at regional scales. Atmospheric scientists have suggested that air-parcel dynamics and interactions with characteristics of the earth's surface that occur at the "mesoscale" (e.g., region) are key processes determining regional climate patterns. Collectively, interactions among human, ecological, and physical forces both define and structure regions and require interdisciplinary study. At a minimum, these issues require that local or site-specific studies be conducted in a manner that incorporates regional influences, designed to be "scalable" to a regional perspective.

Question: How does spatial heterogeneity within an ecosystem influence its dynamics, and what are the critical ways ecosystems are linked together at the larger scale?

Scalable research and management activities require the identification of fundamental units that can be combined functionally and structurally into larger functional units. Inherent in the identification of fundamental units is a thorough understanding of the important spatial and temporal patterns associated with the aquatic system under study. Spatial heterogeneity in aquatic systems results from chemical, physical, and biological structures, which in turn regulate the associated ecosystem processes and patterns. As an example, thermal stratification in lakes controls biotic habitat, biological structure, and biogeochemical cycles. The timing and coherence of the stratification varies among lake systems. Geomorphic characteristics of streams interact with streamflow to create a variety of biotic habitats and control biological structure and function. Several important paradigms in riverine ecology (river continuum, riparian corridor, hyporheic corridor, and river discontinuity theories) are based on the relationship of ecological processes to geomorphic characteristics of streams and their watersheds. As a further example, in boreal peatlands, the landscape patterns reflect the areas of upwelling of base-rich ground water.

Spatial heterogeneity occurs at the micro- to meso- (intermediate) scale. At the microscale, currents can control the biological and geochemical processes occurring in the layer of algae, fungi, and bacteria that covers rocks in a streambed. Interactions occurring at the mesoscale include benthic invertebrates and predatory invertebrates or fish, and ecosystem responses to episodic disturbances such as floods and fires. To develop useful concepts describing the structure and functioning of aquatic systems, it is necessary to identify fundamental units (such as the streambed), based on an understanding of the spatiotemporal patterns.

The heterogeneity within an ecosystem influences the linkages to other adjacent ecosystems. Aquatic systems can be viewed from a four-dimensional framework (Ward and Stanford, 1991). Aquatic systems are affected by vertical linkages to the ground water; lateral linkages to the watershed; longitudinal linkages across the aquatic system, and temporal variation in physical, chemical, and biological processes. In addition, significant transformation may occur as these waters pass through boundaries between the surface water and source water, such as the hyporheic zone (where ground water and surface water interconnect) and the riparian zone (land surface adjacent to a stream). The relative importance of these ecotones and linkages will vary with location in the drainage basin, characteristics of the system under study, and historic land management activities. This four-dimensional framework is critical in understanding aquatic and wetland systems and in formulating scalable research and management, yet few studies have attempted to incorporate it into a holistic ecological view. In most flowing aquatic systems, repeatable four-dimensional units (defined as "reaches") can be identified and form the fundamental basis for scaling up to a watershed or region (Frissell et al., 1986). They are hierarchical, readily expanded, and collapsed in scale depending on the research question being addressed.

Gregory et al (1991) stressed the importance of the four-dimensional perspective for understanding the linkages between land and water in the structuring and functioning of stream ecosystems. Streams and their riparian zone structure and function result from geomorphic, hydrologic, topographic, edaphic, and biotic processes occurring within the watershed. Gregory and colleagues defined reaches by channel and valley floor geomorphic features and as having distinct topographic, edaphic, and disturbance regimes. These reaches had distinct longitudinal, lateral, vertical, and temporal characteristics.

Ecological Significance of the Species

For understanding many biological phenomena, such as evolution, genetics, and the growth and development of organisms, the species, defined as genetically related organisms that can produce offspring, is the fundamental unit. As discussed previously, the ecosystem is another fundamental unit governing biological processes. The definition of boundaries of an ecosystem can be designated to facilitate study of particular processes, whereas the definition of a species is more exact. Questions arise as we try to conceptually relate species and ecosystems.

Species, Niches and Ecosystem Function

Aquatic ecologists strive to understand the controlling relationships within an ecosystem. Two relationships that are currently under active research are (1) grazing on microorganisms by microheterotrophs (referred to as "the microbial loop"), and (2) complex, multiple pathways linking primary producers and top predators (sometimes referred to as "top-down or bottom-up" controls). These are not distinct issues, because the nature of the microbial loop in a lake ecosystem could influence the importance of higher trophic-level organisms in regulating the ecosystem. Although questions may be posed at the level of trophic interactions, in order to examine these interactions in detail in any particular aquatic ecosystem, the questions become focused on which species is doing what, how fast, and when?

> *Question: How do characteristics and habitat requirements of individual species fit together to control ecosystem-scale processes, such as productivity and resilience?*

Aquatic ecosystems often contain many species within each trophic level. The magnitude of the challenge presented to the researcher by this diversity varies with trophic group and aquatic environment. For some groups, species-level identification may be very difficult to obtain. Even the knowledge of species may seem to be insufficient because it does not carry with it detailed knowledge of habitat requirement or ecological function.

Individual benthic invertebrate species have been categorized according to their ecological function as scrapers (on periphyton), shredders (of leaves), predators (on other invertebrates), etc. Therefore, a list of the distribution and abundance of benthic invertebrates can convey a picture of ecosystem interactions in a stream and constrain the hypotheses about ecosystem processes. For example, many species of mayflies graze on algae growing on rocks in the streambed (periphyton), whereas many species of stoneflies shred and ingest leaves. The presence of both mayflies and stoneflies in a stream indicates significant quantities of organic material, either produced in the stream by periphyton or entering the stream from trees adjacent to it in the riparian zone.

For algal species, the ecological relevance of species data may be useful, but less refined than for benthic invertebrates. Hutchinson (1961) asked why so many algal species are present in a 1-liter sample of lakewater, a seemingly uniform environment. He referred to this question as the "paradox of the plankton." Whether or not this is a true paradox can be debated. For algal species, some general habitat requirements and ecological functions are known at the division level (i.e., diatoms require silica and heterocystous cyanobacteria can fix atmospheric nitrogen). However, more detailed habitat information is inferred from distributional patterns rather than by direct study of individual algal species. The significance of interannual shifts in algal species, whether within a lake or stream ecosystem, or the presence of particular rare species, may be more difficult to interpret.

Question: Is understanding of important microbial processes in aquatic ecosystems (the microbial loop, heterotrophic degradation of natural organic material or organic contaminants) constrained by lack of species-level information?

For bacteria, the species concept may seem hardly relevant to ecological studies. In the 1970s and 1980s, microbial ecology classified bacteria functionally, rather than by detailed, recognizable, morphological traits. Their habitat requirements were described broadly (e.g., aerobic, anaerobic, facultative, obligate). However, more current laboratory studies of isolates of aquatic bacteria show that genetic characteristics may be significant in showing relatedness among bacterial strains, but that functional characteristics are not unique.

For example, some species of sulfate-reducing bacteria may be capable of reducing ferric iron under certain conditions. Current cutting-edge research is examining genetic characteristics in the context of the individual and populations; the context of communities or ecosystems is probably the next frontier. This is an important complement to another cutting-edge research area with much practical significance: the development of engineered bacterial species to remediate certain compounds in situ. This research area has much promise and is a certain future direction of applied limnology.

Question: How does biodiversity in aquatic ecosystems develop and change? How much time is needed?

Although the vulnerability of ecosystems to species loss through habitat modification and contamination is widely acknowledged, we lack a coherent data base to document changing biodiversity of inland aquatic ecosystems. An understanding of species habitat requirements and interspecies interactions is fundamental. The dynamic changes in biological populations can occur over different time scales, which will control how ecosystems adapt and evolve. Some current federal programs may provide biodiversity data for some groups and for some ecosystems in the future. Yet even if we had that data now, we might not be able to interpret the data in terms of causative factors. For example, trout fishermen and others claim that there has been a loss in diversity of benthic invertebrates in western streams. Without a knowledge of the habitat or species interactions and other factors that would maintain diversity in benthic invertebrate communities, such changes cannot be attributed to habitat degradation, chemical contamination, or overfishing.

Intraspecies Genetic Variation

For both unicellular and multicellular organisms, genetic variations occur within a species. These genetic variations (genotypes) give rise to different physical characteristics (morphotypes). Under certain environmental conditions, some morphotypes may have a greater probability of survival than others; thus, the distribution of genetic characteristics within a population can shift.

The expression of genetic characteristics can also be regulated by environmental conditions, which is referred to as "plasticity." Modern biochemical techniques allow for characterization of genetic variation within a population of a given species, but interpretation of this information in terms of genetic shifts and adaptations to environmental conditions is a challenge.

Question: How much plasticity and genetic variation is there for different species, and does that influence ecosystem stability?

This is a "big" question, in that there is a tremendous gap in knowledge, and a "domino" question, in that answers have the potential to increase understanding of many limnological processes.

Questions related to genetic characteristics may seem esoteric at first, but they have important consequences for water quality issues. The following examples involving phytoplankton illustrate this point. Copper sulfate treatment is used routinely as a toxicant to control the growth of nuisance algal species in drinking water reservoirs. In a study of copper sulfate treatment, 90 percent of the population of the target species *Ceratium hirundinella* was killed by the treatment, and many cell fragments were observed in the reservoir; however, 10 percent of the population remained with no impairment of motility or other functions. If such survival is attributable to intraspecies variation in tolerance to copper, such genetic variation may contribute to the annual recurrence of this nuisance species in the reservoir.

Another example related to water quality is the development of toxic blooms of cyanobacteria (blue-green algae). Studies have shown that within a population, some strains produce neurotoxins and others do not. Toxic blooms occur when neurotoxin-producing strains become dominant. Occurrence of toxic blooms in farm ponds can cause severe illness or death in livestock drinking at the pond. Understanding the circumstances that favor toxin-producing strains would be useful in managing this problem.

Question: Can generalizations be made about genetic variations within particular classes of organisms?

There are many potential directions for research on genetic variation. Tolerance and bioaccumulation of contaminants is an issue for organisms at all trophic levels. If there are general patterns of genetic variation, such as more variation in tolerance in diatoms compared to cladocerans (water fleas), knowledge of these patterns would help in understanding ecosystem response to environmental stress.

Natural Water—A Chemical World

"Water quality" is a phrase used commonly in discussing water resources, but the meaning of water quality can be a source of confusion. Quality has many meanings, the principal ones being "peculiar and essential character" and "degree of excellence." The first refers to objective features of water; the second, to value judgments about the water for particular human uses (Averett and Marzolf, 1987). This distinction acknowledges that scientific inquiry into water quality issues (first meaning) is essential for the appreciation, definition, and management of water's utility for human uses (second meaning). The topics discussed below address water quality issues from the second perspective; specifically, they address the "so-what" question that can be asked about data sets of water quality parameters.

Major Cations, Anions, and Alkalinity

The major cation and anion concentrations vary greatly among inland waters, from saline to very dilute. There are many potentially significant questions. Addressing them could yield important insights and allow for a meaningful interpretation of the extensive water quality data that routinely are collected around the country and archived in computer data bases with no more analysis than computation of average values. Such questions include the following:

Questions: What is the significance of variations within the more typical range for determining species distributions of microorganisms, aquatic plants, vertebrates, or invertebrates? How does the abundance of monovalent or divalent cations affect the electrical properties of the layer around the cell? How does that in turn affect solute transport (e.g., nutrient uptake) at the cell membrane? Are natural organic polyacids, such as fulvic acid, also important in this context, and do they cross the cell membrane?

Effects of Many Different Constituents

The major nutrients for growth required by autotrophic microorganisms are nitrogen and phosphorus. These nutrients are dilute in the aquatic medium relative to major cations and anions; autotrophic microorganisms therefore must be able to assimilate these and other required solutes effectively at low concentrations. Microorganisms as well as multicellular organisms are also sensitive to toxicants at low concentrations and can release organic constituents to the medium that have specific functions (e.g., vitamins).

Question: Which of these natural and anthropogenic trace constituents are critical in controlling inland aquatic ecosystems?

Chemical regulation of aquatic ecosystems can occur through limitations on the concentration or bioavailibility of particular chemical species or through the toxic effects of naturally present constituents at elevated concentrations or of xenobiotic compounds. Iron is a metal that is required by microorganisms and may be limiting in some aquatic ecosystems because of its inherently low solubility at the neutral and higher pH values of many natural waters.

Ecotoxicology should be studied and applied in the broadest sense, including the effects of concentrations far below lethal doses. How do molecules released by anthropogenic activities interfere with the capacity of individuals and (more important) entire populations to adapt to a changing environment? Environmental estrogens are an example of compounds that may not be directly toxic but may limit the ability of populations to reproduce and survive. A fundamental research question that would lead to better water quality standards for contaminants in the future involves the mechanism of contaminant uptake and biological magnification by organisms. For many important hydrophobic contaminants, for example, a simple hydrophobic partitioning model does not adequately explain uptake.

Questions: How reliable are present data for trace metals in fresh waters? What applications of current analytical capabilities can be used to provide new ways of studying chemical (also physical and biological) processes in inland aquatic environments? Can new in situ methods be devised?

In many water quality problems, hydrologic and chemical processes are tightly linked, and aquatic systems inherently are temporally dynamic and spatially heterogeneous. For some contaminants, a single analysis can be laborious and expensive for a low concentration that is nonetheless sufficient to have biological effect. Because of this, it is difficult to make the numerous chemical analyses that would be required to link hydrology to chemistry. This limitation may be overcome in the future by going on-line and in situ to enhance resolution in time and in space.

Detrital Organic Material

In aquatic ecosystems, detrital organic carbon often is the largest pool of organic carbon, typically exceeding the biomass of living algae or macrophytes by several orders of magnitude. In many systems, this organic material is allochthonous (originating in the surrounding terres-trial environment) rather than autochthonous (produced within the lake or stream). Degradation of this detrital organic matter has the potential to exert strong controls on the ecosystem, such as the development of reducing conditions in retention zones within stream ecosystems.

There are many different forms of detrital organic carbon in aquatic ecosystems and an array of acronyms for each class. For particles, the distinctions are primarily based on size. CPOM (coarse particulate organic matter) can be composed of twigs, leaves, etc., and FPOM (fine particulate organic matter) can be small bits of shredded leaves. Organisms will have different physiological adaptations to feed on CPOM and FPOM. Another meaningful size distinction is particulate and colloidal, in that different processes control their transport and deposition. The term DOC (dissolved organic carbon) typically is used in a nominal sense for organic matter passing through a 0.4-μm filter and includes colloidal and truly dis-solved forms.

Question: How does the chemical composition of this abundant detrital organic material influence heterotrophic degradation and, in turn, the nature of the eco-system?

These size-based distinctions have their utility but do not address the quality of the material as a substrate for growth by either multicellular or unicellular heterotrophic organisms. For example, DOC includes dis-solved glucose, which would be assimilated readily by microorganisms, and dissolved fulvic acids, which are biologically recalcitrant organic acids persisting for many years. A DOC measurement is the quantitative sum of many different forms of carbon. Trying to understand heterotrophic processes with only DOC measurements is comparable to trying to understand the inorganic geochemistry of natural waters with only measurements of conductivity.

Question: What are the different pathways by which detrital organic carbon is respired in aquatic systems? How does the source of the detrital organic carbon determine the quality of the substrate?

Importance of Long-Range Transport of Constituents

Question: What are the mechanisms controlling interactions of watersheds with overlying airsheds?

The transport and deposition of atmospheric acidic constituents, SO_x and NO_x, affects the weathering rates of soils and nutrient conditions on land and in surface waters, with consequent effects on biological processes and ecosystems. For example, approximately one-third of the present nitrogen input to the Chesapeake Bay is estimated to arise from atmospheric deposition in the watershed. Another example is tropospheric ozone: ozone production in regional airsheds such as the Baltimore-Washington area involves photochemical reactions between NO_x and hydrocarbons from anthropogenic and terrestrial sources to yield atmospheric ozone concentrations many miles from urban sources that exceed allowable limits.

Chlorinated organic substances such as PCBs (polychlorinated biphenyls), DDT, PAHs (polycyclic aromatic hydrocarbon), and dioxin are transported to aquatic systems primarily by the air route, and may have important effects on the organisms in these systems. Mercury is also transported atmospherically; however, the available data base for metals in freshwater ecosystems may be limited to recent years because of potential contamination of samples collected earlier.

The interactions of sulfur, nutrients, synthetic organic substances (including but not limited to pesticides), and many metals during transport through the vegetation canopy, across soil surfaces, through the highly reactive surficial zone of soils, and downward into aquifers is a challenging research area of environmental concern that requires an integration of chemical and hydrologic concepts and field measurements.

A Changing Climate

Question: What is the relative importance of direct physical changes compared to indirect biogeochemical changes in determining responses to changing climate?

The effect of global climate change on regional hydrology and, in turn, on vegetation and on wetlands is a very important concern requiring substantial attention. Changes in hydrologic regimes may alter the extent or timing of biogeochemical processes that control the flux of solutes, particularly nutrients. Such biogeochemical changes may have a greater effect on the receiving lakes and streams than direct physical effects of warmer water temperatures, for example.

Humans: A Keystone Species

Humans play a significant role in all aquatic systems. They alter the physical, chemical, and biological structure and function of aquatic systems through direct interactions (e.g., harvesting fish, discharging chemicals, thermal discharges from industry) and indirectly through altering watersheds, airsheds, and in-stream processes (e.g., dams, reservoirs, introduction of exotic species). Although much alteration of aquatic systems has occurred, few follow-up studies have fully explored the ecosystem response to those changes. Humans have been viewed by ecologists as primarily "drivers" or "receivers" in relation to ecological systems and not as an integrated part of the aquatic system. Basic research is needed to understand the fundamental role that humans play as a keystone species in aquatic systems.

Fundamental Associations Between Humans and Aquatic Systems

Question: Is there a common paradigm between human and aquatic sciences that can be used to relate limnology and sociology?

High-quality fresh water will become a limited resource in the next century. Human-environmental interactions are causing the greatest rate of change in land, water and air resources that the earth has seen in more than 10,000 years (Silver and DeFries, 1990; Turner et al., 1990). Projected growth and redistribution of human population centers will result in increased urbanization of rural areas. Land-use changes, waste generation, and hydrologic alteration will significantly modify aquatic systems. These extensive human effects on aquatic systems and associated watersheds have been proceeding without a commensurate understanding of limnological consequences or integrated (including ecological, social, political, and regulatory) management strategies. Our understanding is limited by the complexity of the problem and the lack of appropriate spatial and temporal information. Complicating the situation is the lack of societal and institutional mechanisms for addressing these problems and of appropriate mechanisms for transferring "scientific knowledge" into adaptive holistic resource management. New mechanisms (e.g., technology transfer or training), tools (e.g., integrated models), and multidisciplinary approaches are needed to solve these problems.

Questions: Are human associations more than "drivers" or "receivers"? How do human metrics relate to the linkages between humans and aquatic systems? What ecological metrics relate to the linkages between humans and aquatic systems?

Defining the Roles Humans Play as the Keystone Species

Interactions between natural (e.g., environmental) and human (e.g., socioeconomic) sciences are not well understood. Integrated, interdisciplinary approaches have been applied only infrequently to problem solving. Ecologists tend to model ecological systems, while social or economic scientists tend to model human systems or human values. These different views are linked sometime later, rather than building an integrated model early on as part of a focused effort. Extrapolating from two separate approaches cannot compare with a truly integrated approach taking all relevant factors into account. Site-specific to local information must be developed into an integrated regional perspective using scalable research activities. What is needed is a coherent, integrated theory that fundamentally links human and aquatic systems, and this theory must be a fundamental component of future aquatic research.

Questions: What ecological and human metrics are important in quantifying or defining the keystone species role in aquatic systems? Is the ecological metric structure, function, flux, rate, timing, or variability? How important are human metrics such as perceptions, myths, socioeconomic status, cultural values, education, and governmental policy in defining the association with aquatic systems?

In order to stimulate the development of integrated human and aquatic research, a number of fundamental needs must be considered. These needs relate to the availability of information in a usable form, availability of knowledge and expertise, ability to develop a common language for effective communication, and identification of a common currency for model validation. We must move from viewing humans as peripheral to viewing humans as a fully integrated part of any aquatic system.

Rambo (1983) developed a general systems theory for humans that might prove useful in integrated research. These systems exist in "a complex, dynamic relationship with multi-causal, multi-directional exchanges of energy, material, and information. Each system is open to external influences through diffusion, migration, and colonization" (Turner et al., 1990). Changes in the system may be sudden or gradual and adaptive, with evolution expressed as survival of species and choices of individuals institutionalized as social norms.

This theory incorporates nature's constraints on human behavior and human behavior's feedbacks to the environment. It focuses on connectivity and mutual causality among natural and human components. It positions humans in the natural environment and allows us to understand what we are doing and why. Natural sciences are focused on the quantification of energy, materials, and information with an emphasis on interconnections, dynamics, and exchanges with the external environment. Organisms are linked through the exchange of energy according to the laws of thermodynamics and materials through biogeochemical cycles. A dynamic equilibrium prevails with changes occurring through evolution and succession.

SIGNIFICANCE OF BASIC RESEARCH

There is a greater awareness in the water resource management community that the piecemeal approaches of the past are insufficient. The new ecosystem assessment bandwagon is in some respects a "let's be smart about it" approach, where the relevant knowledge is applied and critical missing information is obtained. Obviously, greater knowledge and understanding of aquatic ecosystems will contribute to the success of these efforts and of this ecosystem-based approach.

A simple example of a remediation effort for an acid mine drainage stream in the Rocky Mountains illustrates how the questions identified above are relevant to addressing important water resource and environmental issues. In the 1800s and early part of this century, there was a mining boom in the Rocky Mountains, and this boom spurred the development of the western states. Many of these mines are now abandoned, but their legacy includes an almost uncountable number of acidic, metal-enriched water discharges into mountain streams, with elevated metal concentrations in larger rivers. The ecological consequences of the acid mine drainage are not subtle. In the headwater streams, the streambed is typically covered with hydrous iron oxides, known as yellowboy; in the larger rivers, fish populations have high body burdens of metals, which limit their survival. In the Arkansas River, which drains the Lake County mining district in the Colorado Rockies, trout do not survive beyond four years because of accumulated metals in their tissues.

In many of these areas, there are public and legal incentives to clean up such sites. For example, in the summer of 1994, 268 members of Volunteers of Colorado came to the Pennsylvania mine site on Peru Creek in Colorado to help restore the creek by shoveling manure into two large plastic-lined pits. These pits will become artificial wetlands to treat the mine effluent. The pits are located adjacent to a liming facility that was built in the 1980s and proved to be a failure. Such treatments are being tried in many areas to ameliorate both coal and metal mining wastes. Although the volunteers at Peru Creek were enthusiastic and hopeful, from a scientific perspective can we be sure that this approach will work in either the short or the long term?

Many of the issues, challenges, and unanswered questions raised in the preceding section come into play in a more specific way in this example of remediating acid mine drainage. Questions related to temporal and spatial dynamics are clearly important. If wetlands are effective, they essentially accumulate and store metal contaminants alongside the stream. What are the interactions between the stream and its watershed that may influence the long-term stability of the wetland, and are there hydrologic connections between the watershed and the stream that would develop as different routes of metal transport? For the treatment to be effective it must also work during the winter, when the snow is meters deep in the valley, and better understanding of wetland and stream ecosystems during the winter would be useful in this regard. Because of the abundant iron oxides accumulated on the streambed, there may be substantial diel variations in the stream driven by photoreduction of the iron oxides.

Questions related to the occurrence of particular species are also relevant. Should the success of the treatment be evaluated in terms of the return of ecosystem function, as measured by rates of primary production by periphyton, or should success be based on the return to the remediated stream of species of algae, benthic invertebrates, and fish normally found in pristine streams in the region? The chemical considerations relate to both the reduction of contaminants in the effluent and any new chemical species that may be released by the organic-rich artificial wetland.

This example also highlights the importance of understanding the changing human dimension and its relationship to water resources and aquatic ecosystems. The miners who caused many of these problems were doing well just to survive in the harsh mountain environment. It was a "boom-or-bust" industry and there was not much emphasis on planning for the future. The environmental ethic of stewardship that now motivates some members of the public, especially the 268 volunteers, may be a relatively new ethic for Western culture.

ACKNOWLEDGMENTS

We acknowledge contributions of ideas and comments to this chapter from our colleagues S. Spaulding, R. Runkel, A. Covich, D. Macalady, Dieter Imboden, Lynn Roberts, Laura Sigg, Alan Stone, Werner Stumm, Bernhard Wehrli, M. Gordon Wolman, and Alexander Zehnder.

REFERENCES

American Society of Limnology and Oceanography. 1994. Regional Assessment of Freshwater Ecosystems and Climate Change in North America Symposium report, October 24–26, 1994, Leesburg, Va.

Averett, R. C., and G. R. Marzolf. 1987. Water quality. Environ. Sci. Technol. 21(9):827

Edmondson, W. T. 1991. The Uses of Ecology: Lake Washington and Beyond. Seattle: University of Washington Press.

Environmental Protection Agency (EPA). 1980. Urban Storm Water and Combined Sewer

Overflow Impact on Receiving Water Bodies, Proceedings of National Conference, Orlando, Fla. EPA-600/9-80-056. Washington, D.C.: EPA.

Environmental Protection Agency (EPA). 1983. Results of the Nationwide Urban Runoff Program. Executive Summary PB84-185545. Springfield, Va.: National Technical Information Service.

Frissell, C. A., W. J. Liss, C. E. Warren, and M. C. Hurley. 1986. A hierarchical framework for stream habitat classification: Viewing streams in a watershed. Environ. Manage, 10:199–214.

Goldman, C. R. 1981. Lake Tahoe: Two decades of change in a nitrogen deficient oligotrophic lake. Proc. Int. Assoc. Theoret. Appl. Limnol. 21:45–70.

Goldman, C. R. 1985. Lake Tahoe: a microcosm for the study of change. Bull. Ill. Nat. Hist. Sur. 33:247–60.

Gregory, S. V., F. J. Swanson, W. A. McKee, and K. W. Cummins. 1991. An ecosystem perspective of riparian zones: Focus on links between land and water. BioScience 41(8):540–551.

Hecky, R. E., and P. Kilham. 1988. Nutrient limitation of phytoplankton in freshwater and marine environments: A review of recent evidence on the effects of enrichment. Limnol. Oceanogr. 33:796–822.

Hutchinson, G. E. 1961. The paradox of plankton. Am. Naturalist 95:137–145.

Lazaro, T. R. 1990. Urban Hydrology: A Multidisciplinary Perspective. Lancaster, Pa.: Technomic Publishing Company.

Loehr, R. C., D. A. Haith, M. F. Walter, and C. Martin. 1979. Best Management Practices for Agriculture and Silviculture. Ann Arbor, Mich.: Ann Arbor Science Publishers.

Novotny, V., and G. Chesters. 1981. Handbook of Non-Point Pollution: Sources and Management. New York: Van Nostrand Reinhold.

Omernik, J. M. 1976. The Influence of Land Use on Stream Nutrient Levels. EPA-600-76-014. Corvallis, Ore.: Environmental Protection Agency.

Rambo, A. T. 1983. Conceptual Approaches to Human Ecology. Research Report No. 14. Honolulu: East-West Center.

Repetto, R. 1986. World Enough and Time: Successful Strategies for Resource Management. New Haven: Yale University Press.

Silver, C. S., and R. S. DeFries. 1990. One Earth One Future: Our Changing Global Environment. Washington, D.C.: National Academy of Sciences.

Straub, C. P. 1989. Practical Handbook of Environmental Control. Boca Raton, Fla.: CRC Press.

Turner, B. L., W. C. Clark, R. W. Kates, J. F. Richards, J. T. Mathews, and W. B. Meyer. 1990. The East as Transformed by Human Action. New York: Cambridge University Press.

Van der Leeden, F., F. L. Troise, and D. K. Todd. 1990. The Water Encyclopedia. Chelsea, Mich.: Lewis Publishers.

Ward, J. V., and J. A. Stanford. 1991. Research directions in stream ecology. Pp. 121–132 in Advances in Ecology: Research Trends, J. Menon, ed. Trivadrum, India: Council of Scientific Research Integration.

The Role of Major Research Centers in the Study of Inland Aquatic Ecosystems

Thomas M. Frost
Trout Lake Station
Center for Limnology
University of Wisconsin–Madison
Boulder Junction, Wisconsin

Elizabeth Reid Blood
J. W. Jones Ecological Research Center
Newton, Georgia

SUMMARY

Major research centers have made substantial contributions to the basic understanding of inland aquatic ecosystems. Such centers function by providing access to important natural ecosystems and by fostering interdisciplinary investigations of a wide range of environmental phenomena. This paper provides seven representative examples of aquatic research centers and summarizes their contributions.

INTRODUCTION

Much of the present understanding of inland aquatic ecosystems can be traced to contributions from major research centers. The success of research centers can be linked to a few of their key features. Many of these centers provide or facilitate access to sites with one or a group of important aquatic habitats. Such centers often provide logistical support for the routine collection of fundamental information on habitats, providing long-term records on system condition. Other types of centers function primarily by fostering interactions among a group of researchers rather than by providing access to a particular location. Perhaps most important, collaborations at many of these different centers frequently lead to interactions between investigators with different disciplinary perspectives often not combined in smaller-scale investigations of aquatic ecosystems. A few research sites also have a capability for large-scale or multiple experimental manipulations. Finally, several research centers have shown how basic science and management perspectives can be combined.

EXAMPLES OF CONTRIBUTIONS FROM RESEARCH CENTERS

A wide variety of research centers have made important contributions to the basic understanding of aquatic ecosystems. Examples, discussed below, are Hubbard Brook Experimental Forest, Experimental Lakes Area, University of Wisconsin–Madison, Woods Hole Marine Biological Laboratory, Lake Washington, Coweeta Experimental Forest, and H. J. Andrews Experimental Forest.

Hubbard Brook Ecosystem Study

The Hubbard Brook Experimental Forest (HBEF) is a striking example of the contributions that can be made by a research center. The HBEF was established by the U.S. Forest Service in 1955 to investigate the management of watersheds in New England. The research program at the forest, developed largely through the efforts of G. E. Likens and F. H. Bormann, has made substantial contributions to the general understanding of both aquatic and terrestrial ecosystems (e.g., Likens et al., 1977; Bormann and Likens, 1979; Likens, 1985a). Work has focused on processes occurring in the forests and streams in the area and on Mirror Lake near the base of HBEF. Fundamental techniques for assessing ecosystem nutrient cycling were developed by combining the perspectives of hydrologists and biogeochemists (Likens et al., 1977). Insights gained into nutrient cycling at HBEF through whole-watershed clear-cutting experiments and other harvesting programs illustrate the usefulness of ecosystem-scale manipulations (Likens et al., 1977). Detection of the acid deposition phenomenon in North America through precipitation records collected over an extended period has demonstrated the importance of long-term monitoring (Likens et al., 1972, 1984). Research has helped to delineate the role of streams in processing nutrients and organic matter (Fisher and Likens, 1973; Meyer and Likens, 1979). Documented responses to varied timber harvesting techniques at Hubbard Brook have also illustrated the ways in which research with a fundamentally basic science perspective can shed important light on the impact of land-use practices (e.g., Bormann and Likens, 1985; Likens, 1985b). More recently, the Hubbard Brook Experimental Forest has been incorporated as a site within the National Science Foundation's Long-Term Ecological Research (LTER) program, facilitating the continued evaluation of aquatic and terrestrial ecosystem processes at the site (Franklin et al., 1990).

Experimental Lakes Area

The Experimental Lakes Area (ELA), located in the lake district of northwestern Ontario, provides a second prime example of the wealth of information that can arise from a research center. ELA was originally established in the early 1960s by the Canadian government under the leadership of W. E. Johnson and J. R. Vallentyne to investigate problems associated with eutrophication (Johnson and Vallentyne, 1971). Research there has since been expanded, under the leadership of D. W. Schindler, to combine a whole-system experimental perspective with long-term monitoring to provide a basic understanding of aquatic ecosystems (e.g., Schindler, 1973, 1987, 1988).

Some of the first projects at ELA provided important support for management programs to combat eutrophication by confirming the importance of phosphorus as a limiting nutrient in most inland aquatic ecosystems (Schindler, 1974). Subsequent work expanded the focus of the site to consider the effects of acid deposition (Schindler et al., 1985) and other contaminants (Schindler, 1987). Insights gained into the microbial factors mitigating the effects of lake acidification illustrate the importance of combining different disciplinary perspectives in understanding aquatic ecosystem processes (Schindler et al., 1986). The breakdown of sulfate ions under anaerobic conditions with an accompanying removal of acid ions was shown to provide a substantially greater resistance to the effects of sulfuric acid in deposition than had been expected (Cook et al., 1986). In a parallel study, bacterially mediated transitions among different forms of nitrogen were also shown to be strongly influenced by pH in some lakes (Rudd et al., 1988). More recently, long-term data collected on unmanipulated lakes at ELA have provided evidence for climatic warming trends in boreal lakes (Schindler et al., 1990), illustrating some of the potential effects of the climatic changes predicted by several global circulation models and the general importance of extended monitoring of inland aquatic systems.

Wisconsin

The University of Wisconsin–Madison is perhaps one the best examples of long-continuing efforts at a research center focused on inland aquatic ecosystems (Beckel, 1987). E. A. Birge initiated his lake studies there in 1875, beginning with a focus on lakes near the university campus in Madison (Frey, 1963). For more than 60 years, C. Juday was his collaborator on most of these efforts. Expanding on their work around Madison, they were attracted by the potential for comparative studies in the extensive lake district in the Northern Highland of Wisconsin. They established a field station on Trout Lake in 1925 and organized regular summer research by a large group of collaborators. Their research continued until the early 1940s.

Following a hiatus, A. D. Hasler led a second major period of research with an emphasis on experimental studies (Hasler, 1964). This research, involving several large-scale experiments, was conducted around Madison and in the Northern Highland area. Particularly important was the neutralization of a naturally acidic lake by using a now classic design in which the basin was divided into similar manipulated and unmanipulated sections (Hasler, 1964). Researchers involved in several more recent whole-ecosystem manipulations trace their inspiration to these early large-scale experiments by Hasler and his coworkers (e.g., Likens, 1985c; Schindler, 1988; Carpenter and Kitchell, 1994).

Research in Wisconsin continues with major ongoing efforts in the Madison area, primarily on Lake Mendota, and in the Northern Highland. By evaluating current conditions relative to the range of conditions in the past, recent work on Mendota has illustrated the utility of long-term data in understanding system processes (Brock, 1985; Kitchell, 1992). This work also provides another example of the potential gains when nontraditional fields are combined in research projects, in this case fisheries biology and limnology (Vanni et al., 1990). Substantial shifts in the lake's water clarity were ultimately attributable to an unusual die-off of one of its dominant fish. Lake Mendota has also been the site of a major joint collaboration between lake managers and basic scientists (Kitchell, 1992).

Other efforts in Wisconsin include a substantial expansion of projects at the Trout Lake Station in the Northern Highland. The station has served as the staging facility for one of the sites in the Long-Term Ecological Research program (Magnuson et al., 1984) and for a whole-lake acidification experiment (Brezonik et al., 1993). LTER work here has provided another example of the utility of combining disciplines, in this case ground water geology and limnology, by demonstrating how a relatively small portion of the lake's hydrologic inputs could play a critical role in its overall nutrient budget because of the high concentration of minerals in ground water (Hurley et al., 1985).

Regional evaluations of long-term records illustrate how variability, often perceived as an impediment to system understanding, can provide useful insights into the processes controlling system features (Kratz et al., 1987). Related projects have further demonstrated how evaluating patterns in data collected during the same period from sets of lakes within a region can indicate how different system properties are controlled by forces that operate on fundamentally different spatial scales (Kratz et al., 1991). Additional northern work has continued, using a large-scale experimental approach to evaluate the effects of lake acidification (Brezonik et al., 1993) and the role of food web structure on basic lake processes (Carpenter and Kitchell, 1993). The latter project has been conducted at the original site of some of the early whole-lake manipulations with the support of the University of Notre Dame's Environmental Research Center on the Michigan-Wisconsin border.

Ecosystem Center at Woods Hole

The Ecosystem Center at Woods Hole Marine Biological Laboratory provides an example of a center that supports a multidisciplinary group of investigators working at different field sites. Under the leadership of J. Hobbie and B. Peterson, the group has used a combination of large-scale manipulations and long-term monitoring to examine the basic limnology of Arctic tundra ponds (Hobbie, 1980). This group has demonstrated that the addition of phosphorus can convert an Arctic river from a heterotrophic to an autotrophic ecosystem (Peterson et al., 1985). Their work has shown the importance of tundra lakes and streams as conduits for the transfer of major quantities of carbon dioxide to the atmosphere (Kling et al., 1991) and helped to develop the use of stable isotope techniques for examining food webs (Peterson and Fry, 1987). Research in other locations has documented the importance of sulfur cycling in the transfer of energy in coastal salt marshes (Howarth and Teal, 1979). The center's more recent efforts have also led to the establishment of an Arctic site within the LTER network (Franklin et al., 1990) and of a near-shore site within the National Science Foundation's Land-Margin Ecosystems program.

Lake Washington

Another example of a different type of research center is Lake Washington near Seattle, Washington. Population development in the Seattle area led to an increased loading of nutrients into Lake Washington through the 1950s with a concomitant deterioration of water quality. A program was then established, with substantial guidance from W. T. Edmondson, to divert sewage from the lake. Edmondson and a group of researchers at the University of Washington then documented Lake Washington's responses to these management efforts (Edmondson, 1972), providing strong evidence for the usefulness of sewage nutrient control as a management practice at a critical stage in the eutrophication controversy. Analyses by Edmondson's group documented the positive effects of the sewage diversion program, provided basic understanding of the nature of the eutrophication process, and demonstrated a classic example of the potential of sound management practices for inland aquatic ecosystems. Their monitoring of Lake Washington continued following the initial responses to sewage diversion, revealing another substantial and surprising change in water transparency. From 1976 onward, the lake exhibited higher levels of water clarity than had been recorded at any earlier stage in its history.

These levels were linked with the emergence of previously inconspicuous *Daphnia* species (Edmondson and Litt, 1982) and provide evidence of how changes in a lake's food web can have a strong influence on its water clarity. Edmondson (1991) reviews the details of the Lake Washington story and illustrates the sometimes complex interplay between basic science and lake management issues.

This program illustrates how a single leader with long-term support for research can do a very effective job of fostering fundamental ecosystem understanding. It also illustrates the precarious nature of such programs; since Edmondson's retirement in the mid-1980s, the University of Washington has chosen not to continue its important history of limnological research and has hired no replacement for him.

Coweeta Experimental Forest

Like the program at HBEF, research at Coweeta Experimental Forest illustrates how a research center can generate important general information about both aquatic and terrestrial ecosystem processes (Swank and Crossley, 1988). In addition to providing fundamental information on the control of forest type over the amount of water exiting a watershed, work at Coweeta has documented how forest conditions influence the amount of organic materials transported by streams (Wallace et al., 1982; Tate and Meyer, 1983). Other research has shown how watershed conditions affect the basic chemistry and biology of streams (Webster and Patten, 1979; Meyer and Tate, 1983) and how the organisms occurring within streams influence the processing of organic matter (Meyer et al., 1988; Perlmutter and Meyer, 1991). The removal of all aquatic insects produced substantial changes in the transport of materials by streams, demonstrating the importance of insects in fundamental stream processes and further illustrating the usefulness of a large-scale experimental approach (Cuffney et al., 1984).

H. J. Andrews Experimental Forest

One of the major synthetic perspectives in river ecology, the River Continuum Concept, concerns how the proportions of fundamental ecosystem processes change systematically with distance downstream from headwaters to the ocean. This model can be traced in part to interactions arising at the H. J. Andrews Experimental Forest (HJAEF) along with the Stroud Water Research Center (Vannote et al., 1980). Multidisciplinary perspectives at HJAEF have helped develop an understanding of how geomorphology exerts a fundamental influence on basic ecosystem processes (Swanson et al., 1988). Additionally, work at HJAEF has indicated how coarse woody debris can influence stream processes, such as the rates at which water is transported, the decomposition of organic materials, and the rates of biogeochemical processes (Harmon et al., 1986). Other research has shown how processes in the riparian zone surrounding a stream have a wide range of effects that can best be understood from an ecosystem perspective (Gregory et al., 1991). Finally, research at HJAEF also demonstrated how forest management practices can have critical influences on stream function, including the usefulness of these habitats for the breeding success of fishes (Gregory et al., 1987).

CURRENT STATUS AND FUTURE POSSIBILITIES FOR RESEARCH CENTERS

Most of the centers described in this paper continue to remain quite active, and many are currently associated with the National Science Foundation Long-Term Ecological Research program. This program also includes additional sites that emphasize lake studies, such as one focused on the Dry Valley Lakes in Antarctica; several stream study programs (Meyer et al., 1993); and a coastal site, the Virginia Coast Reserve.

The extensive contributions from research centers make a strong case for a program that will further the establishment and continuation of such centers. New centers should be based on several basic principles: (1) a broad multidisciplinary foundation; (2) the collection of long-term baseline data on a number of systems within an area; (3) an opportunity to conduct large-scale, whole-ecosystem manipulations; and (4) a solid infrastructure to facilitate data collection and experimental work on a variety of scales. New centers should be planned with a recognition of geographic and system limitations on the present set of research sites. For example, the historic sites are quite limited geographically and do not effectively represent water bodies in the southern part of the continent, nor do they provide sufficient baseline data on reservoirs.

Research centers should be encouraged to develop interactions among major research sites and opportunities for graduate student training, including programs for short- and long-term visiting scientists and semester-long programs for graduate students that involve visits to a series of research centers. A good model is the highly successful Organization for Tropical Studies Tropical Ecology program, which has fostered a range of research throughout tropical ecosystems. Overall, the contributions from research centers are quite substantial. They have played a critical role in developing the fundamental understanding of inland aquatic ecosystems. The opportunities for research at these centers should be continued and expanded.

REFERENCES

Beckel, A. L. 1987. Breaking New Waters: A Century of Limnology at the University of Wisconsin. Transactions of the Wisconsin Academy of Sciences, Arts, and Letters. Special Issue. Madison: Wisconsian Academy of Sciences, Arts, and Letters.

Bormann, F. H., and G. E. Likens. 1979. Pattern and Process of a Forested Ecosystem. New York: Springer-Verlag.

Bormann, F. H., and G. E. Likens. 1985. Air and watershed management and the aquatic ecosystem. Pp. 436–444 in An Ecosystem Approach to Aquatic Ecology: Mirror Lake and Environment, G. E. Likens, ed. New York: Springer-Verlag.

Brezonik, P. L., J. G. Eaton, T. M. Frost, P. J. Garrison, T. K. Kratz, C. E. Mach, J. H. McCormick, J. A. Perry, W. A. Rose, C. J. Sampson, B. C. L. Shelley, W. A. Swenson, and K. E. Webster. 1993. Experimental acidification of Little Rock Lake, Wisconsin: chemical and biological changes over the pH range 6.1 to 4.7. Can. J. Fish. Aquat. Sci. 50:1101–1121.

Brock, T. D. 1985. A Eutrophic Lake: Lake Mendota, Wisconsin. New York: Springer-Verlag.

Carpenter, S. R., and J. F. Kitchell. 1993. The Trophic Cascade in Lakes. Cambridge, England: Cambridge University Press.

Cook, R. B., C. A. Kelly, D. W. Schindler, and M. A. Turner. 1986. Mechanisms of hydrogen ion neutralization in an experimentally acidified lake. Limnol. Oceanogr. 31:134–148.

Cuffney, T. F., J. B. Wallace, and J. R. Webster. 1984. Pesticide manipulation of a headwater stream: Invertebrate responses and their significance for ecosystem processes. Freshwater Invert. Biol. 3:153–171.

Edmondson, W. T. 1972. Nutrients and phytoplankton in Lake Washington. Pp. 172–193 in Nutrients and Eutrophication, G. E. Likens, ed. Special Symposium No. 1. Gloucester Point, Va.: American Society of Limnology and Oceanography.

Edmondson, W. T. 1991. The Uses of Ecology: Lake Washington and Beyond. Seattle: University of Washington Press.

Edmondson, W. T., and A. H. Litt. 1982. Daphnia in Lake Washington. Limnol. Oceanogr. 27:272–293.

Fisher, S. G., and G. E. Likens. 1973. Energy flow in Bear Brook, New Hampshire: An integrative approach to stream ecosystem metabolism. Ecol. Monogr. 43:421–439.

Franklin, J. F., C. S. Bledsoe, and J. T. Callahan. 1990. Contributions of the long-term ecological research program. BioScience 40:509–523.

Frey, D. G. 1963. Wisconsin: The Birge and Juday years. Chapter 1 in Limnology in North America, D. G. Frey, ed. Madison: University of Wisconsin Press.

Gregory, W. V., G. A. Lamberti, D. C. Erman, K. V. Koski, M. L. Murphy, and J. R. Sedell. 1987. Influence of forest practices on aquatic production. Pp. 234–255 in Streamside Management: Forestry and Fishery Interactions, E.O. Salo and T.W. Cundy, eds. Proceedings of a Symposium. Seattle: Insititute of Forest Resources, University of Washington.

Gregory, S. V., F. J. Swanson, W. A. McKee, and K. W. Cummins. 1991. An ecosystem perspective of riparian zones. BioScience 41:540–551

Harmon, M. E., J. F. Franklin, F. J. Swanson, P. Sollins, S. V. Gregory, J. D. Lattin, N. H. Anderson, S. P. Cline, N. G. Aumen, J. R. Sedell, G. W. Lienkaemper, K. Cromack, Jr., and K.W. Cummins. 1986. Ecology of coarse woody debris in temperate ecosystems. Pp. 133–302 in Advances in Ecological Research, A. MacFadyen and E. D. Ford, eds. Orlando, Fla.: Academic Press.

Hasler, A. D. 1964. Experimental limnology. BioScience 14(7):36–38.

Hobbie, J. E., ed. 1980. Limnology of Tundra Ponds. U.S./IBP Synthesis Series V. 13. Stroudsburg, Pa.: Dowden, Hutchinson, and Ross.

Howarth, R. W., and J. M. Teal. 1979. Sulfate reduction in a New England salt marsh. Limnol. Oceanogr. 24:999–1013.

Hurley, J. P., D. E. Armstrong, G. J. Kenoyer, and C. J. Bowser. 1985. Groundwater as a silica source for diatom production in a precipitation-dominated lake. Science 227:1576–1579.

Johnson, W .E., and J. R. Vallentyne. 1971. Rationale, background, and development of experimental lake studies in Northwestern Ontario. J. Fish. Res. Bd. Can. 28:123–128.

Kitchell, J. F. 1992. Food Web Management: A Case Study of Lake Mendota. New York: Springer-Verlag.

Kling, G. W., G. W. Kipphut, and M. C. Miller. 1991. Arctic lakes and streams as gas conduits to the atmosphere: Implications for tundra carbon budgets. Science 251:298–301.

Kratz, T. K., T. M. Frost, and J. J. Magnuson. 1987. Inferences from spatial and temporal variability in ecosystems: Long-term zooplankton data from lakes. Am. Nat. 129:830–846.

Kratz, T. K., B. J. Benson, E. R. Blood, G. L. Cunningham, and R. A. Dahlgren. 1991. The influence of landscape position on temporal variability in four North American ecosystems. Am. Nat. 138:355–378.

Likens, G. E., ed. 1985a. An Ecosystem Approach to Aquatic Ecology: Mirror Lake and Its Environment. New York: Springer-Verlag.

Likens, G. E. l985b. The aquatic ecosystem and air-land-water interactions. Pp. 430–435 in An Ecosystem Approach to Aquatic Ecology: Mirror Lake and Its Environment, G. E. Likens, ed. New York: Springer-Verlag.

Likens, G. E. 1985c. An experimental approach for the study of ecosystems. J. Ecol. 73:381–396.

Likens, G. E., F. H. Bormann, and N. M. Johnson. 1972. Acid rain. Environment 14:33–40.

Likens, G. E., F. H. Bormann, R. S. Pierce, J. S. Eaton, and N. M. Johnson. 1977. Biogeochemistry of a Forested Ecosystem. New York: Springer-Verlag.

Likens, G. E., F. H. Bormann, R. S. Pierce, J. S. Eaton, and R. E. Munn. 1984. Long-term trends in precipitation chemistry at Hubbard Brook, New Hampshire. Atmos. Environ. 18:2641–2647.

Magnuson, J. J., C. J. Bowser, and T. K. Kratz. 1984. Long-term ecological research on north temperate lakes (LTER). Verh. Int. Verein. Limnol. 22:533–535.

Meyer, J. L., and G. E. Likens. 1979. Transport and transformation of phosphorus in a forest stream ecosystem. Ecology 60:1255–1269.

Meyer, J. L., and C. M. Tate. 1983. The effects of watershed disturbance on dissolved organic carbon dynamics of a stream. Ecology 64:33–44.

Meyer, J., T. Crocker, D. D'Angelo, W. Dodds, S. Findlay, M. Oswood, D. Repert, and D. Toetz. 1993. Stream Research in the Long-Term Ecological Research Network. Seattle: Long-Term Ecological Research Network Office.

Meyer, J. L., C. M. Tate, R. T. Edwards, and M. T. Crocker. 1988. The trophic significance of dissolved organic carbon in streams. Pp. 269–278 in Forest Hydrology and Ecology at Coweeta, W. T. Swank, and D. A. Crossley, Jr., eds. New York: Springer-Verlag.

Perlmutter, D. G., and J. L. Meyer. 1991. The impact of a stream-dwelling harpacticoid copepod upon detritally associated bacteria. Ecology 72:2170–2180.

Peterson, B. J., and B. Fry. 1987. Stable isotopes in ecosystem studies. Ann. Rev. Ecol. Syst. 18:293–320.

Peterson, B. J., J. E. Hobbie, A. E. Hershey, M. A. Lock, T. E. Ford, J. R. Vestall, V. L. McKinley, M. A. J. Hular, M. C. Miller, R. M. Ventullo, and G. S. Volk. 1985. Transformation of a tundra river from heterotrophy to autotrophy by addition of phosphorus. Science 229:1383–1386.

Rudd, J. W. M., C. A. Kelly, D. W. Schindler, and M. A. Turner. 1988. Disruption of the nitrogen cycle in acidified lakes. Science: 240:1515–1517.

Schindler, D. W. 1973. Experimental approaches to limnology—An overview. J. Fish. Res. Bd. Can. 30:1409–1413.

Schindler, D. W. 1974. Eutrophication and recovery in experimental lakes: Implications for lake management. Science 184:897–899.

Schindler, D. W. 1987. Detecting ecosystem responses to anthropogenic stress. Can. J. Fish. Aquat. Sci. 44 (Suppl. 1):6–25.

Schindler, D. W. 1988. Experimental studies of chemical stressors on whole lake ecosystems (Edgardo Badi Memorial Lecture). Verh. Int. Verein. Theoret. Angew. Limnol. 23:11–41.

Schindler, D. W., K. H. Mills, D. F. Malley, D. L. Findlay, J. A. Shearer, I. J. Davies, M. A. Turner, G. A. Linsey, and D. R. Cruikshank. 1985. Long-term ecosystem stress: The effects of years of experiments. Science 250:967–970.

Schindler, D. W., M. A. Turner, M. P. Stainton, and G. A. Linsey. 1986. Natural sources of acid neutralizing capacity in low alkalinity lakes of the Precambrian Shield. Science 232:844–846.

Schindler, D. W., K. G. Beaty, E. J. Fee, D. R. Cruikshank, E. R. DeBruyn, D. L. Findlay. G. A. Linsey, J. A. Shearer, M. P. Stainton, and M. A. Turner. 1990. Effects of climatic warming on lakes of the Central Boreal Forest. Science 250:967–970.

Swank, W. T., and D. A. Crossley, Jr., eds. 1988. Forest Hydrology and Ecology at Coweeta. New York: Springer-Verlag.

Swanson, F. J., T. K. Kratz, N. Caine, and R. G. Woodmansee. 1988. Landform effects on ecosystem patterns and processes. BioScience 38:92–98.

Tate, C. M., and J. L. Meyer. 1983. The influence of hydrologic conditions and successional states on dissolved organic carbon export from forested watersheds. Ecology 64:25–32.

Vanni, M. J., C. Luecke, J. F. Kitchell, Y. Allen, J. Temte, and J. J. Magnuson. 1990. Effects on lower trophic levels of massive fish mortality. Nature 344:333–335.

Vannote, R. L., G. W. Minshall, K. W. Cummins, J. R. Sedell, and C. E. Cushing. 1980. The river continuum concept. Can. J. Fish. Aquat. Sci. 37:130–137.

Wallace, J. B., D. H. Ross, and J. L. Meyer. 1982. Seston and dissolved organic carbon dynamics in a southern Appalachian stream. Ecology 63:824–838.

Webster, J. R., and B. C. Patten. 1979. Effects of watershed perturbations on stream potassium and calcium dynamics. Ecol. Monogr. 49:51–72.

Bringing Biology Back into Water Quality Assessments

G. Wayne Minshall
Department of Biological Sciences
Idaho State University
Pocatello, Idaho

SUMMARY

For some time now, the quality of the nation's inland waters has been evaluated largely on the basis of chemical and toxicological criteria. However, more recent theories reflect the idea that the native biota, exposed to the full suite of environmental conditions in nature, more accurately reflect the suitability of that environment for survival and long-term persistence. This paper examines the reasons for the resurgence of interest in the biological assessment of water quality and highlights some important considerations in the application of this approach to inland aquatic ecosystems.

Biological integrity is a concept central to successful bioassessment because it identifies the essential factors to be measured and provides a reference against which the degree of environmental disturbance or stress, either natural or anthropogenic, can be evaluated. Major anthropogenic stresses on the integrity of inland aquatic ecosystems include livestock grazing; forestry; agriculture; mining and smelting; urban usage; manufacturing; impoundment and diversion; and lake-, marsh-, and stream-bottom alteration. Measurement of the biological health of aquatic ecosystems is a complex issue involving multiple spatial and temporal scales and methodological and logistical considerations. Nevertheless, direct assessment of the status or ecological health of aquatic organisms and communities is essential for proper resource management of inland waters and their sustained diversity and productivity.

INTRODUCTION

Over the past few decades, water quality has been defined primarily in chemical terms. More recently, however, water management agencies have been increasingly aware of the need to bring biology back into the water quality equation. In some cases, chemical monitoring has actually exceeded the ability to detect biological impacts of chemical contaminants, so that large sums of money have been spent to remove contaminants that do not even affect aquatic organisms. On the other hand, reliance on chemical criteria or laboratory-derived toxicological information taken out of the environmental context has often allowed levels of toxicants or other materials that are harmful to aquatic populations.

The return of biology to environmental assessments has brought with it a need for knowledge about whole-organism biology, the study of which has become increasingly neglected in academic institutions over the past several decades. At the same time, many exciting developments, in fields ranging from molecular biology to landscape ecology, have potential application to the study and management of inland aquatic resources. This paper reviews the historical basis for the application of biological methods to water quality assessment and discusses factors that need to be considered in evaluating the biological integrity of inland aquatic ecosystems.

HISTORICAL BACKGROUND

Modern bioassessment of inland aquatic ecosystems has given rise to several terms and concepts regarding protection or restoration of aquatic environments (Steedman, 1994). Foremost among these are the ideas of integrity, which relates to whether biological systems are intact or restorable, and health, management, and sustainability, which relate to modification of sites by human activity.

The idea of biological, and subsequently ecological, integrity is traceable at least as far back as the writings of Aldo Leopold (1949), but its emergence as a formal ecosystem concept did not occur until the mid-1970s (e.g., Cairns, 1977a,b). The Water Quality Act Amendments of 1972 (P.L. 92-500) formalized the term "biological integrity" under the directive to restore and maintain the "chemical, physical, and biological integrity of the nation's waters."

Initially, the primary focus was on chemical and physical aspects of the environment and on toxicity tests performed in the laboratory on both individual contaminants and complex mixtures of waste effluents from industry and other sources. The idea of biological integrity gradually evolved to include naturalness, sustainability, and ecosystem balance, structure, and function (Jackson and Davis, 1994). Karr and Dudley (1981) defined biological integrity as the "ability of an aquatic ecosystem to support and maintain a balanced, adaptive community of organisms having a species composition, diversity, and functional organization comparable to that of natural habitats within a region." Others have refined, clarified, and extended this concept to specific applications in a series of articles (e.g., Karr, 1991, 1993; Kay, 1991; King, 1993; Steedman and Haider, 1993; Polls, 1994; Steedman, 1994). Still others have addressed the choice of indicators (Keddy et al., 1993), monitoring considerations (Munn, 1993), and descriptions of national programs to measure ecological integrity (EPA, 1990; Marshall et al., 1993; Woodley, 1993; Jackson and Davis, 1994). Because of increasing public awareness of environmental problems, beginning in the mid-1960s and continuing to the present, key ecological issues have been codified as catch phrases, such as ecological or ecosystem health, management, and sustainability.

• *Ecological health* generally is regarded as a condition in which natural ecosystem properties are not severely restricted, the ability for progressive self-organization is present, the capacity for self-repair when stressed is preserved, and minimal external support for management is needed (Steedman and Regier, 1990; Karr, 1993). Sites modified by human activity may be considered ecologically healthy "when their management neither degrades the sites for future use, nor results in degradation beyond their borders" (Steedman, 1994). Problems associated with deterioration of ecosystem health must be addressed at a landscape scale of resolution since significant cumulative and interactive effects otherwise might be overlooked.

• *Ecosystem management* may be defined as "the skillful, integrated use of ecological knowledge at various scales to produce desired resource values, products, services, and conditions in ways that also sustain the diversity and productivity of ecosystems" over the long term (Avers, 1992). In practice, it means blending the needs of people and environmental values in such a way as to achieve healthy, productive, and sustainable ecosystems. The extent to which these goals are attained can be determined only through biological assessment and monitoring of resource conditions.

• *Ecosystem sustainability* is "the ability to sustain diversity, productivity, resilience to stress, health, renewability, and/or yields of desired values, resource uses, products, or services from an ecosystem while maintaining the integrity of the ecosystem over time" (Overbay, 1992). Its return to the resource management equation has come about through the National Environmental Protection Act, the Endangered Species Act, and numerous other federal laws passed during the 1960s and 1970s (e.g., Overbay, 1992) and sustained by federal court decisions.

Ecosystem management and sustainability are likely to have a major influence on research and management of many inland aquatic ecosystems in the United States for years to come. These concepts are especially relevant to federal agencies with large land holdings and broad responsibilities for the terrestrial and aquatic resources occupying them, such as the Bureau of Land Management, the National Forest Service, the National Park Service, and the Refuge Management branch of the Fish and Wildlife Service. Both the concepts and the recognition of their responsibility to implement them are new to these agencies; it is still unclear how (and at what rate) they will move toward instituting formal ecosystem management policies. However, it is clear that biological assessments and continued monitoring will be important to ecosystem management in inventorying aquatic biological resources and their status, in assessing the effects of various management practices on them, and in determining suitable management strategies through research and adaptive management practices to ensure ecosystem sustainability.

Federal Legislative Initiatives

Historically, three major federal legislative actions are responsible for current efforts to increase the use of biological measures and a more meaningful ecological perspective: (1) the Water Quality Act, (2) the National Environmental Policy Act, and (3) the Endangered Species Act.

The Water Quality Act of 1965 (P.L. 89-234) and the related Clean Water Act (Federal Water Pollution Control Act Amendments of 1972, Clean Water Act of 1977, Water Quality Act of 1987) were enacted in response to widespread surface water degradation and a growing public environmental awareness and concern. The implications of this legislation for inland aquatic science range from classroom to courtroom, and its implementation provides substantial opportunities for involvement in all aspects of water science.

The National Environmental Policy Act of 1969 (NEPA; P.L. 91-190) was responsible for interjecting an ecological perspective into subsequent federal legislation and actions, particularly as they relate to natural resource-oriented projects. NEPA set forth a national policy to protect and improve the national environment by requiring detailed consideration of proposals for federal legislation, construction (e.g., dam construction, channel alteration, draining of wetlands), or resource extraction (e.g., water diversion, logging, or livestock grazing) likely to significantly affect the quality of the air, land, and water environments. Among other things, the law required the identification of (1) any adverse environmental effects that cannot be avoided should the proposal be implemented; (2) alternatives to the proposed action, including total abandonment and mitigation of damages; (3) the relationship between local short-term human uses of the environment and maintenance and enhancement of long-term productivity (i.e., sustainability); and (4) any irreversible and irretrievable commitments of resources that would be involved in the proposed action. This act consistently has been upheld and expanded by federal court decisions. Increasingly, NEPA and its legal interpretations have had far-reaching implications for the management of inland aquatic resources at the ecosystem and landscape scales. NEPA has resulted in a more holistic and long-range view of past, present, and future management actions on natural resources in an ecosystem context and has called for a greater and more thorough knowledge of resource states under different management treatments.

The Endangered Species Act of 1973 (ESA; P.L. 93-205) protects all species (except pests) of plants and animals in danger of extinction. Twelve percent of all animal species live in inland waters, and many species are restricted to limited geographic ranges. As freshwater habitats have been destroyed, altered, or polluted, biodiversity and ecosystem integrity have declined. The listing of federally recognized threatened or endangered freshwater species is an important means of tracking total biological integrity (Covich, 1993). The Endangered Species Act has served to emphasize the importance of identifying and preserving the diversity of inland aquatic organisms and their habitats, and of assessing long-term trends in their conditions.

Several recent developments stemming from these legislative acts have brought the biological aspects of water quality to the forefront: (1) the initiation of several large federal monitoring and assessment programs that emphasize the measurement of water quality in biological rather than solely chemical or physical terms; (2) legal mandates to institute biological criteria into state water quality standards in the next few years; and (3) comprehensive assessment of the status of resources throughout the Columbia River Basin and how to manage these resources. These directives and comprehensive programs at the state and national levels will severely overload existing resource management personnel, a situation that is unlikely to be alleviated at the current rate of qualified graduates entering the work force.

Federal Monitoring and Assessment

Specific federal programs of monitoring and assessment have been instituted by the Environmental Protection Agency (EPA) and the U.S. Geological Survey (USGS). Presumably, the newly instituted National Biological Service (NBS) also will emphasize biological assessments through wetland surveys, inventories of biological resources, and the like, unless these responsibilities are abrogated by the new Congress.

At the moment, the premier U.S. federal program involving bioassessment is the National Water Quality Assessment (NAWQA) Program of the USGS (Gurtz, 1994). This program is designed to integrate chemical, physical, and biological data to assess the status of, and trends in, national water quality. It consists of 60 study units (major river basins and large aquifers) located throughout the country (Gurtz, 1994) that represent major natural and human-impacted conditions that influence water quality. Data collection began in 1991. Biological data include (1) analysis of aquatic organism tissues for a wide array of chemical contaminants; (2) characterizations of algal, macroinvertebrate, and fish communities; and (3) characterizations of vegetation growing in and along streams. Physical data include streamflow and characterizations of in-stream, bank, and floodplain habitats (Meador and Gurtz, 1994). NAWQA has developed nationally consistent biological sampling methods so that results are comparable across different river basins and geographic regions.

The Environmental Monitoring and Assessment Program (EMAP) of the EPA has comparable goals, but it uses a more extensive set of sampling sites, including lakes and wetlands (Hunsaker et al., 1990; Paulsen et al., 1991). This program has spent much time developing its philosophical and conceptual underpinnings and has lagged behind NAWQA in making available a standard set of methods and in initiating a full-fledged data collection program. However, a series of pilot studies focusing on lakes was conducted in 1991 (Larsen and Christie, 1993).

An Intergovernmental Task Force on Monitoring Water Quality led by representatives of NAWQA and EMAP (Gurtz and Muir, 1994) is working to develop a national water quality survey that would demonstrate effective collaboration among federal agencies. This group has chosen to focus on biological aspects of water quality to better understand the condition of the nation's stream communities and to identify opportunities and barriers to cooperative partnerships (M. Gurtz, USGS, personal communication, 1994). The national survey will initially aim to characterize reference conditions, but the long-term goal is to include all streams regardless of their condition.

The Clean Water Act mandates state development of criteria to measure water quality conditions based on biological assessments of natural ecosystems. The general authority for biological criteria comes from Section 101, which establishes as the objective of the act the restoration and maintenance of the chemical, physical, and biological integrity of the nation's waters. This section also includes an interim water quality goal for the protection and propagation of fish, shellfish, and wildlife. Propagation includes the full range of biological conditions necessary to support reproducing populations of all forms of aquatic life and other life that depend on aquatic systems (EPA, 1990). Sections 303 and 304 provide specific directives for the development of biological criteria. Section 303 requires states to adopt protective water quality standards that consist of uses, criteria, and antidegradation measures. Section 303(c)(2)(B), enacted in 1987, requires states to adopt numerical criteria for toxic pollutants specified by EPA. The section further requires that states adopt criteria based on biological assessment and monitoring methods, consistent with information published by EPA under Section 304(a)(B). Section 304 directs

EPA to develop and publish water quality criteria and information on methods for measuring water quality, including biological monitoring and assessment methods to determine (1) the effects of pollutants on aquatic community components (e.g., plants, plankton, fish) and community attributes (e.g., diversity, productivity, stability) in any body of water, and (2) the factors necessary to restore and maintain the ecological integrity of all navigable waters (EPA, 1990).

Development and use of biological criteria also will help states to meet the intent of several other legislative acts that require an assessment of risk to the environment (including resident aquatic communities) to determine the need for regulatory action (EPA, 1990). Some examples of the latter are the Comprehensive Environmental Response, Compensation, and Liability Act of 1980; the Federal Land Policy and Management Act of 1976; the Fish and Wildlife Conservation Act of 1980; NEPA; the Resource Conservation and Recovery Act first enacted in 1976; and the Wild and Scenic Rivers Act passed in 1968.

Under the Clean Water Act, states were required to begin instituting narrative biological criteria into state water quality standards during 1991-1993; numeric criteria and full implementation are scheduled to occur within a few years (EPA, 1990). These requirements also apply to federal agencies responsible for the management of large tracts of public land (e.g., the U.S. Forest Service and Bureau of Land Management), especially in the western United States. Narrative biological criteria are general definable statements of conditions or attainable goals of biological integrity and water quality for a given use designation; numeric criteria establish specific values based on measures such as species richness, presence or absence of indicator taxa (taxonomically related groups), and trophic composition.

Need for Cooperation Among Federal Agencies

A good example of cooperation among federal agencies in addressing these aspects using biological assessment is the cooperative survey of the Apalachicola-Chattahoochee-Flint River Basin recently initiated by NAWQA and the NBS (NAWQA Information Sheet, April 6, 1994). This river basin, one of the largest in the eastern Gulf Coast Plain, was known for its rich diversity of at least 45 species of unionid mussels, but these populations have either declined or died out. Mussels are sensitive indicators because they are sessile and are dependent on good water quality, physical habitat conditions, and populations of host fish. The life cycle of unionid mussels is closely linked to fish because mussel larvae are obligate parasites on fish before becoming free-living adults. Conservation efforts to protect or restore declining mussel populations require information on both mussel and fish populations in watersheds with differing land uses and water quality conditions. The NAWQA program, with its interest in monitoring water quality, and the NBS program, with its interest in identifying and protecting populations of aquatic organisms, both require information on the distribution and abundance of aquatic organisms in different environmental settings.

Recent awareness of the rapidly declining status of 76 anadromous fish stocks in the Columbia River Basin (Nehlsen et al., 1992), together with documentation of declining freshwater habitat conditions (Sedell and Everest, 1991), has resulted in intensive efforts by several federal agencies to head off potential extensive curtailment of their resource extraction activities throughout the entire river catchment. Of immediate concern is the fact that protections offered for threatened and endangered fish under the ESA could result in severe curtailment or alteration of U.S. Forest Service and Bureau of Land Management activities. Ideally, improved management of aquatic and riparian ecosystems on lands administered by these two agencies, combined with improvements in hydropower operations, hatchery practices, and fish harvest management, can prevent additional stocks from becoming extinct and preclude the need to extend the protections of the Endangered Species Act to other at-risk anadromous fish stocks (U.S. Forest Service-U.S. Bureau of Land Management, 1994).

In addition, both agencies are required by the Clean Water Act of 1976 (33 USC 1251, 1329) to ensure that activities occurring on lands they administer comply with requirements concerning the discharge or runoff of pollutants. A reasoned response to this new information on serious declines in anadromous fish stocks and aquatic habitat conditions is crucial to the two agencies' success in meeting the "continuing compliance" obligations of NEPA, ESA, the National Forest Management Act of 1976 (NFMA), the Federal Land Policy and Management Act of 1976 (FLPMA), and other environmental laws. By using the latest scientific information on chemical, physical, and biological integrity, the agencies will be better able to ensure the long-term viability of anadromous fish species and the continuing production of goods and services from public lands. Interim and longer-term management strategies are being examined in several geographically specific environmental impact statements as required under NEPA; also under development is a comprehensive ecosystem management plan for the interior Columbia River Basin (Science Integration Team, 1994).

STRESSES ON BIOLOGICAL HEALTH OF INLAND WATERS

The biological integrity of inland aquatic ecosystems is being assaulted in many ways (Power et al., 1988; Resh et al., 1988; Covich, 1993). Numerous anthropogenic disturbances affect inland waters and their associated riparian ecosystems:

• Livestock grazing contributes to increased inorganic sediments, nutrients, and organic matter; breakdown of stream banks; and removal of riparian vegetation.

• Forestry and logging practices, including extensive road building, introduce sediments and logging slash; remove large woody debris; increase runoff and streambed scour; and erode stream banks.

• Agricultural practices add sediments, nutrients, and toxicants; deplete streams through irrigation withdrawal; channelize streams; drain wetlands; and destroy riparian habitats. Pesticides applied to forest and agriculture lands often reach waterways.

• Mining and smelting operations release heavy metals and other poisonous substances to water bodies via surface, subsurface, and aerial pathways.

• Urban usage removes water for domestic consumption; adds sewage and many complex household and other chemicals; converts stream channels into concrete-lined gutters; and contributes fertilizers, herbicides, and pesticides.

• Manufacturing and processing operations release chemicals and heated water and, along with motorized vehicles, contribute airborne pollutants that reach waterways.

• Fish management practices use poisons to remove unwanted species and introduce exotic species.

• Impoundment for flood control, electric power generation, navigation, and recreation drowns rivers, changes flow patterns, alters nutrient and sediment loads and temperatures, and thereby destroys the habitat and impedes or blocks the movement of native aquatic fauna.

• Diking, channelization, and removal of woody debris for navigation, flow "enhancement," flood control, or fish passage all speed up the flow of water; destroy habitat; disrupt in-stream processing of organic matter and nutrients; and prevent interchange of nutrients, organic matter, and sediments within the riparian environment.

• Production of electricity by coal-fired or nuclear reactor steam plants depletes water by evaporation and by diversion from natural water bodies and may increase temperature, trace elements, and other chemicals.

Nutrients from many of the above activities, particularly nitrogen and phosphorus, cause the accelerated enrichment (cultural eutrophication) of lakes and streams. This can result in large-scale fish kills and the elimination of desirable fish species, production of foul odors, uncontrolled growth of algae and toxic bacteria, and obnoxious accumulations of filamentous algae and vascular plants.

Not only do these activities affect the ecological integrity of inland aquatic ecosystems, but the effects of each type of disturbance may be synergistic among types and cumulative in space and/or time (Sidle, 1990). Although viewed as relatively local, they often have large-scale, far-reaching effects. Some large-scale stresses affecting aquatic ecosystems, whether natural or human induced, are rapid and dramatic. Examples include certain recent cases of massive deforestation, urbanization, development of crop- and pasturelands, forest fires, plant disease outbreaks, and insect infestations. Other disturbances occur over extended periods of time and, hence, often are not recognized as such until the situation becomes extremely difficult or impossible to reverse. These include acidification, some types of logging and mining, livestock grazing, fire suppression, irrigation, and potentially, global climate change (Minshall, 1992, 1993; Covich, 1993). Global climate change could profoundly alter riparian ecosystems through its effect on terrestrial vegetation, thermal and hydrologic regimes, nutrient cycles, and so on (Firth and Fisher, 1992). Fast or slow, disturbances of riparian ecosystems may result in changes in water temperature or runoff, channel straightening, scouring or sedimentation, loss of physical habitat, alteration of food base, and waterlogging or drying of riparian soils.

Challenges of Assessing Biological Integrity

Although legislation calls for maintaining biological integrity, measuring the biological health of inland waters is extremely complex; this complexity results not only from the need to account for natural variations in time and space, but also from the need to consider individual species as well as interactions among organisms in a particular aquatic community.

Importance of Scale (Space and Time)

There is no single correct scale for the study, assessment, or management of aquatic-riparian ecosystems (O'Neill et al., 1986; Levin, 1992; Johnson et al., 1993); rather, the appropriate scale depends on the scientific question or management problem being addressed. The importance of various environmental factors and the interpretation of measurements taken on aquatic ecosystems vary with scale (O'Neill et al., 1986; Minshall, 1988). Further, since ecosystem boundaries vary with scale, the spatial boundaries also must be correlated with the temporal framework appropriate for a particular disturbance (O'Neill et al., 1986).

Most ecosystems extend over comparatively large areas and persist for long periods of time. It is thus difficult to devise large-scale, single-value measurements of ecosystem integrity. However, the hierarchical structure of ecosystems results in a series of scaled interactions that can act as natural integrators of local processes. For example, measurement of community metabolism of a river segment or lake can serve to integrate the status of conditions from a myriad of spatial patches and compartments within these ecosystems. This natural integration is especially evident at the scale of entire water catchments (King, 1993) where, for example, the ecological health of an entire forest may be reflected in the condition of the stream flowing through it. Use of scales of assessment close to the scale of the entire ecosystem will increase the likelihood that observed changes will be of consequence for the entire ecosystem. These larger-scale integrated measures are invaluable for detecting changes in loss of ecosystem integrity, but they may have to be supplemented with finer-scale measurements to determine cause and effect (King, 1993). For example, measurements of primary production at the level of patches or compartments within river segments or lakes are necessary to determine the relative importance of each and to isolate the specific factors responsible for any differences. In lakes, measurements performed on the plankton are better for lakewide assessments of phosphorus availability, whereas measurements utilizing attached algae such as *Cladophora* permit more localized assessments (Cairns et al., 1993).

As noted above, there is a range of biological information that can be used to evaluate water quality. Studies at the population and community levels of organization emphasize species populations and interactions within and among them, such as competition. In this approach, the physical environment is seen as external to the system of organisms and biotic interactions (King, 1993). Population and community studies emphasize biotic interactions, whereas ecosystem studies focus on the processing and transfer of matter and energy in which the environment is an integral (as opposed to external) part of the system (O'Neill et al., 1986; King, 1993). Study of landscapes commonly addresses patterns of distribution within and among ecosystems, thus generally implying spatial scales of relatively broad extent.

Geology, topography, and climate all influence the characteristics of a river basin or watershed ecosystem (e.g., Minshall et al., 1985) and thus act at the scale of the landscape (Omernik, 1987; Hughes and Larsen, 1988). Landscape patterns (such as regional or river basin) influence many ecological phenomena in inland aquatic ecosystems (Hughes et al., 1986; Karr, 1991). For example, streamflow characteristics vary with the type of soils and underlying geology; the topographic relief; and the form, amount, and timing of precipitation. The resulting flow regime in turn influences a variety of ecologically relevant features, including channel form, substratum size composition and stability, woody debris, and the nature of the food base. Patterns of stream discharge and disturbance regimes show a strong geographic separation (Minshall, 1988; Poff and Ward, 1989), implying the operation of landscape-level phenomena. Differences in flow regimes, coupled with climatically mediated thermal characteristics and geologically determined substratum characteristics, are expected to contribute to correspondingly different geographic and other landscape-level patterns in biotic structure and function (Vannote et al., 1980; Minshall et al., 1985; Poff and Ward, 1990).

Patterns of disturbances, both natural and human induced, resulting in a mosaic of patches of different ages and composition also may enter into the natural pattern of variability (White and Pickett, 1985). For example, the River Continuum Concept proposes that there are a number of features of even pristine stream-riparian ecosystems that change progressively throughout a river basin, therefore requiring a landscape perspective for proper interpretation (Vannote et al., 1980). The influence of riparian vegetation, the annual amount of terrestrial leaf litter in the channel, the availability of dissolved organic matter, and the modal size of particulate organic matter all generally decrease with distance from the headwaters of a stream system. The relative contributions of photosynthesis and community metabolism and the composition of macroinvertebrate functional feeding groups also change gradually and in a predictable fashion along the so-called river continuum. In addition, the effects of disturbances vary along a river system; some (especially if widely dispersed) become dissipated with increasing stream size, whereas others may act cumulatively. For example, the effects of moderate amounts of logging or farming within a basin or the entrance of low levels of sewage or nutrients from a point source may be dissipated through biological means and physical dilution as the water proceeds downstream. The effect of an isolated incident, such as the building of a single low-head dam, may be imperceptible, whereas the erection of many such dams can cumulatively have numerous adverse effects on many aspects of the downstream riverine ecosystem.

Landscape is the scale of many forestwide and forestwide-regionwide land uses (e.g., logging, mining, livestock grazing) and their associated management practices that affect aquatic-riparian ecosystems. It also is the focus of many larger-scale problems for resource managers (e.g., drought, acid rain, forest and range fires, disease, and pest outbreaks). The catchment is an appropriate landscape unit for examining stream-riparian ecosystem responses to disturbances on the order of years to decades (Minshall, 1993). However, for events occurring at intervals of 10^2 to 10^4 years, such as wildfire, the focus becomes the entire forest, which itself may cover numerous catchments. Landscape-scale events may affect aquatic ecosystems at various lower levels of resolution because of the hierarchical nature of these systems (Harris, 1980; Frissell et al., 1986; Pringle et al., 1988).

The scale of an ecological system refers to its spatial and temporal dimensions (Allen and Hoekstra, 1992). The concept of ecological integrity is scale dependent (King, 1993). Maintenance of ecological integrity implies perpetuation of some normal state or norm of operation within some prescribed range of variation. Thus, measuring or observing ecosystem integrity, or its loss, requires observations over sufficient space and time to identify the range of variation (King, 1993). Aquatic ecosystems require certain spatial and temporal bounds for the maintenance of their structure and function. For example, many stream ecosystems are dependent on terrestrial leaves and other forms of allochthonous detritus for food and habitat. Thus, severing the connection between the stream and riparian vegetation, because of road building, logging, or livestock grazing, can disrupt the integrity of the stream ecosystem. On a larger scale, attempts to manage or rehabilitate a river without regard to actions in the upstream portions of the river and its tributaries (such as land-use practices that alter the amount, kind, and timing of allochthonous detritus inputs) will be largely ineffectual.

A minimum extent may be required for some processes to operate or interaction to take place. For example, nutrient cycling in streams occurs in a spiraling fashion along the water course, with important excursions into the hyporheic and floodplain zones (Newbold et al., 1982a,b; Green and Kauffman, 1989; Triska et al., 1989). Efforts to assess, protect, or restore biological integrity that do not use the appropriate scale are destined to fail. Failure to observe the system at the appropriate spatial-temporal scale can make inferences about ecological integrity of ecosystems difficult or impossible (e.g., Frost et al., 1988). Furthermore, setting spatial or temporal boundaries of a system smaller than the minimum required for persistence and interaction can affect system function and actually may lead to a loss of ecosystem integrity (King, 1993). For example, it is now clear that survival of native anadromous fish populations over much of the Pacific Northwest is dependent on resource management decisions made throughout the entire Columbia River Basin rather than on a stream-by-stream basis, as done in the past.

Ecosystems in particular must be defined simultaneously in terms of space and time, since ecological dynamics occur over a broad spectrum of space-time scales (O'Neill et al., 1986; Allen and Hoekstra, 1992). For example, lake and stream ecosystem responses occur at scales ranging from millimeters and seconds to hundreds of kilometers and millions of years (Frost et al., 1988; Minshall, 1988, 1993). Small-scale events, such as the rolling of a rock or a 5°C change in temperature, recur with relatively high frequency; larger-scale events, such as a major flood, forest fire, or volcanic eruption, are progressively more rare. Phosphorus released by feeding zooplankton is taken up immediately by algae (Lehman and Scavia, 1982), whereas acid additions to a whole lake take several years to show effects on adult lake trout (Schindler et al., 1985). Recovery of species abundance on individually disturbed rocks in an otherwise intact streambed occurs within a few days to weeks (Robinson and Minshall,

1986), whereas recovery from the effects of a catchment-wide fire or a massive dam failure takes tens to hundreds of years (Minshall et al., 1983, 1989; Richards and Minshall, 1992). The demise of salmonid populations in the Pacific Northwest, referred to earlier, is symptomatic of a region- or basinwide loss of ecological integrity within the present century.

Importance of Seasonal Variations in Time

Variations in activity, condition, distribution, and abundance (hence, recruitment and/or mortality) of aquatic organisms across seasons are common. This is to be expected in strongly seasonal (temperate) environments, but such variations are found even in the tropics (Covich, 1988). Temporal variation in biotic responses should be accounted for in biological assessments of environmental conditions or determinations of change, but frequently it is not. Consideration of seasonal differences is especially important when comparing data from different locations or for the same area over time. For assessment of long-term trends, samples must be collected at the same general time within a season. However, in comparative studies among sites or years, sampling times should be determined on the basis of cumulative temperatures (e.g., number of degree-days) rather than specific calendar date. Temperature often is the primary factor responsible for seasonal variations in the temperate parts of the world. Even when light is the primary factor responsible, temperature is a reasonable surrogate because both are strongly influenced by the amount of solar radiation reaching the surface of a water body. This implies the need for continuous long-term records of temperature from the waters in question and/or the development of reliable regressions between air temperature or solar radiation and water temperature for specific localities and habitats of interest.

Nonequilibrium Nature of Aquatic Systems and the Role of Disturbance

Recently, there has been a paradigm shift from a belief in the dominance of equilibrium processes in ecology to one that emphasizes the importance of nonequilibrium processes (e.g., Harris, 1986; Botkin, 1990; Reice, 1994). Previously, the dynamics within ecological levels of organization, from populations through ecosystems, were viewed as being controlled primarily by processes that were density dependent and tended toward equilibrium conditions. The present view, whose implications have yet to be fully appreciated by most ecologists and resource managers, is that these same dynamics are controlled largely by processes that are density independent and of a nonequilibrium type. Consequently, they are believed to be heavily dependent on random (stochastic) forces and hence to be disturbance driven and variable rather than constant. Though reality probably lies somewhere between these two extremes, the latter currently dominates ecological thinking. Nonequilibrium theory suggests that the development of ecological systems is nonlinear, heavily influenced by discontinuous catastrophic disturbances, and hence largely unpredictable and with multiple developmental pathways (Kay, 1991).

Whether aquatic ecosystems are perceived as equilibrium or nonequilibrium actually may depend on the spatiotemporal scale being considered (O'Neill et al., 1986) and on the magnitude and time since disturbance. For example, in a year-to-year and section-by-section context, most natural stream ecosystems may be perceived to be open and nonequilibrium in character. They receive substantial inputs (environmental influences) from outside their boundaries and exert comparable influences (outputs) on adjacent ecosystems. They also are dynamic and continually changing. Many lakes possess similar features but usually to a lesser extent (e.g., Wetzel, 1983). However, when viewed in broader contexts, aquatic systems may exhibit several levels of stable behavior and show substantial spatial homogeneity (i.e., attributes associated with equilibrium conditions) (e.g., Harris, 1986; Frost et al., 1988; Minshall, 1988, 1993).

Disturbance and the resultant change in conditions have long been recognized as an important factor affecting the structure and dynamics of ecological systems at various levels of organization. More recently, emphasis has shifted from viewing disturbance as rare and unpredictable to treating it as a natural process that occurs at different spatial and temporal scales with varying degrees of predictability (e.g., Pickett and White, 1985; Resh et al., 1988; Townsend, 1989; Fisher, 1990). Postdisturbance ecosystem development should be expected to exhibit the characteristics of self-organizing, nonequilibrium systems (Kay, 1991). The development of such systems is expected to proceed in irregular spurts from one steady state to another. Each spurt results in the system moving further from equilibrium and becoming more organized. For example, in ecosystem succession, each of the stages (seres) corresponds to a transient steady state, and the displacement of a previous seral stage by the next is a spurt that results in increased organization (Kay, 1991). Because change (both natural and human induced) is implicit in the modern, nonequilibrium view of ecosystems, its consideration is important in developing and applying the concept of ecosystem integrity to inland waters. Also, since ecosystem integrity is a scale-dependent concept, measuring or observing integrity or its loss in inland aquatic ecosystems requires observations over sufficient temporal extent to identify and characterize their patterns (King, 1993).

Tools for Measuring Biological Integrity

Tools for measuring biological integrity can be divided into two main groups: (1) those that measure "structural" integrity, meaning those that analyze the building blocks of aquatic communities; and (2) those that measure "functional" integrity, meaning tools that assess processes occurring in aquatic communities (Cairns, 1977a,b; Karr, 1991, 1993). Collectively, these structural and functional measures help restore a biological focus in water quality measurements.

Structural Attributes

An unusual change in one or more structural characteristics is interpreted as evidence of ecological stress. Impairment of biological structure of aquatic communities may be indicated by the absence of pollution-sensitive taxa, dominance by any particular taxon and low overall taxa richness, or changes in community composition relative to the reference condition (Plafkin et al., 1989).

Fundamental measurements of ecosystem structure are (1) the number of species or other taxonomic units present, (2) the number (or mass) of individuals per species, and (3) the particular kinds of species present. Historically, aquatic ecologists have done a fairly good job of measuring structural aspects, although they are hampered by the inability to identify all of the species in a biotic community because of the lack of accurate, up-to-date taxonomic keys and comprehensive systematic treatises at the species level, particularly those of a regional nature. A still greater problem is the limited availability of qualified biologists capable of using even the existing taxonomic keys.

The particular structural attributes analyzed to determine biological integrity vary with the level of developmental complexity (plants, invertebrates, fish) and ecological organization (population, community, ecosystem) of the focal organism or group (e.g., Johnson et al., 1993). The most commonly used groups for determining structural responses vary with the type of aquatic habitat under investigation and the scale appropriate to the question being addressed. In lentic waters, microscopic algae and invertebrates (particularly rotifers, cladocerans, and copepods) are most commonly emphasized. These may be separated further into those suspended in the water (i.e., plankton) and those growing attached to natural or introduced substrata (i.e., aufwuchs on sticks or microscope slides, respectively). In lotic waters, the most commonly studied groups are attached algae (periphyton), macroscopic invertebrates living on the stream bottom (benthos), and fish.

The suitability of a particular group is determined by the specific environmental stressor, the generation time of the organisms, and their mobility and dispersal capabilities. For example, fish may be better indicators of physical habitat conditions in streams and benthic macroinvertebrates more indicative of water quality conditions (C. Yoder, personal communication, 1994). Algae have generation times of hours or days, macroinvertebrates of months or a year or more, and fish of several years. Consequently, the algae may have already recovered from an environmental impact before an investigator arrives on the scene, whereas fish populations may not recover quickly enough to show the effects of multiple episodes within a season or year. Fish are more mobile than attached algae or benthic invertebrates and thus may avoid the effects of a pollutant or other disturbance entirely. High mobility also makes it difficult to obtain an accurate population estimate for fish. However, small relatively immobile organisms, such as algae or snails, may have higher dispersal capabilities than larger more mobile forms because they are often carried long distances by air currents or birds. Frequently, only a single group is used in a bioassessment but the use of several groups is preferred because they provide a better coverage of trophic levels and functional groups, sensitivity to a broader range of stresses, and a greater array of response times.

At the population level, the presence and abundance of one or more key species or so-called indicator organisms may be used to determine the condition of the aquatic environment. However, this approach has been effective only in the limited cases where the monitored species responded clearly to specific types of water quality. To be useful as an indicator organism, an individual species must have a narrow range of tolerance for suitable environmental conditions that are known and related to some attribute of interest to humans. Few species satisfy these requirements because tolerance of a narrow range of conditions means that the organism may not be found widely in space or time and therefore will be of limited general utility as an indicator of environmental stress (Warren, 1971). Individual species often are not satisfactory for detecting gradual increases or decreases in pollution and are not sensitive to low levels of pollution. In addition, individual species may flourish or languish for reasons (such as competitors or predators, substratum or food conditions, or current velocity) that have nothing to do with pollution.

Other measures at the population level may be more responsive to ecological dynamics than is species abundance. These are built on the properties of individuals (size, growth rate, content of particular components such as fats or certain enzymes) or populations (birth and death rates, population growth rate, age-frequency distribution). For example, a decline in birth rate may signal a significant change in environmental conditions and therefore provide a possible assessment approach for detecting human-induced impacts (W. T. Edmondson, personal communication, 1994). Similar methods could be applied to insects and other macroinvertebrates in streams.

The effects of toxic substances on individual species have been studied widely under laboratory conditions, but rarely are toxicity tests applied to intact or partially isolated systems in nature (Rand and Petrocelli, 1985). Mortality generally is the criterion in such tests, but a sublethal condition is just as important in controlling populations (Edmondson, 1993). Properties other than mortality (such as those given above) also vary with toxicity. Neoplastic lesions in fish have been correlated with pollutants (Russell and Kotin, 1957; Tyler et al., 1991). Chironomid head capsule morphology (Warwick, 1988) and diatom shell structure (W. T. Edmondson, personal communication, 1994) can be distorted when the organisms live in conditions containing sublethal toxic material. In Lake Orta in north Italy, copper-rich waste discharged from a rayon factory for many years caused elimination of the entire biota. The sedimentary record for the lake shows an initially normal diatom community; at progressively shallower levels, there are increasing proportions of distorted diatoms, gradual elimination of the various species, and finally no biota. Whether other metals would have the same effect as copper or whether different species of diatoms respond differently to different metals is not known.

Population guilds and communities generally have proven more satisfactory than individual indicator organisms for the assessment of water quality. For example, paleolimnologists working on acid rain have been able to determine the pH of lakes on the basis of groups of diatoms found in bottom sediments. In a slightly different approach, planktonic-diatom biomass estimates were examined for the sediments of three lakes with contrasting types of lake disturbance: acidification (Gaffeln, southwestern Sweden), point-source eutrophication (Lough Augher, northern Ireland), and catchment agriculture (Akassjon, Northern Sweden) (Anderson, 1994). At Gaffeln, biomass declined steadily with acidification until extinction of the planktonic diatoms occurred. In Akassjon, maximum biomass coincided with the maximum areal extent of arable land. At Augher, nutrient input from a creamery resulted in steady biomass increases. Multivariate statistical analyses of species groupings have proven to be more useful than individual indicator organisms in a number of studies, particularly those involving pollution (e.g., Clarke, 1993).

Various methods are employed to obtain samples of the biota for use in water quality assessment. These vary with the type of habitat (e.g., lotic or lentic), location (e.g., open water or benthic), size (microscopic or macroscopic), type of substratum (e.g., mud or rock), and degree of mobility of the organisms being sampled. Application of these methods requires skill and training. Planktonic forms may be trapped, pumped, enclosed, or netted from specific or composite locations within the water column (Lind, 1985; Wetzel and Likens, 1991). Attached microscopic forms can be scraped or brushed to free them from the substratum and then removed by suction for placement in a sample container or collection onto a filter (Lind, 1985; Porter et al., 1993). Benthic macroinvertebrates are collected by a variety of dredges, nets, corers, baskets, and other devices depending on water depth, substratum type, and current velocity (Lind, 1985; Wetzel and Likens, 1991; Cuffney et al., 1993a,b). Fish commonly are collected by electrofishing, seining, or gill netting, or are observed directly by snorkeling or scuba diving (Meador et al., 1993b). Selection of methods is best made based on experience and reference to reputable published sources (e.g., APHA, 1989; MacDonald et al., 1991; Meador et al., 1993a, in addition to those cited above). In addition, important aspects of statistical design and data analysis must be incorporated into the study plan, but these are beyond the scope of this paper (e.g., Green, 1979; MacDonald et al., 1991; Norris and Georges, 1993; Resh and McElravy, 1993).

Bioassessment of pollution in inland waters has been going on for a long time (e.g., Hynes, 1960; Karr, 1991; Cairns and Pratt, 1993), and a number of different measures and indices have been proposed. However, the recent renewed interest in biological integrity and its assessment has resulted in a reevaluation of those metrics that seem most effective, in an attempt to come up with standardized protocols (e.g., Karr et al., 1986; Plafkin et al., 1989; Karr, 1991; Resh and Jackson, 1993; Kerans and Karr, 1994). After representative samples of the biota are obtained, the organisms are identified and enumerated, their biomass and condition are determined, and relevant metrics are calculated. A biological "metric" is an absolute or derived measure that is sensitive to ecological conditions. The number of species present in a community is one of the simplest and most reliable metrics. A variety of other metrics have been examined for use in bioassessment, although additional testing and refinement still are needed. Some are sensitive to certain types of pollution, such as organic matter or inorganic sediments, whereas others are more general in their response. Some of the most popular metrics for periphyton, macroinvertebrates, and fish are listed in Table 1. Adjustments in the list are necessary to accommodate regional differences in distribution and assemblage structure and function (e.g., Miller et al., 1988).

Richness is the number of different kinds of individuals (usually species or genera) in the total community or a specified assemblage. The Pinkham and Pearson community similarity index incorporates abundance and species composition (Plafkin et al., 1989). The quantitative similarity index for taxa compares two communities in terms of presence or absence of taxa and relative abundance (Shackleford, 1988). The Hilsenhoff biotic index summarizes the tolerances of macroinvertebrates to organic pollution (Hilsenhoff, 1987, 1988; Plafkin et al., 1989). Dominance measures assume that a highly skewed species-abundance distribution reflects an impaired community. The Hydropsychidae:Trichoptera ratio includes the mildly pollution-tolerant hydropsychids but excludes the pollution-intolerant arctopsychids from the family total. Functional feeding group designations are based on the manner in which food is obtained (see Merritt and Cummins, 1984, for details). Scrapers rasp or chew food growing attached to a surface such as rock, wood, or living aquatic vascular plants. Filterers employ self-constructed nets or specialized anatomical

TABLE 1   Commonly Used Biological Metrics in the Bioassessment of Water Quality and Physical Habitat Conditions

Attached algae (periphyton)[a]
  Soft-bodied forms
    Taxa richness (number of genera)
    Dominant phylum
    Specific indicator taxa of different levels and causes of pollution
  Diatoms
    Shannon diversity index
    Pollution tolerance index (after Lange-Bertalot, 1979)
    Siltation index (relative abundance of motile forms)
    Similarity index (Whittaker and Fairbanks, 1958)

Macroinvertebrates[b]
  Structure
    Taxa richness
    Richness of taxa within the generally pollution-sensitive orders
      Ephemeroptera-Plecoptera-Trichoptera) (= EPT index)
    Pinkham-Pearson index
    Quantitative similarity index for taxa
  Community balance
    Hilsenhoff biotic index
    Proportion of individuals in the dominant taxon
    Dominants-in-common at each site for five most abundant taxa
    Proportion of Hydropsychidae individuals of the total number of Trichoptera
  Functional feeding group components
    Proportion of scraper abundance to sum of scrapers plus filterers
    Proportion of shredder abundance to total individuals
    Quantitative similarity index for functional feeding groups
Fish[c]
  Species composition
    Total number of fish species
    Species richness and composition of selected groups (e.g., darters, suckers, sunfish)
    Presence of pollution-intolerant species
  Trophic composition
    Proportion of individuals as omnivores
    Proportion of individuals as insectivores
    Proportion of individuals as top carnivores
  Abundance and health
    Number of individuals in sample
    Proportion of individuals with disease or anomalies

[a]From Bahls (1993).
[b]From Barbour et al. (1992).
[c]From Karr et al. (1986) and Leonard and Orth (1986).

structures to remove food particles from the water. Shredders feed on leaves, needles, sticks, and other coarse-sized particles (particularly of terrestrial origin).

Table 2 compares values of macroinvertebrates from five unaffected

TABLE 2   List of Macroinvertebrate Metrics for Five Representative Affected and Unaffected Sites from the Snake River Plain and Northern Basic and Range Ecoregions

| | Snake River Plain (SRP) | | | | | | | | | | Northern Basin and Range (NBR) | | | | | | | | | |
| | Affected | | | | | Unaffected | | | | | Affected | | | | | Unaffected | | | | |
METRIC	1	2	3	4	5	1	2	3	4	5	1	2	3	4	5	1	2	3	4	5
EPT/Chironomidae + oligachaeta (b)	1	3	3	1	3	3	3	3	3	3	1	1	1	1	1	5	3	3	5	5
Taxa richness (a,b)	5	1	3	1	1	3	5	5	5	5	1	1	1	1	1	5	5	5	3	3
EPT Richness (a,b)	3	1	1	1	1	3	5	5	5	5	1	1	1	1	1	5	5	5	1	3
HBI (a,b)	1	1	1	1	1	3	3	5	3	3	1	1	1	1	1	5	5	5	3	5
BCI	1	1	1	1	1	3	5	5	5	3	1	1	1	1	1	5	5	5	3	3
EPT/Chironomidae (b)	1	3	1	1	1	5	3	3	1	1	1	1	1	1	1	5	3	1	3	5
% dominance (a,b)	3	3	3	1	1	5	5	5	1	3	1	3	1	1	1	5	5	3	5	5
Hydropsyche/trichoptera	1	5	1	5	1	5	5	5	5	5	5	5	5	5	5	3	5	5	5	5
H'diversity (a,b)	3	1	3	1	1	5	5	5	3	5	1	1	1	1	1	5	5	5	3	5
Simpson's index (a,b)	3	3	3	1	1	5	5	5	3	5	1	1	1	1	1	5	5	5	3	5
Scrapers/filterers (a)	1	1	1	1	1	1	3	1	5	3	3	1	1	1	1	5	3	5	5	1
Density	na	1	5	3	na	1	5	3	3	5	1	5	1	1	1	1	na	1	5	3
% scraper (a)	1	1	1	1	1	3	3	3	3	3	1	1	1	1	1	5	5	3	5	1
% filterers	3	1	3	5	1	5	3	1	5	3	5	3	3	5	1	5	3	5	5	1
% shredders	3	1	3	1	1	5	3	3	3	3	1	1	3	1	1	3	3	5	3	5
% EPT	1	3	1	1	1	5	5	5	3	5	1	1	1	1	1	5	5	3	5	5
% Chironomidae + oligochaeta (b)	3	3	1	1	5	5	3	3	3	3	1	1	1	1	1	5	3	3	5	5
% Chironomidae (b)	1	3	1	1	1	5	1	3	5	1	1	1	1	1	1	5	3	3	5	5
PCA Score	19	11	15	7	7	25	31	29	25	29	10	12	10	10	10	50	42	38	36	46
Total Score	35	35	31	25	23	69	65	65	61	59	27	25	25	25	21	81	71	69	67	67

NOTE: Summed scores for all metrics and metrics determined important through principal components analysis where (a) = SRP and (b) = NBR; "na" indicates that data are not available for this metric.

SOURCE: Robinson and Minshall (1995).

reference streams (i.e., located in relatively undisturbed catchments) and five affected streams (catchments heavily grazed by livestock) in two different ecoregions in southern Idaho (Robinson and Minshall, 1995). Values are presented for each of the individual metrics examined in the study. Total scores (i.e., summed index values) based on all 18 metrics and on the 8 to 10 most responsive metrics identified through principal components analysis (PCA) are given at the bottom of the table. Some metrics are more responsive (and consistent) than others to the effects of livestock grazing. For example, in the Snake River Plain (SRP) ecoregion, the metrics Ephemeroptera-Plecoptera-Trichoptera (EPT) richness, Hilsenhoff biotic index (HBI), and biotic condition index (BCI) show consistent responses to grazing; in the Northern Basin and Range (NBR) region, EPT/Chironomidae, taxa richness, EPT richness, HBI, BCI, percent dominance, Shannon-Weiner diversity (H'), Simpson's index, and percent EPT show the most consistent responses. Nevertheless, most metrics clearly show lower values for the affected sites than for the unaffected ones. Also, the metrics differ in their effectiveness between ecoregions. For example, of the 18 metrics, 9 showed a large and consistent difference between affected and unaffected streams in the NBR but only 4 did so in the SRP. Often a variety of different scores are seen for a given metric, even within the unaffected streams, presumably reflecting differences in environmental effects.

Examination of the total metric scores (Figure 1) showed substantially lower values for the affected streams in both ecoregions, thereby quantifying the effects of livestock grazing. In using the full 18 metrics, the condition of both the grazed and the ungrazed treatment types appears comparable between the two ecoregions. However, in using the subset of metrics selected by PCA, the unaffected streams in NBR are shown to be in better condition and the affected streams in worse condition than those in SRP,

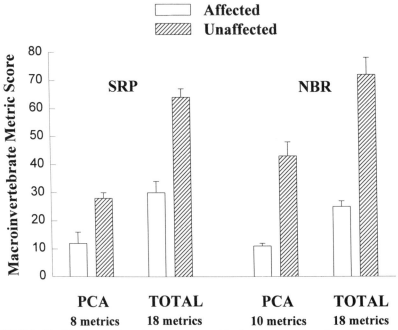

FIGURE 1   Health of macroinvertebrate communities in streams affected and unaffected by livestock grazing in two Idaho ecoregions: the Snake River Plain (SRP) and Northern Basin and Range (NBR).
SOURCE: Based on data reported in Robinson and Minshall (1995). © 1995 by CRC Press.

a fact that supports on-site impressions of stream conditions in the two regions.

Functional Attributes

Functional integrity involves processes such as photosynthesis and community respiration, nutrient transfer, energy flow (secondary production), and decomposition. Abnormal rates of activity and accumulation or depletion of materials are indications of disruptions of an ecosystem's functional integrity. For example, during the course of a whole-lake acidification experiment, the entire nitrification process was halted at a pH of less than 5.7, possibly due to the loss of all nitrifying bacteria (Rudd et al., 1988). High rates of ecosystem metabolism commonly are associated with eutrophication; lowered or negative rates of metabolism may indicate active decomposition of excess supplies of organic matter trapped in the sediments. To distinguish abnormal conditions from normal ones, the natural range of variability must be known for each particular location (e.g., Frost et al., 1988).

Functional measures of ecosystem integrity are often considered less sensitive to environmental factors, including pollutants, than measures of structure. Also, system function may give a very different impression of the effects of environmental stress than structural responses, particularly species composition (Frost et al., 1995). However, these apparent anomalies simply may occur because functional measures operate at longer temporal or larger spatial scales than structural measures or because their responses are less well known. It often is assumed that ecosystems are functionally resilient to alteration of structure due to compensatory responses, but this assumption has not been tested adequately for aquatic ecosystems. A compensatory functional response occurs when rates and amounts of ecosystem processes remain unchanged in the face of changes in structural composition, such as alteration of species dominance or loss of species richness. Studies of compensation and complementarity in ecosystem function are few and limited largely to lakes (Schindler, 1987; Howarth, 1991; Frost et al., 1995). The loss of species from ecosystems produces inconsistent results. In some cases, such losses are accompanied by no apparent compensation (e.g., primary production remains unchanged); in others, considerable alteration of ecosystem processes occurs (e.g., Kitchell and Carpenter, 1993). Understanding the reasons for such apparent inconsistency is an important area for future research.

Measurement of ecosystem function has been avoided because methods dealing with it have been lacking or are thought to be more difficult and time-consuming to employ. Although it often is easier to evaluate aspects of system structure, this is not always the case. For example, it generally is easier to make a measurement of open-water community metabolism in a lake than to collect, identify, count or weigh, and determine the appropriate structural metrics for even one group of organisms. In addition, freshwater ecologists now have the fundamental tools needed to begin assessing some functional aspects of ecosystem integrity (e.g., metabolism chambers, nutrient uptake techniques, leaf pack and litter bag decomposition). Although methods are now available to begin adding functional measures to the assessment of ecological integrity in inland waters, many are still crude and simplistic. Considerable refinement in methodology and the development of additional approaches and techniques are sorely needed. In addition, further efforts are needed to develop practical, cost-effective techniques for routine bioassessment and to train field technicians in their use.

All aquatic life is dependent, either directly or indirectly, on the photosynthetic fixation of radiant and chemical energy into primary organic compounds. Part of this fixation occurs directly in the aquatic environment through primary production (i.e, "self-produced," or autochthonous); the remainder occurs on land, with the organic matter entering the aquatic environment in the form of leaves and other terrestrial plant products (i.e., "produced outside," or allochthonous). Thus, primary production is an important measure of both the autotrophic capacity of a water body and the availability of food resources for the various consumer levels in the food web. In addition, when used in conjunction with measurement of community respiration, it provides useful insights into the relative importance of autochthonous versus allochthonous carbon resources to energy flow in an ecosystem. Primary production in inland waters is measured by determining the amount of carbon dioxide taken up, the amount of higher-energy carbon compounds produced, or the amount of oxygen evolved in the presence of light. Measurement techniques all involve determination of changes in the amount of reactant or of one or more products in either "open systems" (i.e., the water in all or a portion of a basin or channel) or "closed systems" (i.e., sealed bottles, spheres, recirculating chambers, and the like containing isolated portions of water and components of the ecosystem).

Although the measurement of primary production is a time-honored procedure in aquatic ecology, there is still considerable room for improvement. Difficulties include accurate correction for diffusion losses and gains across the air-water interface in open system measurements and nutrient depletion and the lack of adequate circulation (or flow) in closed systems. The potential for other artifacts due to enclosure in containers of various sizes and shapes is likely but has yet to be examined. Additional problems arise in interpreting the meaning of whole-system values when the contributions of various compartments or subsystems remain undetermined and in extrapolating from closed-system measurement of ecosystem fragments to realistic intact-system estimates.

Community respiration is the combined sum of the respiration by autotrophs and heterotrophs (e.g., bacteria, fungi, invertebrates, vertebrates). It measures the total consumptive process of the ecosystem of fixed carbon, whether autochthonous or allochthonous in origin. The procedure involves measurement of the amount of oxygen consumed, organic carbon compounds used, or carbon dioxide produced using techniques identical to those for primary production. Measurements are made in the dark in order to eliminate the confounding and counteracting effect of photosynthesis. Respiration also occurs during the daytime and is believed be higher than at night, but techniques for routine measurement of daytime respiration outside the artificial conditions of the laboratory are not available.

A special case of measuring utilization of organic carbon compounds involves determination of the rate of mineralization of dead plant matter. This material may be exposed directly in the water in the form of loosely aggregated packs of tree leaves or other materials, or enclosed in mesh bags (Petersen and Cummins, 1974; Allard and Moreau, 1986; Cummins et al., 1989). Control of mesh size has been used to exclude or enclose selected decomposer groups (Minshall and Minshall, 1978), but in many cases, inhibition of decomposition due to the creation of anaerobic conditions may occur (Cummins et al., 1989). The usual measure is the amount of weight lost (after correction for leaching and handling) per time interval, adjusted for temperature. Stressed aquatic environments generally show rates of decomposition reduced from those of unstressed ones. In particular, increased levels of acid, chlorine, chlorine plus ammonia, and salt have been shown to lower litter decomposition (Reice and Wohlenberg, 1993).

Another important process in aquatic ecosystems is biogeochemical cycling. In lakes and wetlands, cycling and uptake rates commonly are measured in vertical tubes or other types of containers (Harris, 1986; Wetzel and Likens, 1991). In streams, where the cycling does not occur in place but extends longitudinally downstream in a sort of corkscrew fashion, this process is termed "spiraling." In flowing waters, several aspects can be measured, such as rate of downstream movement, rate of cycling, turnover length, and turnover time (Newbold et al., 1982a,b; Minshall et al., 1992), which may provide insights into the effect of various human and anthropogenic impacts. In addition, nutrient uptake rates are measured relatively easily by releasing known amounts of a nutrient (such as nitrogen or phosphorus) at an upstream location, allowing the nutrient-enriched water to flow over a representative stretch of stream, and determining the amount removed at some downstream point (Munn and Meyer, 1990; Stream Solute Workshop, 1990). Nutrient uptake rates can provide insights into an important aspect of nutrient cycling or spiraling and enable the measurement of various environmental stresses on this aspect of ecosystem functioning.

In addition to measures of ecosystem processes such as energy flow and nutrient cycling, measurement of functional integrity should include genetic and evolutionary aspects of the biota (Regier, 1993). Biological systems are in continual states of adjustment (adaptation) to their environment and evolve over time. Thus, the scale of their response and the patterns subsequently produced can be expected to be the product of selection by long-term evolution. Failure to address evolutionary aspects adequately has led to major misconceptions regarding ecosystem properties and processes such as succession (Hagen, 1992; Colinvaux, 1993). The dilution, isolation, and extinction of genetic pools are bound to be major problems in inland waters both now and in the future (Noss and Cooperrider, 1994). Awareness of this problem is just becoming widespread and is restricted mainly to fish and mollusks (Williams and Miller, 1990; Nehlsen et al., 1992; Bogan, 1993), but effects on other aquatic groups are expected to be equally severe (e.g., Zwick, 1992).

Methods for measuring these long-term fitness features of ecosystem functional integrity have only recently begun to be developed for freshwater organisms. They are often more difficult to apply to field conditions and require larger numbers of samples than other types of processes because of the sensitive, tedious, and specialized laboratory analyses involved (Funk and Sweeney, 1990; Robinson et al., 1992) or the detailed life history, functional, and behavioral information required (e.g., Resh et al., 1994; Usseglio-Polatera, 1994). Nonetheless, it is important to address genetic and evolutionary components in the assessment of the ecological integrity of inland waters.

Two general types of tools are available to do this: (1) those that permit determination of the genetic makeup, particularly the extent of allelic heterozygosity and gene polymorphism, of populations within the ecosystem; and (2) those that assess the occurrence of various measurable traits expected to evolve in particular environments or be selected for under different types and frequencies of environmental stress. Most natural environments are predominantly nonequilibrium, populated by organisms whose populations have relatively high levels of heterozygosity (Hedrick, 1986). Genetic diversity will decline as populations near extinction. Species traits that are likely to contribute to fitness, and hence are likely to be selected for or against under particular environmental conditions, include (1) physiological adaptations to generally unfavorable physical conditions, (2) adaptations for defense, (3) food harvesting and somatic development, (4) reproduction, and (5) tactics for escape in space or time (Southwood, 1988). These traits are postulated to vary in a predictable manner in relation to the degree of stress (adversity) in the environment and the frequency of disturbance or extent of temporal heterogeneity (Southwood, 1977, 1988; Hildrew and Townsend, 1987). Environmental disturbances to which a population has not adapted over evolutionary time may adversely affect its genetic diversity and threaten its long-term survival.

Tools that permit determination of the degree of genetic heterozygosity rely on biochemical measures of gene frequencies and polymorphisms. One common biochemical approach is to use gel electrophoresis to identify loci associated with various enzymes (Shaw and Prasad, 1970; Harris and Hopkinson, 1976). The loci are then scored, based on relative distance of the bands from the origin and determinations of allelic frequencies and heterozygosities. Finally, population polymorphisms and population heterozygosities are calculated from the scored loci (Ayala, 1982). Measurement of species traits important to long-term fitness uses relevant morphological, physiological, and behavioral features of a population (e.g., size, body form, reproductive capacity, mode of respiration, dispersal ability, feeding method). This procedure involves the selection of traits, quantification of the extent to which the different life history stages of individual populations possess those traits, derivation of species-trait and species-habitat-type matrices, and evaluation of relationships between the two matrices using appropriate statistical analyses (Doledec and Statzner, 1994; Usseglio-Polatera, 1994).

Bioassessment procedures that incorporate multiple measures (metrics) of the responses of population aggregates ("communities") are recommended because different measures are sensitive to different types of water quality impairment. Often individual metric scores are summed in the belief that a collective "signal" is easier to discern than individual ones (Plafkin et al., 1989; Karr, 1991, 1993). However, some of the metrics respond in opposite ways, many are biased toward a particular type of pollution (e.g., organic wastes), and not all types of pollution are represented or adequately determined. Therefore, the common practice of summing the results of individual metrics to obtain a single total score tends to conceal valuable information and to produce equivocal results. Additional work is needed to remove uninformative redundancy and develop metrics specific to different types of degradation.

IMPLICATIONS FOR THE FUTURE

Modern water science encompasses a broad array of skills and areas of expertise. Future scientists, teachers, and resource managers will need to be broadly trained in these areas. The complexity and magnitude of the questions facing researchers and resource managers will increasingly require an interdisciplinary approach and the ability to work cooperatively.

The ecological integrity of inland waters is being assailed on many fronts. Direct assessment of the biota is crucial for the protection and management of aquatic resources. Sound understanding of basic biological (ecological) relationships is prerequisite to sound management (Jumars, 1990; Edmondson, 1993). Several large federal monitoring and assessment programs under development emphasize the measurement of water quality in biological, rather than solely chemical-physical, terms. Consequently, the need for training in systematics, basic biology, and ecology of key groups of inland aquatic flora and fauna (e.g., diatom algae, macroinvertebrates, fish) will increase in the future.

Considerations of scale are increasingly a part of the process by which aquatic ecologists approach a variety of ecological issues and problems (King, 1993). The effects of natural and human-caused factors on inland aquatic ecosystems require consideration at multiple spatiotemporal scales that include adequate heterogeneity across landscapes (Covich, 1993). Hierarchy theory commonly is used to address these questions of scale (Allen and Hoekstra, 1992). Thus, ecological integrity will have to address questions of scale and hierarchy; the approach will vary with the particular research question or management problem. However, for the immediate future, the ecosystem and landscape perspectives will be especially important if sustainable biological aquatic resources are to be protected adequately in the face of pressures from the burgeoning human population.

Numerous pressing challenges face the future of inland waters, their study, and their management for sustained benefits and uses. Ecosystem, landscape, and global perspectives will be necessary to provide adequate quantity and high-quality water for human use and natural habitats (Covich, 1993). The remaining natural freshwater habitats that have high biodiversity or endemic species should be protected (e.g., Boon, 1992), and degraded waterways should be restored and their natural linkages reestablished. Basinwide planning and management are needed to protect and restore riverine ecosystems and avoid cumulative effects. Agency personnel and the public must be educated on the new strategies and techniques in aquatic ecosystem management and restoration at the catchment and basin levels. Improved funding is necessary for research and education to enhance information, improve skills, and increase the number of personnel to ensure proper management of sustainable inland aquatic ecosystems.

Most of the methods being applied to the evaluation of structural and functional attributes of inland aquatic ecosystems have been around for a long time. Although there is much to be said for the use of widely accepted, time-tested approaches, there is also the danger that complacency will lead to lack of needed improvement and innovation. Continued refinement of existing methods is necessary. At the same time, many exciting new developments in biology—including genetic markers; molecular, morphological, and behavioral indicators of exposure to toxic substances; and molecular measures of function—are emerging as fertile fields for additional research and for application to the understanding and management of inland aquatic ecosystems (e.g., Petersen and Petersen, 1983; Alvarez et al., 1993; O'Brien, 1994). Computer-based geographical information systems, satellite imagery, and remote sensing are providing valuable techniques for addressing both research and management questions at various levels of resolution in the landscape (Osborne and Wiley, 1988; Johnston et al., 1990; Paris, 1992; Richards and Host, 1994). Rapid technological advances in these and other areas such as data logging and wireless transmission, radiotelemetry, geographical positioning systems, acoustical sounding, electronic surveying and distance measurers, and pressure transducers for remote water-level sensing will increasingly provide powerful tools for addressing important questions relating to inland aquatic resources.

ACKNOWLEDGMENTS

W. T. Edmondson and several committee members (P. Brezonik, E. Gorham, T. Frost, and R. Wetzel) contributed a number of helpful ideas, comments, and references to an early draft of this paper. J. MacDonald is responsible for major improvements in the organization and presentation of the material. J. N. Minshall provided editorial suggestions that improved the presentation. Special recognition is due my colleague C. T. Robinson and graduate students, especially K. N. Myler, T. V. Royer, and E. B. Snyder, for their comments and advice.

REFERENCES

Allard, M., and G. Moreau. 1986. Leaf decomposition in an experimentally acidified stream channel. Hydrobiologia 139:109–117.

Allen, T. F. H., and T. W. Hoekstra. 1992. Toward a Unified Ecology. New York: Columbia University Press.

Alvarez, A. J., E. A. Hernandez-Delgado, and G. A. Toranzos. 1993. Advantages and disadvantages of traditional and molecular techniques applied to the detection of pathogens in waters. Water Sci. Technol. 27:253–256.

American Public Health Association (APHA). 1989. Standard Methods for the Examination of Water and Wastewater. New York: APHA.

Anderson, N. J. 1994. Comparative planktonic diatom biomass responses to lake and catchment disturbance. J. Plankton Res. 16:133–150.

Avers, P. E. 1992. Introduction in Taking an Ecological Approach to Management: Proceedings of the National Workshop. Washington, D.C.: U.S. Forest Service, Watershed and Air Management.

Ayala, F. J. 1982. Population and Evolutionary Genetics: A Primer. Menlo Park, Calif.: Benjamin/Cummings Publishing Company.

Bahls, L. L. 1993. Periphyton Bioassessment Methods for Montana Streams. Helena, Mont.: Department of Health and Environmental Sciences, Water Quality Bureau.

Barbour, M. T., J. L. Plafkin, B. P. Bradley, C. G. Graves, and R. W. Wisseman. 1992.

Evaluation of EPA's rapid bioassessment benthic metrics: metric redundancy and variability among reference sites. Environ. Toxicol. Chem. 11:437–449.

Bogan, A. E. 1993. Freshwater bivalve extinctions (Mollusca: Unionoida): A search for causes. Am. Zool. 33:599–609.

Boon, P. J. 1992. Essential elements in the case for river conservation. Pp. 11–33 in River Conservation and Management, P. J. Boon, P. Calow, and G. E. Petts, eds. London: John Wiley & Sons Ltd.

Botkin, D. 1990. Discordant Harmonies: A New Ecology for the Twenty-First Century. New York: Oxford University Press.

Cairns, J. 1977a. Aquatic ecosystem assimilative capacity. Fisheries 2:5–7, 24.

Cairns, J. 1977b. Quantification of biological integrity. Pp. 171–185 in The Integrity of Water, R. K. Ballentine and L. J. Guarraia, eds. Washington, D.C.: U.S. Environmental Protection Agency, Office of Hazardous Materials.

Cairns, J. C., and J. R. Pratt. 1993. A history of biological monitoring using benthic macroinvertebrates. Pp. 10–27 in Freshwater Biomonitoring and Benthic Macroinvertebrates, D. M. Rosenberg and V. H. Resh, eds. New York: Chapman and Hall.

Cairns, J. C., P. V. McCormick, and B. R. Niederlehner. 1993. A proposed framework for developing indicators of ecosystem health. Hydrobiologia 263:1–44.

Clarke, K. R. 1993. Non-parametric multivariate analyses of changes in community structure. Aust. J. Ecol. 18:117–143.

Colinvaux, P. 1993. Ecology, Vol. 2. New York: John Wiley & Sons.

Covich, A. P. 1988. Geographical and historical comparisons of neotropical streams: biotic diversity and detrital processing in highly variable habitats. J. N. Am. Benthol. Soc. 7:361–386.

Covich, A. P. 1993. Water and ecosystems. Pp. 41–55 in Water in Crisis: A Guide to the World's Fresh Water Resources, P. H. Gleick, ed. New York: Oxford University Press.

Cuffney, T. F., M. E. Gurtz, and M. R. Meador. 1993a. Methods for Collecting Benthic Invertebrate Samples as Part of the National Water-Quality Assessment Program. U. S. Geological Survey Open-File Report 93-406. Reston, Va.: U.S. Geological Survey.

Cuffney, T. F., M. E. Gurtz, and M. R. Meador. 1993b. Guidelines for the Processing and Quality Assurance of Benthic Invertebrate Samples as Part of the National Water-Quality Assessment Program. U. S. Geological Survey Open-File Report 93-407. Reston, Va.: U.S. Geological Survey.

Cummins, K. W., M. A. Wilzbach, D. M. Gates, J. B. Perry, and W. B. Taliaferro. 1989. Shredders and riparian vegetation. BioScience 39:24–30.

Doledec, S., and B. Statzner. 1994. Theoretical habitat templets, species traits, and species richness: 548 plant and animal species in the Upper Rhone River and its floodplain. Freshwater Biol. 31:523–538.

Edmondson, W. T. 1993. Experiments and quasi-experiments in limnology. Bull. Mar. Sci. 53:65–83.

Environmental Protection Agency (EPA). 1990. Biological Criteria: National Program Guidance for Surface Waters. EPA-440/5-90-004. Washington, D.C.: EPA, Office of Water Regulations and Standards.

Firth, P., and S. G. Fisher, eds. 1992. Global Climate Change and Freshwater Ecosystems. New York: Springer-Verlag.

Fisher, S. G. 1990. Recovery processes in lotic ecosystems: Limits of successional theory. Environ. Manage. 14:725–736.

Frissell, C. A., W. J. Liss, C. E. Warren, and M. C. Hurley. 1986. A hierarchical framework for stream habitat classification: Viewing streams in a watershed context. Environ. Manage. 10:199–214.

Frost, T. M., D. L. DeAngelis, S. M. Bartell, D. J. Hall, and S. H. Hurlbert. 1988. Scale in the design and interpretation of aquatic community research. Pp. 229–258 in Complex Interactions in Lake Communities, S. R. Carpenter, ed. New York: Springer-Verlag.

Frost, T. M., S. R. Carpenter, A. R. Ives, and T. K. Kratz. 1995. Species compensation and complementarity in ecosystem function. Pp. 224–239 in Linking Species and Ecosystems, C. G. Jones and J. H. Lawton, eds. New York: Chapman and Hall.

Funk, D. H., and B. W. Sweeney. 1990. Electrophoretic analysis of species boundaries and phylogenetic relationships in some taeniopterygid stoneflies (Plecoptera). Trans. Am. Entomol. Soc. 116:727–751.

Green, D. M., and J. B. Kauffman. 1989. Nutrient cycling at the land-water interface: the importance of the riparian zone. Pp. 61–71 in Practical Approaches to Riparian Resource Management, R. E. Gresswell, B. A. Barton, and J. L. Kershner, eds. BLM-MT-PT-89-001-4351. Washington, D.C.: U.S. Bureau of Land Management.

Green, R. H. 1979. Sampling Design and Statistical Methods for Environmental Biologists. New York: John Wiley & Sons.

Gurtz, M. E. 1994. Design of biological components of the National Water-Quality Assessment (NAWQA) Program. Pp. 323–354 in Biological Monitoring of Aquatic Systems, S. L. Loeb and A. Spacie, eds. Boca Raton, Fla.: Lewis Publishers.

Gurtz, M. E., and T. A. Muir. 1994. Report of the Interagency Biological Methods Workshop. Open-File Report 94-490. Reston, Va.: U.S. Geological Survey.

Hagen, J. B. 1992. An Entangled Bank: The Origins of Ecosystem Ecology. New Brunswick, N.J.: Rutgers University Press.

Harris, G. P. 1980. Temporal and spatial scales in phytoplankton ecology: Mechanisms, methods, models and management. Can. J. Fish. Aquat. Sci. 37:877–900.

Harris, G. P. 1986. Phytoplankton Ecology: Structure, Function, and Fluctuation. New York: Chapman and Hall.

Harris, H., and D. A. Hopkinson. 1976. Handbook of Enzyme Electrophoresis in Human Genetics. Sunderland, Mass.: Sinauer Associates.

Hedrick, P. W. 1986. Genetic polymorphism in heterogeneous environments: A decade later. Ann. Rev. Ecol. Syst. 17:535–566.

Hildrew, A. G., and C. R. Townsend. 1987. Organization in freshwater benthic communities. Pp. 347–372 in Organization of Communities Past and Present, J. H. R. Gee and P. S. Giller, eds. 27th Symposium British Ecological Society, Aberystwyth, 1986. Oxford: Blackwell Press.

Hilsenhoff, W. L. 1987. An improved biotic index of organic stream pollution. Great Lakes Entomol. 20:31–39.

Hilsenhoff, W. L. 1988. Rapid field assessment of organic pollution with a family level biotic index. J. N. Am. Benthol. Soc. 7:65–68.

Howarth, R. W. 1991. Comparative responses of aquatic ecosystems to toxic chemical stress. Pp. 169–195 in Comparative Analyses of Ecosystems: Patterns, Mechanisms, and Theories, J. Cole, G. Lovett, and S. Findlay, eds. New York: Springer-Verlag.

Hughes, R. M., and D. P. Larsen. 1988. Ecoregions: An approach to surface water protection. J. Water Pollut. Control Fed. 60:486–493.

Hughes, R. M., D. P. Larsen, and J. M. Omernik. 1986. Regional reference sites: a method of assessing stream potentials. Environ. Manage. 10:629–635.

Hunsaker, C. T., D. E. Carpenter, and J. Messer. 1990. Ecological indicators for regional monitoring. Bull. Ecol. Soc. Am. 71:165–172.

Hynes, H. B. N. 1960. The Biology of Polluted Waters. Liverpool, England: Liverpool University Press.

Jackson, S., and W. Davis. 1994. Meeting the goal of biological integrity in water-resource programs in the U.S. Environmental Protection Agency. J. N. Am. Benthol. Soc. 13:592–597.

Johnson, R. K., T. Wiederholm, and D. M. Rosenberg. 1993. Freshwater biomonitoring using individual organisms, populations, and species assemblages of benthic macroinvertebrates. Pp. 40–158 in Freshwater Biomonitoring and Benthic Macroinvertebrates, D. M. Rosenberg and V. H. Resh, eds. New York: Chapman and Hall.

Johnston, C. A., N. E. Detenbeck, and G. J. Niemi. 1990. The cumulative effect of wetlands on streamwater quality and quantity: A landscape approach. Biogeochemistry 10:105–141.

Jumars, P. A. 1990. W(h)ither limnology? Limnology and Oceanography 35:1216–1218.

Karr, J. R. 1991. Biological integrity: A long-neglected aspect of water resource management. Ecol. Appl. 1:66–84.

Karr, J. R. 1993. Measuring biological integrity: Lessons from streams. Pp. 83–104 in Ecological Integrity and the Management of Ecosystems, S. Woodley, J. Kay, and G. Francis, eds. Delray Beach, Fla.: St. Lucie Press.

Karr, J. R., and D. R. Dudley. 1981. Ecological perspective on water quality goals. Environ. Manage. 5:55–68.

Karr, J. R., K. D. Fausch, P. L. Angermeier, P. R. Yant, and I. J. Schlosser. 1986. Assessing Biological Integrity in Running Waters: A Method and Its Rationale. Champaign, Ill.: Illinois Natural History Survey Special Publication 5.

Kay, J. J. 1991. A nonequilibrium thermodynamic framework for discussing ecosystem integrity. Environ. Manage. 15:483–495.

Keddy, P. A., H. T. Lee, and I. C. Wisheu. 1993. Choosing indicators of ecosystem integrity: Wetlands as a model system. Pp. 61–79 in Ecological Integrity and the Management of Ecosystems, S. Woodley, J. Kay, and G. Francis, eds. Delray Beach, Fla.: St. Lucie Press.

Kerans, B. L., and J. R. Karr. 1994. A benthic index of biotic integrity (B-IBI) for rivers of the Tennessee Valley. Ecol. Appl. 4:768–785.

King, A. W. 1993. Considerations of scale and hierarchy. Pp. 19–45 in Ecological Integrity and the Management of Ecosystems, S. Woodley, J. Kay, and G. Francis, eds. Delray Beach, Fla.: St. Lucie Press.

Kitchell, J. F., and S. R. Carpenter. 1993. Variability in lake ecosystems: Complex responses by the apical predator. Pp. 111–124 in Humans as Components in Ecosystems, M. J. McDonnell and S. T. A. Pickett, eds. New York: Springer-Verlag.

Lackey, R. T. 1994. Ecological risk assessment. Fisheries 19(9):4–18.

Lange-Bertalot, H. 1979. Pollution tolerance of diatoms as a criterion for water quality estimation. Nova Hedwigia 64:285–304.

Larsen, D. P., and S. J. Christie, eds. 1993. EMAP-Surface Waters 1991 Pilot Report. EPA/620/R-93/003. Washington, D.C.: Environmental Protection Agency.

Lehman, J. T., and D. Scavia. 1982. Microscale patchiness of nutrients in plankton communities. Science 216:729–730.

Leonard, P. M., and D. J. Orth. 1986. Application and testing of an index of biotic integrity in small, coolwater streams. Trans. Am. Fish. Soc. 115:401–414.

Leopold, A. 1949. A Sand County Almanac: And Sketches Here and There. New York: Oxford University Press.

Levin, S. A. 1992. The problem of pattern and scale in ecology. Ecology 73:1943–1967.

Lind, O. T. 1985. Handbook of Common Methods in Limnology. Dubuque, Ia.: Kendall/Hunt Publishing Company.

MacDonald, L. H., A. W. Smart, and R. C. Wissmar. 1991. Monitoring Guidelines to Evaluate Effects of Forestry Activities on Streams in the Pacific Northwest and Alaska. EPA 910/9-91-001. Washington, D.C.: Environmental Protection Agency.

Marshall, I. B., H. Hirvonen, and E. Wiken. 1993. National and regional scale measures of Canada's ecosystem health. Pp. 117–129 in Ecological Integrity and the Management of Ecosystems, S. Woodley, J. Kay, and G. Francis, eds. Delray Beach, Fla.: St. Lucie Press.

Meador, M. T., and M. E. Gurtz. 1994. Biology as an Integrated Component of the U.S. Geological Survey's National Water-Quality Assessment Program. Water-Resources Notes 94–83. Reston, Va.: U.S. Geological Survvey.

Meador, M. T., T. F. Cuffney, and M. E. Gurtz. 1993a. Methods for Sampling Fish Communities as Part of the National Water-Quality Assessment Program. U.S. Geological Survey Open-File Report 93-104. Reston, Va.: U.S. Geological Survey.

Meador, M. T., C. R. Hupp, T. F. Cuffney, and M. E. Gurtz. 1993b. Methods for Characterizing Stream Habitat as Part of the National Water-Quality Assessment Program. Open-File Report 93-408. Reston, Va.: U.S. Geological Survey.

Merritt, R. W., and K. W. Cummins, eds. 1984. An Introduction to the Aquatic Insects of North America, 3rd ed. Dubuque, Ia.: Kendall/Hunt Publishing Company.

Miller, D. L., P. M. Leonard, R. M. Hughes, J. R. Karr, P. B. Moyle, L. H. Schrader, B. A. Thompson, R. A. Daniels, K. D. Fausch, G. A. Fitzhugh, J. R. Gammon, D. B. Halliwell, P. L. Angermeier, and D. J. Orth. 1988. Regional applications of an index of biotic integrity for use in water resource management. Fisheries 13(5):12–20.

Minshall, G. W. 1988. Stream ecosystem theory: A global perspective. J. N. Am. Benthol. Soc. 7:263–288.

Minshall, G. W. 1992. Troubled waters of greenhouse earth: Summary and synthesis. Pp. 308–318 in Global Climate Change and Freshwater Ecosystems, P. Firth and S. G. Fisher, eds. New York: Springer-Verlag.

Minshall, G. W. 1993. Stream-riparian ecosystems: Rationale and methods for basin-level assessments of management effects. Pp. 153–177 in Eastside Forest Ecosystem Health Assessment, Vol. II: Ecosystem Management—Principles and Applications, M. E. Jensen and P. S. Bourgeron, eds. General Technical Report PNW-GTR-318. Portland, Ore.: U.S. Forest Service Pacific, Northwest Research Station.

Minshall, G. W., D. A. Andrews, and C. Y. Manuel-Faler. 1983. Application of island biogeographic theory to streams: Macroinvertebrate recolonization of the Teton River, Idaho. Pp. 279–297 in Stream Ecology: Application and Testing of General Ecological Theory, J. R. Barnes and G. W. Minshall, eds. New York: Plenum Press.

Minshall, G. W., K. W. Cummins, R. C. Petersen, C. E. Cushing, D. A. Bruns, J. R. Sedell, and R. L. Vannote. 1985. Developments in stream ecosystem theory. Can. J. Fish. Aquat. Sci. 42:1045–1055.

Minshall, G. W., J. T. Brock, and J. D. Varley. 1989. Wildfires and Yellowstone's stream ecosystems: A temporal perspective shows that aquatic recovery parallels forest succession. BioScience 39:707–722.

Minshall, G. W., and J. N. Minshall. 1978. Further evidence on the role of chemical factors in determining the distribution of benthic invertebrates in the River Duddon. Arch. Hydrobiol. 83:324–355.

Minshall, G. W., R. C. Petersen, T. L. Bott, C. E. Cushing, K. W. Cummins, R. L. Vannote, and J. R. Sedell. 1992. Stream ecosystem dynamics of the Salmon River, Idaho: An 8th-order system. J. N. Am. Benthol. Soc. 11:111–137.

Munn, R. E. 1993. Monitoring for ecosystem integrity. Pp. 105–115 in Ecological Integrity and the Management of Ecosystems, S. Woodley, J. Kay, and G. Francis, eds. Delray Beach, Fla.: St. Lucie Press.

Munn, N. L., and J. L. Meyer. 1990. Habitat-specific solute retention in two small streams: An intersite comparison. Ecology 71:2069–2082.

Nehlsen, W., J. E. Williams, and J. A. Lichatowich. 1992. Pacific salmon at the crossroads: Stocks at risk from California, Oregon, Idaho, and Washington. Trout (Winter):24–51.

Newbold, J. D., P. J. Mulholland, J. W. Elwood, and R. V. O'Neill. 1982a. Organic carbon spiralling in stream ecosystems. Oikos 38:266–272.

Newbold, J. D., R. V. O'Neill, J. W. Elwood, and W. Van Winkle. 1982b. Nutrient spiralling in streams: Implications for nutrient limitation and invertebrate activity. Am. Nat. 120:628–652.

Norris, R. H., and A. Georges. 1993. Analysis and interpretation of benthic macroinvertebrate surveys. Pp. 234–286 in Freshwater Biomonitoring and Benthic Macroinvertebrates, D. M. Rosenberg and V. H. Resh, eds. New York: Chapman and Hall.

322

Noss, R. F., and A. Y. Cooperrider. 1994. Saving Nature's Legacy: Protecting and Restoring Biodiversity. Washington, D.C.: Island Press.

O'Brien, S. J. 1994. A role for molecular genetics in biological conservation. Proc. Natl. Acad. Sci. USA 91:5748–5755.

Omernik, J. A. 1987. Ecoregions of the conterminous United States. Ann. Assoc. Am. Geogr. 77:118–125.

O'Neill, R. V., D. L. DeAngelis, J. B. Waide, and T. F. H. Allen. 1986. A Hierarchical Concept of Ecosystems. Princeton, N. J.: Princeton University Press.

Osborne, L. L., and M. J. Wiley. 1988. Empirical relationships between land use/cover patterns and stream water quality in an agricultural watershed. J. Environ. Manage. 26:9–27.

Overbay, J. C. 1992. Ecosystem management. Pp. 3–15 in Taking an Ecological Approach to Management: Proceedings of the National Workshop, Watershed and Air Management. WO-WSA-3. Washington, D.C.: U.S. Forest Service.

Paris, J. F. 1992. Remote sensing applications for freshwater systems. Pp. 261–284 in Global Climate Change and Freshwater Ecosystems, P. Firth and S. G. Fisher, eds. New York: Springer-Verlag.

Paulsen, S. G., D. P. Larsen, and P. R. Kaufmann. 1991. EMAP Surface Waters Monitoring and Research Strategy: Fiscal Year 1991. EPA/600/3-91/022. Corvallis, Ore.: Environmental Protection Agency, Environmental Research Laboratory.

Petersen, L. B. N., and R. C. Petersen. 1983. Anomalies in hydropsychid capture nets from polluted streams. Freshwater Biol. 13:181–191.

Petersen, R. C., and K. W. Cummins. 1974. Leaf processing in a woodland stream. Freshwater Biol. 4:343–368.

Pickett, S. T. A., and P. S. White, eds. 1985. The Ecology of Natural Disturbance and Patch Dynamics. New York: Academic Press.

Plafkin, J. L., M. T. Barbour, K. D. Porter, S. K. Gross, and R. M. Hughes. 1989. Rapid Bioassessment Protocols for Use in Streams and Rivers: Benthic Macroinvertebrates and Fish. EPA/444/4-89-001. Washington, D.C.: Environmental Protection Agency.

Poff, N. L., and J. V. Ward. 1989. Implications of streamflow variability and predictability for lotic community structure: a regional analysis of streamflow patterns. Can. J. Fish. Aquat. Sci. 46:1805–1818.

Poff, N. L., and J. V. Ward. 1990. Physical habitat template of lotic systems: Recovery in the context of historical pattern of spatiotemporal heterogeneity. Environ. Manage. 14:629–645.

Polls, I. 1994. How people in the regulated community view biological integrity. J. N. Am. Benthol. Soc. 13:598–604.

Porter, S. D., T. F. Cuffney, M. E. Gurtz, and M. R. Meador. 1993. Methods for Collecting Algal Samples as Part of the National Water-Quality Assessment Program. Open-File Report 93-409. Reston, Va.: U.S. Geological Survey.

Power, M. E., R. J. Stout, C. E. Cushing, P. P. Harper, F. R. Hauer, W. J. Matthews, P. B. Moyle, B. Statzner, and I. R. Wais de Badgen. 1988. Biotic and abiotic controls in river and stream communities. J. N. Am. Benthol. Soc. 7:456–479.

Pringle, C. M., R. J. Naiman, G. Bretschko, J. R. Karr, M. W. Oswood, J. R. Webster, R. L. Welcomme, and M. J. Winterbourn. 1988. Patch dynamics in lotic systems: the stream as a mosaic. J. N. Am. Benthol. Soc. 7:503–524.

Rand, G. M., and S. R. Petrocelli. 1985. Fundamentals of Aquatic Toxicology: Methods and Applications. New York: Hemisphere Publishing Corporation.

Regier. H. A. 1993. The notion of natural and cultural integrity. Pp. 3–18 in Ecological Integrity and the Management of Ecosystems, S. Woodley, J. Kay, and G. Francis, eds. Delray Beach, Fla.: St. Lucie Press.

Reice, S. R. 1994. Nonequilibrium determinants of biological community structure. Am. Sci. 82:424–435.

Reice, S. R., and M. Wohlenberg. 1993. Monitoring freshwater benthic macroinvertebrates and benthic processes: Measures for assessment of ecosystem health. Pp. 287–305 in Freshwater Biomonitoring and Benthic Macroinvertebrates, D. M. Rosenberg and V. H. Resh, eds. New York: Chapman and Hall.

Resh, V. H., and J. K. Jackson. 1993. Rapid assessment approaches to biomonitoring using benthic macroinvertebrates. Pp. 195–233 in Freshwater Biomonitoring and Benthic Macroinvertebrates, D. M. Rosenberg and V. H. Resh, eds. New York: Chapman and Hall.

Resh, V. H., and E. P. McElravy. 1993. Contemporary quantitative approaches to biomonitoring using benthic macroinvertebrates. Pp. 159–194 in Freshwater Biomonitoring and Benthic Macroinvertebrates, D. M. Rosenberg and V. H. Resh, eds. New York: Chapman and Hall.

Resh, V. H., A. V. Brown, A. P. Covich, M. E. Gurtz, H. W. Li, G. W. Minshall, S. R. Reice, A. L. Sheldon, J. B. Wallace, and R. C. Wissmar. 1988. The role of disturbance in stream ecology. J. N. Am. Benthol. Soc. 7:433–455.

Resh, V. H., A. G. Hildrew, B. Statzner, and C. R. Townsend. 1994. Theoretical habitat templets, species traits, and species richness: A synthesis of long-term ecological research on the Upper Rhone River in the context of concurrently developed ecological theory. Freshwater Biol. 31:539–554.

Richards, C., and G. Host. 1994. Examining land use influences on stream habitats and macroinvertebrates: A GIS approach. Water Resources Bulletin 30:729–738.

Richards, C., and G. W. Minshall. 1992. Spatial and temporal trends in stream macroinvertebrate communities: The influence of catchment disturbance. Hydrobiologia 241:173–184.

Robinson, C. T., and G. W. Minshall. 1986. Effects of disturbance frequency on stream benthic community structure in relation to canopy cover and season. J. N. Am. Benthol. Soc. 5:237–248.

Robinson, C. T., and G. W. Minshall. 1995. Effects of open-range livestock grazing on stream communities. Pp. 39–48 in Animal Waste and the Land-Water Interface, K. F. Steele. Boca Raton, Fla.: Lewis Publishers.

Robinson, C. T., L. M. Reed, and G. W. Minshall. 1992. Influence of flow regime on life history, production, and genetic structure of Baetis tricaudatus (Ephemeroptera) and Hesperoperla pacifica (Plecoptera). J. N. Am. Benthol. Soc. 11:278–289.

Rudd, J. W. M., C. A. Kelly, D. W. Schindler, and M. A. Turner. 1988. Disruption of the nitrogen cycle in acidified lakes. Science 240:1515–1517.

Russell, F. E., and P. Kotin. 1957. Squamous papillomas in white croaker. J. Natl. Cancer Inst. 18:857–861.

Schindler, D. W. 1987. Detecting ecosystem responses to anthropogenic stress. Can. J. Fish. Aquat. Sci. 44 (Supplement 1):6–25.

Schindler, D. W., K. H. Mills, D. F. Malley, D. L. Findlay, J. A. Shearer, I. J. Davies, M. A. Turner, G. A. Linsey, and D. R. Cruikshank. 1985. Long-term ecosystem stress: The effects of eight years of acidification on a small lake. Science 228:1395–1401.

Science Integration Team. 1994. Scientific Framework for Ecosystem Management in the Interior Columbia River Basin. Walla Walla, Wash.: Eastside Management Project.

Sedell, J. R., and F. H. Everest. 1991. Historic Changes in Pool Habitat for Columbia River Basin Salmon Under Study for Listing. Corvallis, Ore.: U.S. Forest Service, Pacific Northwest Research Station.

Shackleford, B. 1988. Rapid Bioassessments of Lotic Macroinvertebrate Communities: Biocriteria Development. Little Rock, Ark.: Arkansas Department of Pollution Control and Ecology.

Shaw, C. R., and R. Prasad. 1970. Starch-gel electrophoresis of enzymes: A compilation of recipes. Biochem. Gen. 4:297–320.

Sidle, R. C. 1990. Overview of cumulative effects concepts and issues. Pp. 103–107 in Forestry on the Frontier, Proceedings of the 1989 National Convention, Society of American Foresters. Bethesda, Md.: Society of American Foresters.

Southwood, T. R. E. 1977. Habitat, the templet for ecological strategies. J. Anim. Ecol. 46:337–365.

Southwood, T. R. E. 1988. Tactics, strategies and templets. Oikos 52:3–18.

Steedman, R. J. 1994. Ecosystem health as a management goal. J. N. Am. Benthol. Soc. 13:605–610.

Steedman, R. J., and W. Haider. 1993. Applying notions of ecological integrity. Pp. 47–60 in Ecological Integrity and the Management of Ecosystems, S. Woodley, J. Kay, and G. Francis, eds. Delray Beach, Fla: St. Lucie Press.

Steedman, R. J., and H. A. Regier. 1990. Ecological Bases for an Understanding of Ecosystem Integrity in the Great Lakes Basin. Great Lakes Fishery Commission and International Joint Commission.

Stream Solute Workshop. 1990. Concepts and methods for assessing solute dynamics in stream ecosystems. Journal of the North American Benthological Society 9:95–119.

Steele, K., ed. 1995. Animal Waste and Land-Water Interface. New York: CRC, Inc.

Townsend, C. R. 1989. The patch dynamics concept of stream community ecology. Journal North American Benthological Society 8:36–50.

Triska, F. J., V. C. Kennedy, R. J. Avanzino, G. W. Zellweger, and K. E. Bencala. 1989. Retention and transport of nutrients in a third-order stream in northwest California: Hyporheic processes. Ecology 70:1893–1905.

Tyler, C., J. Sumpter, and R. Walton. 1991. Neoplasms in fish—A consequence of water pollutants? Freshwater Forum 1:89–94.

U.S. Forest Service–U.S. Bureau of Land Management. 1994. Environmental Assessment for the Implementation of Interim Strategies for Managing Anadromous Fish-Producing Watersheds in Eastern Oregon and Washington, Idaho, and Portions of California. 300-094/00016. Washington, D.C.: U. S. Government Printing Office.

Usseglio-Polatera, P. 1994. Theoretical habitat templets, species traits, and species richness: Aquatic insects in the Upper Rhone River and its floodplain. Freshwater Biol. 31:417–437.

Vannote, R. L., G. W. Minshall, K. W. Cummins, J. R. Sedell, and C. E. Cushing. 1980. The river continuum concept. Can. J. Fish. Aquat. Sci. 37:130–137.

Warren, C. E. 1971. Biology and Water Pollution Control. New York: W. B. Saunders Company.

Warwick, W. F. 1988. Morphological deformities in Chironomidae (Diptera) larvae as biological indicators of toxic stress. Pp. 281–320 in Toxic Contaminants and Ecosystem Health, a Great Lakes Focus, M. S. Evans, ed. New York: John Wiley & Sons.

Wetzel, R. G. 1983. Limnology, 2nd ed. Philadelphia: Saunders College Publishing.

Wetzel, R. G., and G. E. Likens. 1991. Limnological Analyses. New York: Springer-Verlag.

White, P. S., and S. T. A. Pickett. 1985. Natural disturbance and patch dynamics: An introduction. Pp. 3–13 in The Ecology of Natural Disturbance and Patch Dynamics, S. T. A. Pickett and P. S. White, eds. New York: Academic Press.

Whittaker, R. H., and C. W. Fairbanks. 1958. A study of plankton copepod communities in the Columbia Basin, southeastern Washington. Ecology 39:46–65.

Williams, J. E., and R. R. Miller. 1990. Conservation status of the North American fish fauna in fresh water. J. Fish Biol. 37(Suppl. A):79–85.

Woodley, S. 1993. Monitoring and measuring ecosystem integrity in Canadian national parks. Pp. 155–176 in Ecological Integrity and the Management of Ecosystems, S. Woodley, J. Kay, and G. Francis, eds. Delray Beach, Fla.: St. Lucie Press.

Zwick, P. 1992. Stream habitat fragmentation—A threat to biodiversity. Biodiversity Conserv. 1:80–97.

APPENDIXES

A

Limnology Programs in U.S. Institutes of Higher Education: A Survey

TABLE A-1 Survey of Limnology Programs at U.S. Institutes of Higher Education: Summary

Survey Question		Number of Schools
Number of faculty members in limnology and related disciplines	0	3
	1	9
	2	5
	3	6
	4	7
	5–10	28
	11–20	5
	>20	3
Students obtaining degrees in limnology-related fields each year	0	3
	1–5	26
	6–10	12
	11–15	6
	16–20	6
	>20	6
Level of introductory limnology course	Freshman/sophomore	0
	Junior/senior/graduate	59
	Graduate only	4
Student interest in limnology	Increasing	38
	Holding steady	17
	Decreasing	2
Schools with access to research or field stations		40
Schools responding to survey		69

SOURCE: Written surveys sent to schools belonging to the Universities Council on Water Resources or housing a Water Resources Research Institute. See Table A-3 for detailed responses.

TABLE A-2 Departments Housing Faculty Working in Limnology and Related Disciplines

Department	Number of Schools with Limnology-Related Faculty Housed in This Type of Department
Agriculture/soil science	6
Biology	42
Botany	12
Chemistry	3
Civil/environmental engineering	32
Ecology	6
Entomology	9
Environmental science/studies	10
Fisheries	17
Forestry	10
Geography	3
Geology/geophysics/geoscience	18
Hydrology/hydrogeology	4
Landscape architecture	1
Life science	2
Marine science and limnology	1
Natural resources	9
Oceanography	3
Public health	2
Rangeland management	3
Urban and regional planning	1
Zoology	15
Zoology and limnology	1

SOURCE: Written surveys of schools belonging to the Universities Council on Water Resources or housing U.S. Geological Survey Water Resources Research Institutes. See Table A-3 for detailed responses.

TABLE A-3 Limnology Programs at U.S. Institutes of Higher Education: Survey Responses

University	Departments Offering Degrees in Limnology-related Fields	Departments Housing Faculty in Limnology-related Fields	Students in Limnology-related Fields Graduating Each Year	Faculty in Limnology and Related Disciplines	Frequency of Introductory Limnology Course	Level of Student Interest in Limnology	Availability of Limnological Research Stations or Labs
Alabama	Biological sciences Geology Geography Civil and environmental engineering	Biological sciences Geology Geography Civil and environmental engineering	10	22	Once per semester (junior/senior/ graduate level)	Increasing	One field station
Arizona State	Zoology Civil engineering	Zoology Civil engineering	na	5	Once per year (senior level)	Increasing	None
Arkansas	Biological sciences Zoology Botany Environmental science Civil engineering	Biological sciences Zoology Botany Environmental science Civil engineering	5–10	3	Once per year (senior/ graduate level)	Holding steady	None
Alaska, Fairbanks	Biology and wildlife Fisheries Marine science and limnology	Biology and wildlife Marine science and limnology	na	3	Once per year (senior level)	Holding steady at high level	Two ecological research stations
Arizona	Hydrology and water resources Wildlife and fisheries science	Hydrology and water resources Wildlife and fisheries science Entomology	6–8	2	Once per year (senior/ graduate level)	Increasing rapidly	None

Institution							
Auburn	Fisheries and allied aquacultures Zoology	Fisheries and allied aquacultures Zoology	5	4	Once per year (senior/graduate level)	Increasing	250 small impound-ments used for fish cultures
California, Berkeley	Civil and environmental engineering Integrated biology Environmental science, planning, and management Landscape architecture	Civil and environmental engineering Integrated biology Landscape architecture Entomology Geology	5	10	Once per semester (junior/senior/graduate level)	Increasing rapidly	Two field stations; other labs can be constructed on field sites
California, Davis	Environmental studies Ecology	Environmental studies Land, air, and water resources Ecology Hydrology	10	5	Once per year (senior/graduate level)	Increasing	Two field research stations
California, Irvine	Ecology and evolutionary biology	Biological sciences	None	1	Every other year (junior/senior level)	na	None
California, Santa Barbara	Biological sciences	Biological sciences Geology Geography	34	5	Once per year (junior/senior level)	Increasing	Once field lab with many aquatic habitats
Cincinnati	Biological sciences	Biological sciences Geology Environmental engineering	3–5	4	Every other year (junior/senior level)	Increasing	None

Continues on next page

TABLE A-3 Limnology Programs at U.S. Institutes of Higher Education: Survey Responses (*Continued*)

University	Departments Offering Degrees in Limnology-related Fields	Departments Housing Faculty in Limnology-related Fields	Students in Limnology-related Fields Graduating Each Year	Faculty in Limnology and Related Disciplines	Frequency of Introductory Limnology Course	Level of Student Interest in Limnology	Availability of Limnological Research Stations or Labs
Clemson	Biological sciences	Biological sciences Environmental systems engineering Aquaculture Fisheries and wildlife	3	5	Once per year (senior level)	Holding steady	Access to Corps of Engineers research facility
Colorado	Environmental, population, and organismic biology	Environmental, population, and organismic biology Civil, environmental, and architectural engineering	2	2	Once per year (junior/senior/ graduate level)	Increasing	One field station
Colorado State	Biology Fishery and wildlife biology	Biology Fishery and wildlife biology	3–4	3	Three courses offered in alternate years (senior/ graduate level)	Increasing	None
Connecticut	Ecology and evolutionary biology	Ecology and evolutionary biology Civil engineering Natural resources management and engineering	8–10	3	Two courses, offered each year (junior/ senior level)	Increasing	Marine science field station

Cornell	Natural resources Ecology and systematics	Ecology and systematics	2–3	6	Every other year (junior/senior/graduate level)	Increasing	One field station
Drexel	Bioscience and biotechnology	Bioscience and biotechnology Environmental studies institute	2–4	1 (plus 3 in related fields)	Once per year (senior/graduate level)	Holding steady	One field station
Duke	Botany Geology Zoology School of the environment	Botany Geology Zoology School of the environment	2	5	Every other year (senior/graduate level)	Increasing	None
East Carolina	Biology	Biology	20	6	Once per year (senior/graduate level)	Holding steady	One field station in development
Georgia	Forest resources Institute of ecology Botany Entomology Microbiology	Forest resources Institute of ecology Botany Entomology Microbiology	10	10	Once per year (junior/senior/graduate level)	Increasing	Access to four field labs
Hawaii	None	Zoology	None	None	Every other year (senior/graduate level)	Increasing	One rainforest lab
Idaho	Fish and wildlife Life sciences Hydrogeology Entomology Civil engineering Bacteriology/biochemistry	Fish and wildlife Life sciences Hydrogeology Entomology Civil engineering Bacteriology/biochemistry	5	4	Once per year (senior level)	Increasing rapidly	Three field stations

Continues on next page

TABLE A-3 Limnology Programs at U.S. Institutes of Higher Education: Survey Responses (*Continued*)

University	Departments Offering Degrees in Limnology-related Fields	Departments Housing Faculty in Limnology-related Fields	Students in Limnology-related Fields Graduating Each Year	Faculty in Limnology and Related Disciplines	Frequency of Introductory Limnology Course	Level of Student Interest in Limnology	Availability of Limnological Research Stations or Labs
Johns Hopkins	Geography and environmental engineering Earth and planetary sciences	Geography and environmental engineering Earth and planetary sciences	5	6	New course (graduate level)	na	na
Kansas	Biology Systematics and ecology	Systematics and ecology Entomology Botany	2	8	Once per year (junior/senior/graduate level)	Increasing rapidly	One large field site with many aquatic systems
Kansas State	Biology	Biology	21 (20 of these are B.S. degrees in fisheries)	2	Every other year (junior/senior/graduate level)	Holding steady	One field station
Kentucky	Biological sciences Geological sciences Toxicology Forestry	Biological sciences Forestry Geological sciences	20	na	Every two to three years (senior level)	na	na
Louisiana	Biological and agricultural engineering Civil and environmental engineering Oceanography and coastal sciences	Biological and agricultural engineering Civil and environmental engineering Oceanography and coastal sciences	100	6	None	Increasing in all fields related to the environment	Natural systems engineering lab; several agricultural experimental stations

Maine	Plant biology Geological sciences Zoology Wildlife Institute of quaternary studies	Plant biology Geological sciences Zoology	0–3	5	Every other year (junior/senior/graduate level)	Increasing	One field site
Maryland	None	None	None	None	None	na	Two estuarine labs
Massachusetts	Biology Environmental sciences Wildlife and fisheries biology Entomology Geology Plant and soil science Civil engineering Environmental engineering	Biology Environmental sciences Wildlife and fisheries biology Entomology Geology Plant and soil science Civil engineering Environmental engineering Chemistry Public health Microbiology	na	1 limnologist; 16 faculty in related areas	Every semester (full-year course, junior/senior/graduate level)	Decreasing	None
Michigan State	Fisheries and wildlife Zoology Botany Resource development	Fisheries and wildlife Zoology Botany Resource development	3–4	7	Once per year (junior/senior/graduate level)	Holding steady	Two field stations

Continues on next page

TABLE A-3 Limnology Programs at U.S. Institutes of Higher Education: Survey Responses (*Continued*)

University	Departments Offering Degrees in Limnology-related Fields	Departments Housing Faculty in Limnology-related Fields	Students in Limnology-related Fields Graduating Each Year	Faculty in Limnology and Related Disciplines	Frequency of Introductory Limnology Course	Level of Student Interest in Limnology	Availability of Limnological Research Stations or Labs
Minnesota, Twin Cities	Civil engineering Ecology, evolution and behavior Fisheries and wildlife Geology and geophysics Geography Horticulture Soil, water and climate Forest resources Entomology Plant science Public health	Agricultural engineering Ecology, evolution and behavior Civil engineering Geology and geophysics Entomology Horticulture Soil, water and climate Fisheries and wildlife Biology Plant science Environmental and occupational health	85	65	Three courses offered twice per year (junior/senior level)	Holding steady	One field station with many aquatic systems, one with experimental stream channels, and one with wetlands
Mississippi State	Wildlife and fisheries	Wildlife and fisheries Biological engineering Biological science	6	3	Once per year (junior level)	Holding steady	One research and extension center
Missouri	Fisheries and wildlife	Fisheries and wildlife Biological sciences Civil engineering	3–5	8	Once per year (senior/graduate level)	Increasing	One field laboratory

Institution			Graduate students		Course frequency	Trend	Field stations
Michigan	Biology Natural resources and environment	Biology Natural resources and environment	15–20	6–10	Once per year (senior/graduate level)	Increasing	One biological research station and many field research preserves
M.I.T.	Civil and environmental engineering	Civil and environmental engineering	5	1–2	Every other year (graduate level)	na	na
Montana	Biological sciences Wildlife biology	Biological sciences Environmental studies Wildlife biology Geology	20	20	Once per year (senior/graduate level)	Increasing	Three large field stations
Montana State	Biology	Biology Environmental engineering Microbiology	10	6	Once per year (junior/senior level)	Holding steady	Field lab in Antarctica
Nebraska	Forestry, fisheries, and wildlife Biological sciences	Forestry, fisheries, and wildlife Biological sciences	20–23	1	Once per year (junior/senior/graduate level)	Increasing	One field station
Nevada, Las Vegas	Biological sciences	Biological sciences Geological sciences/hydrology	5–10	5	Once per year (junior/senior level)	Holding steady	None
New Hampshire	Biology Natural resources	Biology Natural resources Botany	na	3	Once per year (senior level)	na	na
New Mexico Institute of Mining and Technology	Geoscience	Geoscience Biology Chemistry	15 (hydrology)	1	None	na	None

Continues on next page

TABLE A-3 Limnology Programs at U.S. Institutes of Higher Education: Survey Responses (*Continued*)

University	Departments Offering Degrees in Limnology-related Fields	Departments Housing Faculty in Limnology-related Fields	Students in Limnology-related Fields Graduating Each Year	Faculty in Limnology and Related Disciplines	Frequency of Introductory Limnology Course	Level of Student Interest in Limnology	Availability of Limnological Research Stations or Labs
New Mexico State	Interdisciplinary	Fishery and wildlife sciences Civil engineering	10	1	Once per year (senior/graduate level)	Holding steady	None
New York, Syracuse	Environmental and forest biology	Forestry Chemistry	2–3	1	Once per year (senior/graduate level)	Increasing	Many field stations and research sites
North Carolina, Chapel Hill	Environmental science and engineering Biological sciences	Environmental science and engineering Biological sciences Marine science Geology Urban and regional planning	4–6	5	Once per year (senior/graduate level)	Decreasing	None
North Carolina State	Zoology Botany Forestry Fisheries and wildlife Crop science Biomeathematics Ecology Eivil engineering Biological and agricultural engineering	Zoology Botany Forestry Civil engineering Biological and agricultural engineering	15	6	Once per year(senior level)	Increasing	Two reservoirs, one farm pond

							Access to Northern Prairie Science Center
North Dakota State	Zoology	Zoology	13	2	Every other year (junior/senior/graduate level)	Increasing	
Northern Illinois	Not taught at this institution	na	na	na	na	na	na
Ohio State	Zoology Natural resources	Zoology Natural resources Geology Entomology Civil engineering	20	4	Twice per year (junior/senior/graduate level)	Increasing	Field labs on Lake Erie, reservoirs, wetlands
Oklahoma State	Zoology	Zoology Botany Civil and environmental engineering	3	2–3	Every other year (junior/senior/graduate level)	Holding steady	None
Penn State	Environmental resource management	Civil/environmental engineering Biology Forest resources	na	5	Once per year (senior level)	na	None
Puerto Rico, Mayaguez	Biology	Biology	None	1	Once per year (graduate level)	Holding steady	None
Purdue	Forestry and natural resources	Forestry and natural resources Botany and plant pathology Entomology Environmental engineering Biological science	4	1	Once per year (junior/senior/graduate level)	Increasing interest in water quality	One field station; nearby river; experimental watershed

Continues on next page

TABLE A-3 Limnology Programs at U.S. Institutes of Higher Education: Survey Responses (*Continued*)

University	Departments Offering Degrees in Limnology-related Fields	Departments Housing Faculty in Limnology-related Fields	Students in Limnology-related Fields Graduating Each Year	Faculty in Limnology and Related Disciplines	Frequency of Introductory Limnology Course	Level of Student Interest in Limnology	Availability of Limnological Research Stations or Labs
Rutgers	Biology Ecology Environmental sciences	Biology Environmental sciences Natural resources	5 or fewer	4	Every other year (junior/senior level)	Increasing	None
South Dakota State	Wildlife and fisheries sciences Civil and environmental engineering Biology/microbiology Agricultural engineering	Wildlife and fisheries sciences Biology/microbiology Civil engineering Plant science	3–4	8	Once per year (junior/senior level)	Holding steady (some increase in wetland ecology)	One field station
South Florida	Biology Civil engineering	Biology Geology Civil engineering	1	5	Every third year (senior/graduate level)	Increasing	None
Tennessee	Ecology Zoology Fisheries	Ecology Zoology Botany Forestry, wildlife, and fisheries Environmental engineering	3–7	8	Once per year (junior level)	Holding steady	None

Texas, Arlington	Biology Environmental science and engineering	Biology Civil and environmental engineering	4–9	4 (biology department)	Once per year or every other year (junior/senior/graduate level)	Increasing	None
Texas, Austin	Zoology Biological sciences	Zoology Civil engineering	20	5 or fewer	Once per year (junior/senior level)	na	None
Texas A & M	Oceanography Wildlife and fisheries	Wildlife and fisheries Oceonography Rangeland ecology and management	30	5	Once per year (junior/senior level)	Holding steady	None
Texas Tech	Civil engineering Environmental engineering Environmental technology management Range and wildlife management Fisheries and wildlife management	Civil engineering Range and wildlife management Biological sciences	10	13	Once per year (graduate level)	Increasing	One field station
Utah State	Biology Fisheries and wildlife Aquatic ecology Watershed science Civil and environmental engineering	Biology Fisheries and wildlife Range science Forestry Civil and environmental engineering Soil science	13	11	Once per year (junior/senior level)	Increasing	One field laboratory

Continues on next page

TABLE A-3 Limnology Programs at U.S. Institutes of Higher Education: Survey Responses (*Continued*)

University	Departments Offering Degrees in Limnology-related Fields	Departments Housing Faculty in Limnology-related Fields	Students in Limnology-related Fields Graduating Each Year	Faculty in Limnology and Related Disciplines	Frequency of Introductory Limnology Course	Level of Student Interest in Limnology	Availability of Limnological Research Stations or Labs
Vermont	Natural resources	Natural resources	5 or fewer	1	Once per year (junior/senior/ graduate level)	na	na
Virginia Military Institute	None	None	na	na	na	na	na
Virginia Tech	Biology Fisheries and wildlife	Biology Fisheries and wildlife Civil engineering	na	15	Once per year (senior/ graduate level)	Increasing	None
Washington State	Natural resources science Environmental science Environmental engineering	Civil and environmental engineering Natural resources science Zoology-limnology	10–15	6	Every semester (junior/senior/ graduate level)	Increasing	Several lakes and reservoirs
Wisconsin, Madison	Oceanography and limnology graduate program Zoology Botany Geology Water chemistry program Environmental studies	Botany Geology and geophysics Civil and environmental engineering Zoology Environmental studies	5–10	22	Once per year (junior/senior/ graduate level)	Increasing	One field station

Wyoming	Zoology and physiology	Zoology and physiology	15	8	Once per year (junior/senior/ graduate level)	Increasing	One field lab; access to National Park Service research center
Yale	Biology Geology Forestry and environmental studies	Forestry and environmental studies Biology Geology Geophysics	None	0	None	Increasing	11,000 acres of forest land

NOTE: The notation "na" indicates that this question was not answered on the survey.

B

Where Are the Limnologists?
Surveys of Professional
Societies and Journals

TABLE B-1 Demographics of Professional Societies Including
Limnologists in Their Membership

	ASLO	NALMS	NABS	SWS	IAGLR	ESA
Total number of members	4,000	2,500	1,873	3,500	1,075	7,400
Student members (percent of total membership)	25	5	18	5	10	13
Final degree (percent):						
Ph.D.	88[a]	15	39	na	52	79
M.S./M.A.	10	na[b]	43	na	29	na
B.S./B.A.	2	na	17	na	15	na
Other			1		4	
Gender distribution (percent):						
Male	77	na	na	na	na	77
Female	23	na	na	na	na	23
Employment distribution (percent):						
Academia	na	10	49	11	na	61
Government	na	22	34	25	na	18
Private sector	na	15	17	41	na	18

NOTE: ASLO, American Society of Limnology and Oceanography; NALMS, North American Lake Management Society; NABS, North American Benthological Society; SWS, Society of Wetland Scientists; IAGLR, International Association for Great Lakes Research; ESA, Ecological Society of America.

[a]21 percent of ASLO members did not indicate their highest degree; these figures represent those who did indicate their education level.
[b]"na" indicates that this information is not available from the professional society.

TABLE B-2 Home Institutions of American Limnologists Publishing in Selected Limnological Journals

Journal	Year	Total Number of Papers	Papers Covering Topics in Limnology (percent of total papers)	Papers in Limnology with U.S. First Authors (percent of total papers)	Academic — Major University	Academic — Small College	Government	Private
Limnology and	1985	109	50	27	83	7	3	7
Oceanography	1986	129	68	29	86	0	14	0
	1987	113	51	24	70	0	30	0
	1988	128	92	43	89	2	9	0
	1989	134	63	24	81	9	6	3
	1990	183	85	27	88	8	4	0
	1991	179	75	21	66	11	21	0
	1992	167	81	22	76	8	13	3
	1993	142	74	26	84	8	8	0
	1994	139	66	19	77	15	8	0
Journal of	1991	36	97	67	21	63	12	4
the North	1992	34	85	74	12	64	20	4
American								
Benthological								
Society								
Ecology	1990	231	39	13	90	7	3	0
	1991	219	28	12	73	15	12	0
	1992	176	21	8	57	29	14	0
Biogeochemistry	1990	28	29	18	80	20	0	0
	1991	34	32	18	100	0	0	0
Journal of	1991	103	29	17	78	17	5	0
Phycology	1992	110	34	15	76	12	12	0

NOTE: A wide range of other journals also publish papers in limnology and related fields of aquatic science. Examples include *Hydrobiologia, Freshwater Biology, Archiv für Hydrobiologie,* and *Journal of Aquatic Ecosystem Health.*

C

Biographical Sketches of Committee Members and Staff

Patrick L. Brezonik (Chair) is professor of environmental engineering at the University of Minnesota and director of the university's Water Resources Center. His research interests include cycling of mercury and other trace metals in watersheds, lake eutrophication, nutrient dynamics in natural waters and sediments, acid deposition, and organic matter in water. His is a member of the Water Science and Technology Board and served on the National Research Council's Committee on Inland Aquatic Ecosystems and Committee to Review the Environmental Protection Agency's Environmental Monitoring and Assessment Program. Previously, he served as a professor of water chemistry and environmental science at the University of Florida. He obtained a B.S. in chemistry from Marquette University and M.S. and Ph.D. in water chemistry from the University of Wisconsin.

Elizabeth Reid Blood is associate scientist at the J. W. Jones Ecological Research Center in Newton, Georgia. Her research interests include nutrient cycling, the role of macroinvertebrates in decomposition in streams, forest-marsh interactions, and a variety of other ecological subjects. Her prior work experience includes positions as professor in the Department of Environmental Health Sciences at the University of South Carolina, assistant program director for the National Science Foundation's Ecosystem Studies Program, and manager of the Okefenokee Swamp project at the University of Georgia's Institute of Ecology, where she researched the biogeochemistry and hydrology of the swamp. She earned a B.S. in biology and an M.S. in limnology from Virginia Commonwealth University. She earned a Ph.D. in ecology from the University of Georgia.

W. T. Edmondson is professor emeritus of zoology at the University of Washington. His research interests include the eutrophication of lakes, links between lake water quality and land development, and the ecology and taxonomy of Rotifera; he was instrumental in documenting the causes of eutrophication in Seattle's Lake Washington and in guiding policymakers on strategies to reverse damage to the lake. Edmondson is a member of the National Academy of Sciences and has received numerous professional awards, including the G. Evelyn Hutchinson medal of the American Society of Limnology and Oceanography, the August Thienemann–Einar Nauman Medal of the International Association for Theoretical and Applied Limnology, and the National Academy of Sciences' Cottrell Award for Environmental Quality. He received a B.S. degree in biological science and a Ph.D. degree in zoology from Yale University.

Thomas M. Frost is the associate director for Trout Lake Station, Center for Limnology of the University of Wisconsin, Madison. His prior work experience includes positions as an instructor and research associate in the Department of Biology at the University of Colorado, Boulder, and as director of the Lake Valencia Project's North American Field Group in Maracay, Venezuela. He earned a B.S. in biology from Drexel University and a Ph.D. in biology from Dartmouth College.

Eville Gorham is Regent's Professor of Ecology and Botany at the University of Minnesota. He researches the ecology and biogeochemistry of wetlands, acid rain and its effects on ecosystems, the effects of global warming on wetlands, and the history of ecology and biogeochemistry. Previously, he held positions as lecturer at the University of London, senior scientific officer at Britain's Freshwater Biological Association, assistant professor of botany at the University of Toronto, and head of the biology department at the University of Calgary. He is a member of the National Academy of Sciences and fellow of the Royal Society of Canada, and has received numerous professional awards, including the G. Evelyn Hutchinson Medal of the American Society of Limnology and Oceanography. He earned B.Sc. and M.Sc. degrees from Dalhousie University in Nova Scotia and a Ph.D from the University of London.

Douglas R. Knauer is chief of water resources research at the Wisconsin Department of Natural Resources. In this position, he helps the department define its research needs in water resources and coordinates research with state environmental management bureaus, the state legislature, and the University of Wisconsin. In addition, he has led a major study of the biogeochemical fate of mercury in aquatic ecosystems. He earned a B.S. from North Dakota State University and an M.S. from Northern Michigan University.

Diane M. McKnight is research scientist at the U.S. Geological Survey, Water Resources Division. Her major research interest is in biogeochemical processes in natural waters. She is a principal investigator for the National Science Foundation Office of Polar Programs project in Antarctica, where she conducts research on Antarctic lakes. She earned a B.S. in mechanical engineering, an M.S. in civil engineering, and a Ph.D. in environmental engineering from the Massachusetts Institute of Technology.

G. Wayne Minshall is a professor of ecology and zoology at Idaho State University. His research focuses on the ecology of flowing waters, emphasizing aquatic benthic invertebrates, community dynamics, and stream ecosystem structure and function. A special research focus is the effects of wildfires on stream ecosystems. His professional recognitions include the Award of Excellence from the North American Benthological Society. He earned his B.S. from Montana State University and his Ph.D. in zoology from the University of Louisville.

Charles R. O'Melia is a professor in the Department of Geography and Environmental Engineering at Johns Hopkins University. His research interests are in aquatic chemistry, environmental fate and transport of pollutants, predictive modeling of natural systems, and theory of water and wastewater treatment. His professional experience includes positions as assistant engineer for Hazen & Sawyer, Engineers; assistant sanitary engineer at the University of Michigan; assistant professor at the Georgia Institute of Technology; lecturer at Harvard University; and associate professor of environmental science and engineering at the University of North Carolina, Chapel Hill. He is a member of the National Academy of Engineering and the Water Science and Technology Board; he served on the National Research Council's Committee on Wastewater Management in Coastal Urban Areas. He received a B.C.E. from Manhattan College and an M.S.E. and Ph.D. in sanitary engineering from the University of Michigan.

Kenneth W. Potter is professor of civil and environmental engineering and chair of the Water Resources Management Program at the University of Wisconsin, Madison. His teaching and research interests are in hydrology and water resources and include estimation of hydrologic risk, especially flood risk; hydrologic modeling and design; stormwater modeling, management, and design; assessment of human impacts on hydrologic systems; and estimation of hydrologic budgets for surface and ground water. Dr. Potter is a former member of the Water Science and Technology Board and has participated in several National Research Council activities.

He received a B.S. in geology from Louisiana State University and a Ph.D in geography and environmental engineering from the Johns Hopkins University.

Dean B. Premo is president of White Water Associates, Inc., of Amasa, Michigan, an ecological consulting and analytical laboratory services company. His research and teaching interests include biodiversity and ecosystem management, especially as they relate to aquatic and wetland ecosystems and riparian areas. His education and research programs with members of the forest products industry have received national and regional recognition. Dr. Premo is also a consultant to the Environmental Protection Agency's Science Advisory Board and an adjunct faculty member of Michigan Technological University. He received a B.S. in biology and an M.S. and Ph.D. in zoology from Michigan State University.

David W. Schindler is Killam Professor of Zoology at the University of Alberta and was for many years a research scientist at the Freshwater Institute of the Canadian Department of Fisheries and Oceans. His research interests include biological and chemical ecology, biogeochemistry, and experimental manipulation of whole ecosystems. His previous work includes positions as an assistant professor of biology at Trent University and director of the Experimental Lakes Area Project. He is a fellow of the Royal Society of Canada. Among his many awards are the G. Evelyn Hutchinson Medal of the American Society of Limnology and Oceanography, the August Thienemann–Einar Nauman Medal of the International Association for Theoretical and Applied Limnology, and the Stockholm Water Prize. He earned his B.S. from North Dakota State University and his Ph.D in ecology as a Rhodes scholar at Oxford University.

Robert G. Wetzel is the Bishop Professor of the Department of Biological Sciences at the University of Alabama in Tuscaloosa. His research interests include the physiology and ecology of bacteria, algae, and higher aquatic plants; biogeochemical cycling in fresh waters; and functional roles of organic compounds and detritus in aquatic ecosystems. His prior professional experience includes positions as professor at Michigan State University, Erlander National Professor of the Institute of Limnology of Uppsala University in Sweden, and professor at the University of Michigan. Dr. Wetzel is an elected member of the Royal Danish Academy of Sciences and the American Academy of Arts and Sciences. His awards include the G. Evelyn Hutchinson Medal of the American Society of Limnology and Oceanography and the August Thienemann–Einar Nauman medal of the International Association for Theoretical and Applied Limnology. He earned B.Sc. and M.Sc. degrees from the University of Michigan and a Ph.D. from the University of California, Davis.

Project Staff

Jacqueline A. MacDonald is a senior staff officer with the National Research Council's Water Science and Technology Board. She has directed projects on bioremediation of environmental contaminants, cleanup of contaminated ground water, and provision of safe drinking water to small communities. She received a B.A. magna cum laude in mathematics from Bryn Mawr College and an M.S. in environmental science in civil engineering from the University of Illinois.

Anita Hall is an administrative assistant for the Water Science and Technology Board. She was the senior project assistant for the completion of the limnology study and has been on the staff of the National Research Council since 1987.

Gregory K. Nyce was senior project assistant with the National Research Council's Water Science and Technology Board during the course of the study presented in this report. He also served as senior project assistant for the National Research Council's Committees on Wetlands Characterization, Mexico City Water Supplies, and In Situ Bioremediation. He received a B.S. in psychology from Eastern Mennonite College.

D

Other Contributors to This Report

During the course of this project, many individuals other than committee members and those acknowledged in the preface contributed their time to the development of the study and this report. These individuals assisted in planning the study that led to the report, developing recommendations for improving the value of education in limnology for water resources management jobs, writing about limnology programs in their universities, and developing written material for the report. We acknowledge the possibility of overlooking some individual from this list of contributors and apologize if this occurred.

Participants in planning meeting that led to this project:

Judy Meyer, University of Georgia, Chair
Jill Baron, Colorado State University
Ken Bencala, U.S. Geological Survey
Nick Clesceri, Rensselaer Polytechnic Institute
Christopher D'Elia, University of Maryland
Penelope Firth, National Science Foundation
Stuart Fisher, Arizona State University
William Lewis, University of Colorado
John Magnuson, University of Wisconsin
Donald Nielsen, University of California, Davis
Ann Spacie, Purdue University
Richard Sparks, Illinois Natural History Survey
Robert Wetzel, University of Alabama

Attendees at the workshop on perspectives on limnological education from water managers:

Jeff Bode, Wisconsin Department of Natural Resources
Wayne Cheyney, U.S. Bureau of Reclamation

Richard DiBuono, U.S. Army Corps of Engineers
David Dilks, Limno-Tech, Inc.
James Erckman, Seattle Water Department
Tom Fontaine, South Florida Water Management District
Wayne Poppe, Tennessee Valley Authority
Don Porcella, Electric Power Research Institute
Dan Tadgerson, Sault Ste. Marie Tribe of Chippewa Indians
William Taft, Michigan Department of Natural Resources

Contributors of text boxes and other material in Chapters 2, 3, and 4:

Arthur Brooks, University of Wisconsin–Milwaukee
Dennis Cooke, Kent State University
Stuart Fisher, Arizona State University
Ralph Fuhrman, civil and environmental engineer, Washington, D.C.
Charles Hawkins, Utah State University
Dieter Imboden, Swiss Federal Institute of Technology
Library staff of the J. W. Jones Ecological Research Center
William Lewis, University of Colorado
Gene Likens, Cary Arboretum
Bill Mitsch, Ohio State University
Earl Spangenberg, University of Wisconsin–Stevens Point
Heinz Stefan, University of Minnesota
Jack Vallentyne, Great Lakes Laboratory for Fisheries and Aquatic Sciences

We also thank the individuals at the 69 universities who responded to our questionnaire and provided the information found in Appendix A and the staff of limnological societies and journals for providing the information on membership and publication statistics found in Appendix B.

Index